INTERFACE SCIENCE AND TECHNOLOGY
Series Editor: ARTHUR HUBBARD

In this series:

Vol. 1: Clay Surfaces: Fundamentals and Applications
   Edited by F. Wypych and K.G. Satyanarayana

Vol. 2: Electrokinetics in Microfluidics
   By Dongqing Li

Vol. 3: Radiotracer Studies of Interfaces
   Edited by G. Horányi

Vol. 4: Emulsions: Structure Stability and Interactions
   Edited by D.N. Petsev

Vol. 5: Inhaled Particles
   By Chiu-sen Wang

Vol. 6: Heavy Metals in the Environment
   Edited by H.B. Bradl

Heavy Metals in the Environment

INTERFACE SCIENCE AND TECHNOLOGY – VOLUME 6

Heavy Metals in the Environment

Edited by

H.B. Bradl

University of Applied Sciences Trier
Neubrucke, Germany

2005

ELSEVIER
ACADEMIC
PRESS

Amsterdam – Boston – Heidelberg – London – New York – Oxford – Paris
San Diego – San Francisco – Singapore – Sydney – Tokyo

ELSEVIER B.V.
Radarweg 29
P.O. Box 211, 1000 AE Amsterdam
The Netherlands

ELSEVIER Inc.
525 B Street, Suite 1900
San Diego, CA 92101-4495
USA

ELSEVIER Ltd
The Boulevard, Langford Lane
Kidlington, Oxford OX5 1GB
UK

ELSEVIER Ltd
84 Theobalds Road
London WC1X 8RR
UK

© 2005 Elsevier Ltd. All rights reserved.

This work is protected under copyright by Elsevier Ltd., and the following terms and conditions apply to its use:

Photocopying
Single photocopies of single chapters may be made for personal use as allowed by national copyright laws. Permission of the Publisher and payment of a fee is required for all other photocopying, including multiple or systematic copying, copying for advertising or promotional purposes, resale, and all forms of document delivery. Special rates are available for educational institutions that wish to make photocopies for non-profit educational classroom use.

Permissions may be sought directly from Elsevier's Rights Department in Oxford, UK: phone (+44) 1865 843830, fax (+44) 1865 853333, e-mail: permissions@elsevier.com. Requests may also be completed on-line via the Elsevier homepage (http://www.elsevier.com/locate/permissions).

In the USA, users may clear permissions and make payments through the Copyright Clearance Center, Inc., 222 Rosewood Drive, Danvers, MA 01923, USA; phone: (+1) (978) 7508400, fax: (+1) (978) 7504744, and in the UK through the Copyright Licensing Agency Rapid Clearance Service (CLARCS), 90 Tottenham Court Road, London W1P 0LP, UK; phone: (+44) 20 7631 5555; fax: (+44) 20 7631 5500. Other countries may have a local reprographic rights agency for payments.

Derivative Works
Tables of contents may be reproduced for internal circulation, but permission of the Publisher is required for external resale or distribution of such material. Permission of the Publisher is required for all other derivative works, including compilations and translations.

Electronic Storage or Usage
Permission of the Publisher is required to store or use electronically any material contained in this work, including any chapter or part of a chapter.

Except as outlined above, no part of this work may be reproduced, stored in a retrieval system or transmitted in any form or by any means, electronic, mechanical, photocopying, recording or otherwise, without prior written permission of the Publisher. Address permissions requests to: Elsevier's Rights Department, at the fax and e-mail addresses noted above.

Notice
No responsibility is assumed by the Publisher for any injury and/or damage to persons or property as a matter of products liability, negligence or otherwise, or from any use or operation of any methods, products, instructions or ideas contained in the material herein. Because of rapid advances in the medical sciences, in particular, independent verification of diagnoses and drug dosages should be made.

First edition 2005

Library of Congress Cataloging in Publication Data
A catalog record is available from the Library of Congress.

British Library Cataloguing in Publication Data
A catalogue record is available from the British Library.

ISBN: 0-12-088381-3
ISSN: 1573-4285 (Series)

♾ The paper used in this publication meets the requirements of ANSI/NISO Z39.48-1992 (Permanence of Paper).
Printed in The Netherlands.

Working together to grow libraries in developing countries

www.elsevier.com | www.bookaid.org | www.sabre.org

ELSEVIER   BOOK AID International   Sabre Foundation

Preface

Heavy metals in the environment pose a variety of very interesting scientific questions. The fields of work involved cover a wide range of disciplines. Thus, heavy metals are a good example for an interdisciplinary field of work ranging from geology, mineralogy, and geochemistry, if their origin and natural occurrence is concerned, to analytical, physical, and colloid chemistry, when it comes to detection of heavy metals and their interactions with environmental media such as water, groundwater, soil, rock, and air, and biology, ecology, ecotoxicology, and medicine, if one is concerned with their impact on global ecosystems and their effects on human and animal health. Finally, the remediation of heavy metals requires cooperation of several engineering disciplines such as environmental, chemical, and civil engineering.

Of course it is not possible to cover this wide range in sufficient depth in one single book alone. Nevertheless this book aims at giving an overview on the most important topics for the reader interested in the subject. Although this book is not meant to be an introductory textbook, pain was taken to keep the text to a level, which allows graduate students to read and understand it. The first chapter gives some ideas on both natural and anthropogenic sources of heavy metals in the environment. The second chapter introduces analytical methods for their detection, the most important biogeochemical processes regulating their mobility, and their ecotoxicological effects on plants, animals, and humans. In this chapter, detailed information over the behaviour of some selected heavy metals is given as well. The third chapter gives an overview over different strategies for the remediation of heavy metals. In this context, innovative new strategies for the remediation of soil and groundwater contaminated with heavy metals such as permeable reactive barriers are discussed along with approved technologies such as encapsulation, soil washing, solidification, and phytoremediation.

There have been many sources of support during the work on this book. First I would like to thank my contributors, who took pain, work, and patience in preparing their subchapters. Prof. Dr. Doris Stüben, University of Karlsruhe, Germany, gives an overview over Platinum Group Metals. Prof. Dr. Chris Kim, Chapman University, Orange, CA, USA, prepared the subchapter on sorption of heavy metals. Dr. Utz Kramar, University of Karlsruhe, Germany, introduces analytical methods for their detection, and last but not least, Dr. Anthimos Xenidis, National Technical University of Athens, Greece, wrote a subchapter on stabilization and solidification. Their time and effort is greatly appreciated. Parts of this book were prepared during a sabbatical leave at the Environmental Research Centre, University of Karlsruhe, Germany. I would like to express my thanks to all the colleagues there for never-ending support and a good time. Finally, I would like to thank my colleagues and students at the Umwelt-Campus, Birkenfeld, for their help and patience during the work on this book.

Heike B. Bradl
Birkenfeld
November, 2004

Table of Contents

Preface V

CHAPTER 1: SOURCES AND ORIGINS OF HEAVY METALS

1. Introduction 1
2. Heavy Metals in Rocks and Soils 1
2.1. Magmatic Rocks 1
2.2. Sedimentary Rocks 4
2.3. Metamorphic Rocks 5
2.4. Soil Formation 6
    2.4.1. Organic Material 7
    2.4.2. Clay Minerals 8
    2.4.3. Oxides and Hydroxides 11

3. Heavy Metals in Water and Groundwater 12
3.1. Surface Waters 12
3.2. Groundwater 12

4. Heavy Metals in the Atmosphere 15

5. Anthropogenic Sources of Heavy Metals 17
5.1. Agricultural Activities 18
    5.1.1. Phosphatic Fertilizers 18
    5.1.2. Pesticides 19
    5.1.3. Sewage Effluents 19
    5.1.4. Biosolids 21
5.2. Industrial Activities 22
    5.2.1. Mining 22
    5.2.2. Coal and Petroleum Combustion 23
    5.2.3. Indoor and Urban Environments 23
    5.2.4. Solid Waste Disposal 25

References 25

CHAPTER 2: INTERACTION OF HEAVY METALS

1. Analytical Procedures for the Detection of Heavy Metals (U. Kramar) 28
1.1. Sample Preparation 28
    1.1.1. Soils and Sediments 29
    1.1.2. Vegetation 29
    1.1.3. Waters 29
1.2. Digestion Methods 30
    1.2.1. Soils, Sediments, and Building Materials 30
    1.2.2. Vegetation 32
1.3. Analytical Methods 32
    1.3.1. Optical Spectroscopic Methods 32
    1.3.2. Microanalytical Methods 45
2. Biogeochemical Processes regulating Heavy Metal Mobility 46
2.1. Sorption (C. Kim) 47
    2.1.1. Introduction 47
    2.1.2. Adsorption Mechanisms 48
    2.1.3. Utility of X-Ray Absorption Spectroscopy in Determining Sorption
        Mechanisms 56
    2.1.4. Model Approaches for Heavy Metal Sorption (H.B. Bradl) 59
    2.1.5. Geochemical Parameters influencing Adsorption 73
2.2. Redox Reactions 76
2.3. Weathering 77
2.4. Driving Factors 77
    2.4.1. pH and Redox Potential 77
    2.4.2. Complexing Agents 78
    2.4.3. Type and Chemical Speciation of Metal 83

3. Ecotoxicological Effects of Heavy Metals 85
3.1. Pathways of Heavy Metal Access 85
    3.1.1. Respiration 85
    3.1.2. Water 86
    3.1.3. Food 86
3.2. Bioavailability and Bioaccumulation 87
    3.2.1. Definition 87
    3.2.2. Bioavailability in the Soil-Plant System 90
    3.2.3. Bioavailability in the Aquatic System 91

4. Individual Behaviour of Selected Heavy Metals 93
4.1. Arsenic 93
    4.1.1. Chemical and Physical Character of Arsenic 93
    4.1.2. Sources and Applications of Arsenic 94
    4.1.3. Ecotoxicological Effects of Arsenic 96

4.2. Cadmium 98
   4.2.1. Chemical and Physical Character of Cadmium 98
   4.2.2. Sources and Applications of Cadmium 101
   4.2.3. Ecotoxicological Effects of Cadmium 103
4.3 Chromium 104
   4.3.1. Chemical and Physical Character of Chromium 104
   4.3.2. Sources and Applications of Chromium 106
   4.3.3. Ecotoxicological Effects of Chromium 107
4.4. Copper 108
   4.4.1. Chemical and Physical Character of Copper 108
   4.4.2. Sources and Applications of Copper 110
   4.4.3 Ecotoxicological Effects of Copper 111
4.5. Lead 111
   4.5.1. Chemical and Physical Character of Lead 111
   4.5.2. Sources and Applications of Lead 114
   4.5.3. Ecotoxicological Effects of Lead 115
4.6. Manganese 115
   4.6.1. Chemical and Physical Character of Manganese 115
   4.6.2. Sources and Applications of Manganese 117
   4.6.3. Ecotoxicological Effects of Manganese 118
4.7. Mercury 119
   4.7.1. Chemical and Physical Character of Mercury 119
   4.7.2. Sources and Applications of Mercury 121
   4.7.3. Ecotoxicological Effects of Mercury 122
4.8. Molybdenum 124
   4.8.1. Chemical and Physical Character of Molybdenum 124
   4.8.2. Sources and Applications of Molybdenum 125
   4.8.3. Ecotoxicological Effects of Molybdenum 126
4.9. Nickel 126
   4.9.1. Chemical and Physical Character of Nickel 126
   4.9.2. Sources and Applications of Nickel 127
   4.9.3. Ecotoxicological Effects of Nickel 128
4.10. Platinum Group Elements PGE (D. Stüben) 128
   4.10.1. Introduction 128
   4.10.2. Chemical and Physical Character of PGE 129
   4.10.3. Sources and Applications of PGE 131
   4.10.4. PGE Emission by Car Catalytic Converters 134
   4.10.5. PGE in Environmental Matrices 135
   4.10.6. Transformation of PGE and Bioaccumulation
         in the Environment 137
4.11. Zinc (H.B. Bradl) 139
   4.11.1. Chemical and Physical Character of Zinc 139
   4.11.2. Sources and Applications of Zinc 141
   4.11.3. Ecotoxicological Effects of Zinc 142

4.12. Other Heavy Metals 143
    4.12.1. Cobalt 143
    4.12.2. Silver 144
    4.12.3. Thallium 146
    4.12.4. Tin 147

References 148

CHAPTER 3: REMEDIATION TECHNIQUES

1. Introduction 165

2. Physical Remediation Techniques 165
2.1. Soil Washing 165
    2.1.1. Particle-Size Dependent Distribution of Pollutants 167
    2.1.2. Wet Liberation 168
    2.1.3. Classification of Fine Particles 169
2.2. Encapsulation 170
    2.2.1. Slurry Walls 170
    2.2.2. Thin Walls 171
    2.2.3. Sheet Pile Walls 172
    2.2.4. Bored-pile Walls and Jet Grouting 172
    2.2.5. Injection Walls 172
    2.2.6. Artificial Ground Freezing and Frozen Walls 173
2.3. Vitrification 174
2.4. Electrokinetic Techniques 176
    2.4.1. Principle Electrokinetic Transport Processes 177
    2.4.2. Electrode Reactions 178
    2.4.3. Applications 179
2.5. Permeable Reactive Barrier Systems 179
    2.5.1. Permeable Walls 180
    2.5.2. Funnel and Gate Systems 182
    2.5.3. Reactor Technologies for Removal of Heavy Metals 183
    2.5.4. Engineering Methods for Execution of
           Permeable Reactive Barriers 186

3. Chemical Remediation Techniques 191
3.1. Precipitation 192
3.2. Ion Exchange 193
3.3. Flocculation 194
    3.3.1. Colloidal Systems 194
    3.3.2. Flocculation Chemicals 197
3.4. Membrane Filter Processes 198
3.5. Solidification/Stabilization (A. Xenidis) 200

    3.5.1. Introduction 200
    3.5.2. Solidification/Stabilisation Mechanisms 202
    3.5.3. Evaluation of Solidification/Stabilisation Processes 209
    3.5.4. Technology Description 216
    3.5.5. Field Applications 230

4. Phytoremediation of Heavy Metals (H.B. Bradl) 235
4.1. Introduction 236
4.2. Basic Physiological Processes 236
    4.2.1. Processes involving Microorganisms 237
    4.2.2. Plant Processes 239
4.3. Mechanisms of Phytoremediation 241
    4.3.1. Phytoextraction 241
    4.3.2. Phytostabilization 245
    4.3.3. Phytovolatilization 247
    4.3.4. Phytofiltration 248
4.4. Advantages and Limitations of Phytoremediation 249

References 251

Index 263

Heavy Metals in the Environment
H.B. Bradl (editor)
© 2005 Elsevier Ltd. All rights reserved.

Chapter 1

Sources and Origins of Heavy Metals

H. B. Bradl[a]

[a]Department of Environmental Engineering, University of Applied Sciences Trier, Umwelt-Campus Birkenfeld, P.O. Box 301380, 55761 Birkenfeld, Germany

1. INTRODUCTION

There are different sources for heavy metals in the environment. These sources can be both of natural or anthropogenic origin. This chapter gives a general introduction into the different heavy metal sources such as magmatic, sedimentary, and metamorphic rocks, weathering and soil formation, the rock cycle, the origin of heavy metals in surface and groundwater as well as in the atmosphere, and anthropogenic sources stemming from human activities such as industrial production and agriculture [1, 2].

2. HEAVY METALS IN ROCKS AND SOILS

2.1. Magmatic Rocks

Rocks and soils are the principal natural sources of heavy metals in the environment. The primary rocks, which are called magmatic or igneous rocks, crystallize from magma upon cooling down. Magma is defined as molten rock material originating from the earth's mantle, which can be transported to the surface by several geological processes such as volcanism or plate tectonics [3]. Magma contains a large variety of different chemical elements. Heavy metals are incorporated as trace elements into the crystal lattice of the primary minerals, which form during the cooling of the magma. This process is called isomorphic substitution, as the heavy metals substitute other atoms during the crystallization. The amount of isomorphic substitution is determined by the ion radius, the ion charge, and the electronegativity of the main element and of the substituting element. The trace elements occurring in the most common rock forming minerals are given in Table 1.

Table 1
Trace elements of the most common rock forming minerals

Mineral	Trace Element	Resistance to Weathering
Olivine	Ni, Co, Mn, Li, Zn, Cu, Mo	weathers easily
Hornblende	Ni, Co, Mn, Se, Li, V, Zn, Cu, Ga	
Augite	Ni, Co, Mn, Se, Li, V, Zn, Pb, Cu, Ga	
Biotite	Rb, Ba, Ni, Co, Mn, Se, Li, V, Zn, Cu, Ga	
Apatite	Rare earth elements, Pb, Sr	
Anorthite	Sr, Cu, Ga, Mn	
Andesine	Sr, Cu, Ga, Mn	
Oligoclase	Cu, Ga	
Albite	Cu, Ga	
Garnet	Mn, Cr, Ga	intermediate stability
Orthoclase	Rb, Ga, Sr, Cu, Ga	
Muscovite	F, Rb, Ba, Sr, Cu, Ga, V	
Titanite	Rare earth elements, V, Sn	
Ilmenite	Co, Ni, Cr, V	
Magnetite	Zn, Co, Ni, Cr, V	
Turmaline	Li, F, Ga	
Zircon	Hf, U	
Quartz	-	resistant

Modified after Ref. 4.

Magmatic rocks can be classified by their chemical composition on the one hand and their crystal size and texture on the other hand. If magma cools down slowly within the earth's crust, there is enough time for large crystals to be formed, which can easily be recognized with the naked eye. If magma is extruded rapidly onto the earth's surface (e.g. by volcanic activity), crystallization occurs quickly, and consequently, those magmatic rocks are characterized by very fine crystals that can not be seen with the naked eye. Magmatic rocks showing large crystals are called intrusive rocks or intrusiva, while those with fine texture and small crystals are called effusive rocks or effusiva. The principal effusive and intrusive magmatic rocks and their mineral and chemical components are identified in Fig. 1. For each individual intrusive rock, there is an equivalent effusive counterpart, which is identical in chemical composition, yet shows different crystal size and texture (e.g. granite and rhyolithe). During the crystallization phase of a magma body, a process called chemical differentiation takes place. Mineral crystallization is a function of both temperature and pressure conditions, which change constantly during cooling. Different minerals precipitate according to their stability fields at limited ranges of temperature, pressure, and chemical composition conditions.

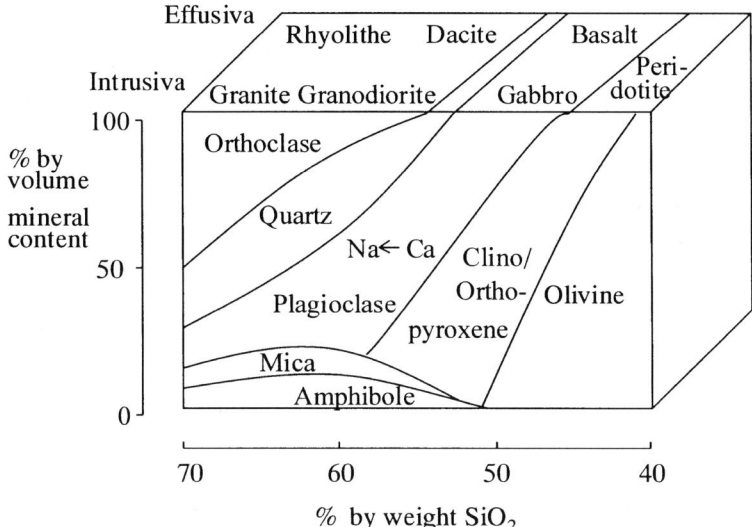

Fig. 1. General classification of magmatic rocks (effusiva and intrusiva) showing principal minerals and their chemical composition.

The magma is depleted of those elements, which have been bound into the crystallizing minerals, while is at the same time enriched in those elements that have not been incorporated into those minerals. Consequently, the chemical composition of the magma body is subject to changes during the cooling down process. Most heavy metals concentrate mainly in the residual magma. Only a few heavy metals form their own mineral or form an important component of a principal mineral.

One example is Cr, which crystallizes as the mineral chromite ($FeCr_2O_4$), or Ni, which occurs in the mineral forsterite ($Mg_2[Ni]SiO_4$) as a substitute for Mg. In the later stages of differentiation, metal concentrations increase, which may lead to the precipitation as their own mineral (e.g. U as uranitite, Be as beryl) or their incorporation into late-stage forming accessory mineral. One example is zircon, which contains such elements as U and rare earths. Most heavy metals concentrate in the hot residual hydrothermal fluids, which are formed in the final stages of magma differentiation. As these fluids infiltrate into the enclosing rock, chemical reactions take place between the enclosing rock and the hydrothermal fluid, and minerals precipitate as ores. Examples are Hs as cinnabar (HgS), As as arsenopyrite (FeAsS), Pb as galena (PbS), Zn as sphalerite (ZnS), Cu as chalcopyrite ($CuFeS_2$), Mo in molybdenite (MoS_2), Fe as pyrite (FeS_2), and U as uranitite (UO_2). Cd can substitute in part for Zn in sphalerite ($Zn[Cd]S$), and As can accompany Fe in pyrite ($Fe[As]S_2$). Most frequently, ore is an assemblage of several minerals so that smelting and processing of one metal often results in the release of other metals in the environment.

2.2. Sedimentary Rocks

Physical weathering causes rock to disintegrate into particles, which are called sediments. Chemical weathering dissolves the rock into ions. These sediment particles are transported to sedimentary basins, in which they deposit. Upon accumulating of new sediment, a process called diagenesis or lithification takes place. The loose particles are then connected to each other by chemical alterations of the pore fluid and slowly, solid rock forms out of the loose particles. Pore fluids are squeezed out of the underlying sediments by compaction, and chemical reactions lead to the formation of pore cements, which bind the loose particles together. Sedimentary rocks or sedimentites are formed slowly over geological periods of time. Sedimentary rocks can be classified into clastic, chemical, and biogenic sedimentites. Table 2 gives a general classification of sediments and sedimentary rocks. Clastic sedimentites can easily be classified by their grain size (e.g. gravel, sand, silt, and clay) and composition, while biogenic and chemical sediments can be classified according to their chemical composition.

The most abundant chemical and biogenic sediments comprise of the mineral calcite, which can be precipitated either by living organisms such as corals or precipitates directly if the chemical conditions for precipitation are given. If evaporation exceeds water inflow into ocean basins, chemical sediments precipitate directly. Among these precipitation products are the minerals halite (or rock salt, $NaCl$), gypsum ($CaSO_4 \cdot 2H_2O$), anhydrite ($CaSO_4$), calcite ($CaCO_3$), phosphorite ($Ca_3(PO_4)_2$), borate salts, and goethite ($FeOOH$). The most important biogenic sediment is coal.

Sedimentary rocks are characterized by two properties, which make them unique and economically important. First, they are of porous structure, which enables them to hold fluids such as water, gas and oil. Second, they are permeable, which enables them to transport fluids. These two abilities make sedimentary rocks so important for water and energy supply. They also may contain ore deposits of many heavy metals if they are penetrated by ore-bearing hydrothermal fluids.

2.3. Metamorphic Rocks

Metamorphic rocks are the third rock type. Any magmatic, sedimentary or metamorphic rock can be subjected to increased temperature and pressure conditions (e.g. during transport of plates into deeper parts of the earth's crust), which leads to chemical alterations. The process, by which metamorphic rocks are generated, is called metamorphosis. Metamorphic rocks are characterized by the reorganization and recrystallization of rock-forming minerals. There are two main groups of metamorphosis, which are of importance. First, there is kinetic or regional metamorphosis, which results in crystalline schists and second, contact or static metamorphosis occurs, if molten magma comes into contact with consolidated rock.

Table 2
General classification of sediments and sedimentary rocks

Sediment	Sedimentary rock	Size and/or Composition
Gravel	Conglomerate	> 2 mm rounded rock and /or mineral detritus; in rock this is cemented by silica (SiO_2), calcite ($CaCO_3$), or iron oxides
Sand	Sandstone	1/16 – 2 mm particles, mainly quartz, cemented as above
Silt	Siltstone	1/256 -1/16 mm particles as given above with cementation as above or by clay-size matrix
Clay	Shale	> 2µm particles mainly clay minerals as weathering or decomposition products of feldspars and other minerals; lithification by compaction and cementation
Accumulated Shells Lime Mud	Limestone	Rock of $CaCO_3$ from shell remains or from chemical precipitation by evaporation
	Dolostone	Limestone or lime mud altered by interaction of Mg-rich waters to the mineral dolomite $MgCa(CO_3)_2$
	Gyprock	Precipitate of $CaSO_4 \cdot 2H_2O$ (gypsum) from evaporation
	Rock Salt	Precipitate of NaCl (halite) from evaporation
	Coal	Accumulation of vegetation to form peat, Lignite, and bituminous coal under increasing compaction, and anthracite with increasing heat and pressure

Modified after Ref. 1.

These processes are important for the origin of ore-bearing deposits. The classification of metamorphic rocks is based upon foliation, layering, texture, and mineral composition. The transition between diagenesis and metamorphosis can be defined at low pressure and temperature conditions (approximately 200 °C and 2 kbar).

2.4. Soil Formation

Consolidated rock is disintegrated by weathering processes (e.g. by the influence of temperature, water, ice etc.). Soluble minerals are solved by interaction of carbonic acid and water, while relatively insoluble minerals such as quartz (SiO_2) disintegrate as finer particles. The end result of weathering is soil. Soil is one of the key elements for all terrestrial ecosystems. It provides the nutrient-bearing environment for plant life and is of essential importance for degradation and transfer of biomass. Soil is a very complex heterogeneous medium, which consists of solid phases (the soil matrix) containing minerals and organic matter and fluid phases (the soil water and the soil air), which interact with each other and ions entering the soil system [5]. Soil formation or pedogenesis is influenced by the factors climate, soil organisms, topography, type of parent rock, and time [6]. During soil development, distinct layers or horizons develop, which comprise a soil profile. Fig. 2 shows soil horizons that develop in a temperate humid climate.

The first layer is called O-horizons and consists mostly of decomposed organic matter such as leaves, twigs, and other humic substances. The underlying A horizon is composed of mineral and organic matter and is subject to leaching by rainwater filtration. Clay-size particles and dissolved chemical elements (such as Fe, Ca, Mg, and heavy metals) are transferred into the underlying B horizon, a process, which is called eluviation. In the B horizon, this material accumulates (alluviation). The B horizon is characterized by a large content of clay minerals and Fe oxyhydroxides, which are able to absorb heavy metals. This horizon is also the main source of plant nutrients. Finally, the C horizon underlying the B horizon is composed of partially weathered parent rock, which is followed by the unaltered parent rock. According to the type of parent rock, climate, time, and soil organisms, soil composition may vary to a large extent. Its chemical composition mirrors the chemistry of the parent rock. Natural or geogenic background concentrations of heavy metals vary significantly from one area to another. Table 3 gives some average natural concentration of heavy metals in selected rocks.

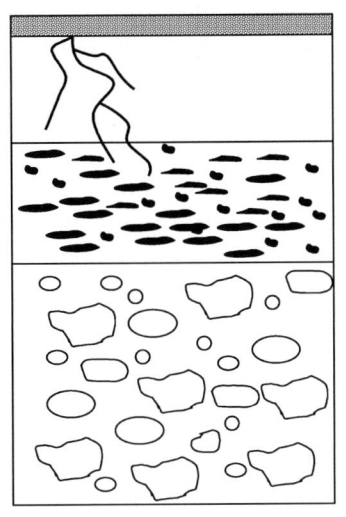

Soil horizons

O - Composed mostly of organic matter

A - Composed of mineral and organic matter

B - composed of earth materials enriched in clay, Fe oxyhydroxides, $CaCO_3$, and other constituents leached from the overlying horizons.

C - Composed of partially weathered parent rock.

Fig. 2. Soil horizons that develop in a temperate humid climate (modified after Ref. 7).

If a soil is derived from e.g. basalt, which is enriched in Cr, Co, and Ni, then this soil can be expected to contain higher concentrations of those elements than a soil derived from granite. There are three main components of soils, which are the key elements for the characteristic properties of a soil. These components are organic material, clay minerals, and oxides of Fe, Al, and Mn.

2.4.1. Organic Material

All soils contain organic material in form of living organisms, organic decomposition products, and humic substances. Their content may vary according to the individual soil type and greatly influences the chemical reactions occurring in soils. Humic substances are acid, yellow to blackish polyelectrolytes of intermediate atomic weight [5]. They are formed by secondary synthesis reactions, in which microorganisms play an important role [8]. Humic acids display a variety of functional groups such as carboxy groups, phenolic hydroxy groups, carbonyl, ester, chinone, and methoxy groups [9, 10]. Atomic weights of humic acids vary between 20 000 and 100 000. The typical elementary composition of humic substances is 44 – 53 % c, 3.6 – 5.4 % H, 1.8 – 3.6 % N, and 30 – 47 % O [6].

2.4.2. Clay Minerals

Clay minerals are weathering products of feldspars and display several important characteristics. The most important of these properties are a large specific surface, negative surface charge, and the resulting ability to adsorb cations. The term „clay" can be used both as a rock term and a particle size term. When used as a rock term, clay means a natural, earthy, fine-grained material which is composed largely of a group of crystalline material, the clay minerals. As a particle size term, clay is used for the category containing the smallest particles. In soil science and mineralogy, generally the fraction < 2µm is used as the maximum size for the clay fraction [11].

Table 3
The average natural contents in selected rocks of heavy metals (Values in ppm unless otherwise noted)

Metal	Granite	Basalt	Shale	Ocean Clay	Lime-stone	Deep-Sea Carbonate	Streams (ppb)
Al (%)	7.2	8.2	8.0	8.4	0.42	2.0	50
As	2	2.2	13	13	1	1	2
Be	3	0.7	3	2.6	0.X	0.X	0.001
Cd	0.13	0.21	0.3	0.03	0.3	0.0X	0.01
Co	4	47	20	74	0.1	7	0.1
Cr	10	185	100	90	11	11	1
Cu	20	94	50	250	4	30	7
Fe (%)	1.42	8.6	5.1	6.50	0.38	0.9	40
Hg (ppb)	0.03	0.09	0.4	0.03	0.04	0.0X	0.07
Mn (%)	0.045	0.18	0.09	0.07	0.11	0.1	7
Mo	1	1.5	2.627	0.4	3		0.6
Ni	10	145	60	230	20	30	0.3
Pb	17	7	20	30	9	9	1
Sb	0.22	0.6	1.5	1	0.2	0.15	0.07
Sc	7	27	16	19	1	2	0.004
Se	0.05	0.05	0.6	0.17	0.08	0.17	0.06
Sn	3	1.5	6	1.5	0.X	0.X	0.04
Ti (%)	0.12	1.14	0.60	0.46	0.04	0.08	3
Tl	2.3	0.21	1.4	0.8	0.0X	0.16	-
V	50	225	140	120	20	20	0.9
Zn	50	118	85	200	20	35	20

Assumed shale equivalent (volatile-free, carbonate-free basis): see Ref. 12.
Average pelagic (ocean) clay: see Ref. 12.
Stream water: see Ref. 13.
Basalt: average see Ref. 14, 15.
Granite: Low Ca, see Ref. 14.
Soil and natural Vegetation: see Ref. 16 – 18.

Table 3 (continued)
The average natural contents in selected rocks of heavy metals (Values in ppm unless otherwise noted)

Metal	Cultivated Soil	Uncultivated Soil	Vegetation Ash Natural	On Mineralized Terrain
Al (%)		1.1 – 6.5	0.1 – 3.9	
As	5.5 – 12	6.7 – 13		
Be	1– 1.2	0.76 – 1.3		
Cd		0.1-0.13	0.95 - 20	
Co	1.3 – 10	1 – 14	0.65 – 400	> 50
Cr	15 – 70	11 – 78	2.2 - 22	
Cu	9.9 – 39	8.7 – 33	50 – 270	50 – 60 to 100 – 200 Sometimes > 1000
Fe (%)	1.4 – 2.8	0.47 – 4.3	0.08 – 0.93	
Hg (ppb)	30 – 69	45 - 160		
Mn (%)	0.099 – 0.74	0.006 – 0.11	0.05 – 1.4	
Mo		0.2 – 5	0.76 – 7.6	
Ni	1.8 – 18	4.4 – 23	0.81 – 130	> 100 Sometimes 2000
Pb	2.6 – 27	2.6 – 25	24 - 480	
Sb		2.0		
Sc	2.8 – 9	2.1 - 13		
Se	0.28 – 0.74	0.27 – 0.73	0.01 – 0.42	
Sn		3 – 10		
Ti (%)	0.17 – 0.40	0.17 – 0.66	0.07 – 0.12	
V	20 – 93	15 – 110	2.6 - 23	
Zn	37 – 68	25 – 67	170 – 1800	500 – 1000 Wood

Modified after Ref. 1.

Clays are comprised of fine-grained alumosilicates of mostly monocline or tricline symmetry that may contain sodium, calcium, potassium, and other ions. The basic crystalline structure of clay minerals consists of two main structural units. The first unit are layers of (Si, Al) O_4 – tetraeders, which are bonded over the oxygen atoms in one plain, and the second structural element consists of octaeders, in which the central ion (mostly aluminium, but also Fe^{3+}, Fe^{2+} or Mg^{2+}) is surrounded by OH^- - ions. The clay crystal lattice is formed by layers of these tetraeders (T) and octaeders (O). According to the succession of these layers, there are two-layer minerals, whose structures are formed by a regular series of tetraeder and octaeder layers (TOTOT....). Another possibility is the formation of three-layer minerals, where one silicate layer has the structure TOT TOT TOT (Fig. 3). The most important clay mineral groups, which are common in soils, are kaolins, smectites and illites. The kaolin minerals belong to the two-layer clay minerals. They are characterized by the approximate chemical composition 2 $H_2O \cdot Al_2O_3 \cdot 2\ SiO_2$. The most common kaolin mineral is kaolinite, which consists of a single silica tetrahedral sheet and a single alumina octahedral sheet, which form the kaolin unit layer. Smectites belong to the three-layer minerals and are composed of units consisting of two silica tetrahedral sheets with a central alumina octahedral sheet. A widely-used smectite is montmorillonite, which occurs naturally as Na-montmorillonite and Ca-montmorillonite [19]. As the lattice has an unbalanced charge because of isomorphic substitution of alumina for silica in the tetrahedral sheet, and of iron and magnesium for alumina in the octahedral sheet, the attractive force between the unit layers in the stacks is weak and cations and polar molecules are able to enter between the layers and cause the layers to expand. Illite is a more general term used for mica like clay, whose basic structural unit is similar to that of montmorillonite.

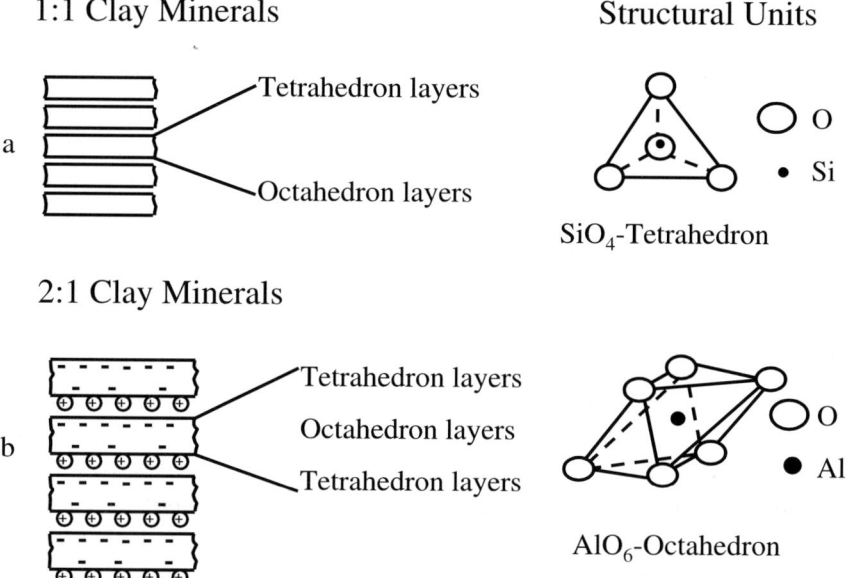

Fig. 3. Layer structures of two-layer minerals (1:1 clay minerals) (a) and of three-layer minerals (2:1 clay minerals) (b); T, O = tetraeder resp. octaeder layer; (reprinted with permission from Ref. 20).

As there is a large substitution of silica for alumina in the tetrahedral sheet illites usually are characterized by a charge deficiency, which is balanced by potassium ions, which bridge the unit layers. As a consequence, illites are non-expandable clay minerals. There are also other clay mineral groups like chlorites and the mixed-layer clays which consist of mixtures of the unit layers on a layer-by-layer basis, e.g. illite-smectite, smectite-chlorite, illite-chlorite and many others [21].

The attraction between silicate layers of three-layer minerals is formed by the socalled interlayer cations. Two-layer minerals like kaolinite or halloysite have no additional ions between their silicate layers. The silicate layers bear an electric charge because of isomorphic substitution. This fact leads to a permanent negative charge excess of the silicate ions because the charge of the surrounding structure of oxygen and hydroxyl ions remains unchanged. Often Al^{3+} is incorporated into the tetraeders instead of Si^{4+} and Fe/Zn^{2+} is incorporated into the octaeders. Additional cations (mostly K^+, Na^+, Ca^{2+}, and Mg^{2+}) are intercalated between the silicate layers to compensate this negative layer charge. These cations are called interlayer cations. The charge on the inner surfaces of the swellable three-layer minerals is caused by the substitution of Al ions in the tetraeder layers and is always negative. It is completely compensated by the exchangeable interlayer cations. Internal surfaces of montmorillonites can go as high as 97% of the total area [22]. There are also negative charges on the outer surface, which are caused by the alumosilicate layers. Their charge density is varying according to the different substitution for the different clay mineral types. The negative charges generated by substitution are independent from the surrounding milieu and are therefore called permanent charges.

There are also variable charges which depend on the pH. These charges are caused by the amphoteric properties of some functional groups like hydroxyl groups on the sides and edges of the clay minerals. These groups can be charged positively or negatively according to the pH of the surrounding solution. Hydroxyl groups tend to dissolve protons at higher pH, while they absorb protons in acid pH. Therefore such surfaces usually bear positive charges at low pH and negative values at higher pH.

2.4.3. Oxides and Hydroxides

Oxides, hydroxides, and oxyhydroxides of Fe, Mn, and Al play an important role in the chemical behaviour of soils. They display large surface areas, very small particle sizes $< 2\mu m$, and occur as coatings of other soil particles, as pore fillings, and as concretions. The most common Fe and Al minerals are aluminium oxide ($Al_2O_3 \cdot 3H_2O$, bauxite) and iron hydroxide (FeOOH, goethite), which remain as the residual decomposition product, laterite. Bauxite is the principal ore of Al remains. Oxides and hydroxides precipitate from soil solutions, sometimes with the help of specialized bacteria such as *Thiobacillus ferrooxidans* and *Metallogenum sp*. During precipitation, metal cations such as Co, Cr, Cu, Mn, Ni, V, and Zn, as well as HPO_3^{4-} and AsO_4^{3-} can be coprecipitated. If chemical soil conditions such as pH and redox conditions change, oxides and hydroxides can be resolved, and heavy metal ions adsorbed by these substances can be released [5]. Oxides and hydroxides usually occur in combination with humic substances and clay minerals.

3. HEAVY METALS IN WATER AND GROUNDWATER

3.1. Surface Waters

Water chemistry of surface waters such as streams, rivers, springs, ponds, and lakes, is greatly influenced by the kind of soil and rock the water flows on or flows through. The main physical, chemical, and biological parameters, which influence water composition, are temperature, pH, redox potential, adsorption and desorption processes from inorganic or organic suspended matter or bottom sediments, cation exchange, dilution, evaporation, and organisms present.

For example, water flowing over limestone ($CaCO_3$) will develop a pH of about 8, while water flowing through granite, which consists mainly of quartz (SiO_2) and feldspars, will develop a more acid pH of about 6. If pyrite (FeS_2) is present, oxidation of the mineral will cause the generation of acid waters, which can affect heavy metal solubility and may lead to an increased mobility of those metals [23]. A drop of pH to 5 or lower is reported to cause serious problems for the aquatic ecosystems [24]. The mobilized heavy metals are dispersed downstream and immobilized by adsorption onto clay minerals and Fe and Mn oxyhydroxides or absorbed onto algae at a lower trophic level in the food web. These heavy metals may accumulate to critical levels in the food web and will cause damages to organisms on a higher trophic level [1].

3.2. Groundwater

Groundwater is a very important direct source of drinking water from wells drilled into the aquifer and for water used for agricultural purposes, mainly irrigation. Fig. 4 explains some of the basic hydrogeological terms connected with groundwater [25]. A rock body, which contains pores and is able to transport groundwater, is called aquifer. As all pores in the aquifer are filled with water, the term "saturated zone" is also used in contrast to the unsaturated zone beneath the surface soil, where the pores are also filled with soil air. The aquifer is limited to the bottom by a relatively impermeable stratum, which is called aquitard. Sometimes, the term "aquifuge" is also used. Water moves vertically in the unsaturated zone, but in the aquifer, it moves horizontally along a natural gradient.

Permeability is a rock specific constant, which is defined according to Darcy's Law:

$$Q = k_f \cdot grad\ I \cdot A \tag{1}$$

$$v_f = k_f \cdot grad\ I \tag{2}$$

with: Q: quantity [m³/s], I: hydraulic gradient, A: area of flow [m²], v_f: filter velocity [m/s], k_f: permeability coefficient [m/s]. Maintaining groundwater quality is of utmost importance for human society, as tainted groundwater contaminates both terrestrial and aquatic food chains, and is a direct threat to human health.

Basic Hydrogeological Terms

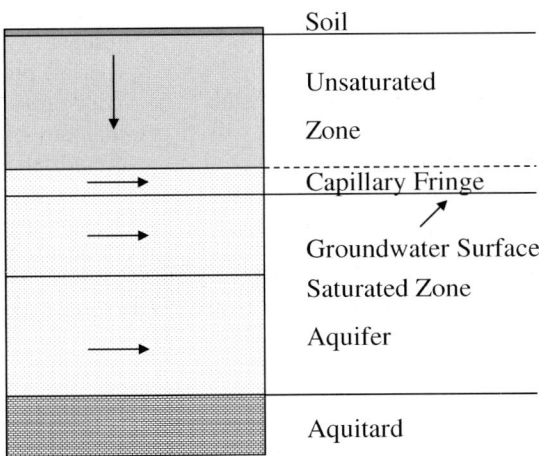

Fig. 4. Some hydrogeological terms connected with groundwater. Black arrows denote water flow direction.

The main organic and inorganic groundwater contaminants and their sources are shown in Table 4. Heavy metals are mainly introduced into groundwater by agricultural and industrial activities, landfilling, mining, and transportation. There are various possibilities for the fate and transport of heavy metals in soil and groundwater (Fig. 5).

Solved metal ions Me^{n+} can be taken up by plants, can be sorbed unto mineral phases, or can be bound unto particulate organic matter via complexation/sorption mechanisms. These colloids are very mobile in soil and groundwater systems and will thus increase heavy metal mobility. Solved metal ions Me^{n+} can also be precipitated to form large immobile crystals. Dissolved organic matter also plays an important role for heavy metal mobility.

Table 4
Typical sources of inorganic and organic substances for groundwater contamination

Source	Inorganic contaminants	Organic Contaminants
Agricultural areas	Heavy metals, salts (Cl^-, NO_3^-, SO_4^{2-})	Pesticides
Urban areas	Heavy metals (Pb, Cd, Zn) salts	Oil (petrol) products, biodegradable organics
Industrial sites	Heavy metals, metalloids, Salts	Polycyclic aromatic hydrocarbons (PAH), chlorinated hydrocarbons (trichloroethylene and tetrachloroethylene), hydrocarbons (benzene, toluene, xylene), oil (petrol) products
Landfills	Salts (Cl^-, NH_4^+), heavy metals	Biodegradable organics and xenobiotics
Mining disposal Sites	Heavy metals, metalloids, salts	Xenobiotics
Dredged sediments	Heavy metals, metalloids	Xenobiotics
Hazardous waste sites	Heavy metals, metalloids	Concentrated xenobiotics
Leaking storage Tanks	-	Oil products
Line sources (motorways, railways, sewerage systems, etc.)	Heavy metals (Cd, V, Pb) salts	PAHs, oil products, pesticides

Modified after Ref. 2.

Soil and dissolved organic matter may incorporate heavy metals via sorption processes. The resulting soil-DOC-Me complexes are subject to leaching and transport into the groundwater. At the soil surface, heavy metals may be released into the atmosphere by surface erosion and colloid loss.

4. HEAVY METALS IN THE ATMOSPHERE

Heavy metals are carried in the atmosphere as gases, aerosols, and particulates. Sources of heavy metals are mineral dusts, sea salt particles, extraterrestrial matter, volcanic aerosols, forest fires, and industrial sources such as emissions from transportation, coal combustion, and fugitive particulate emissions [26]. There are two major types of aerosols. Primary aerosols are directly emitted into the atmosphere from the earth's surface, while secondary aerosols are formed by chemical reactions in the atmosphere, which involve gases, pre-existing aerosols, and water vapour [27]. Table 5 lists estimated emissions of Cd, Pb, Cu, and Zn into the atmosphere. Ref. [28] provides a comprehensive study on origin and chronological development of atmospheric global emissions and the global cycles of heavy metals.

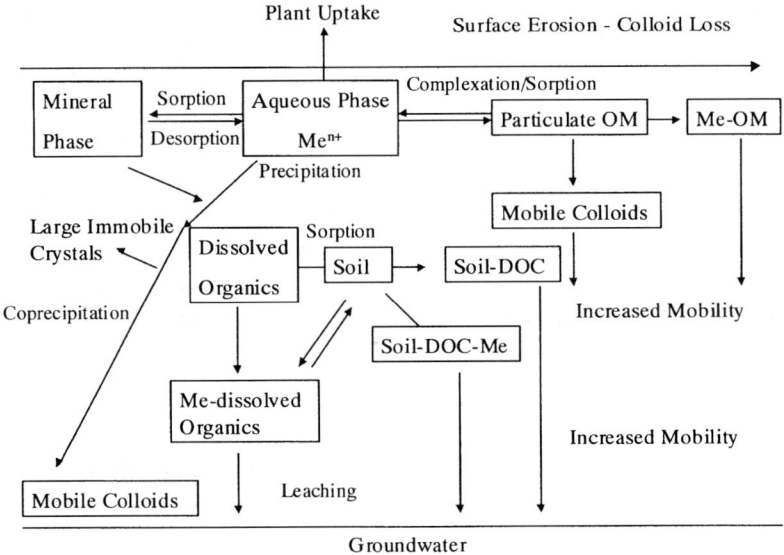

Fig. 5. Transport mechanisms of heavy metals in soil and groundwater (modified after Ref. 2).

Increase in cadmium atmospheric emissions since the year 1900 is caused primarily by ore processing and waste incineration, while lead emissions reflect the effect of volatilization of gasoline additives. Nickel is a natural component of oil and is released into the atmosphere during the combustion process (Fig. 6). Typical enrichment products in fly ash, which were originally bound in sulfide minerals occurring in coal and some ores, are molybdenum, copper, zinc, cadmium, and lead as well as volatile phases such as arsenic, selenium, antimony, and mercury [29, 30].

The substances released into the air (emissions) are spread (transmission) and effect humans, animals, and plants (immission). The term "emission" is used for all solid, liquid, and gaseous pollutants, which are released into the air [31]. The term "immission" refers to the release of solid, liquid, and gaseous pollutants, which permanently or temporarily remain close to the earth's surface. The term "transmission" denotes all "processes, during which the spatial location and the distribution of pollutants in the atmosphere changes because of the forces of movement or due to additional physical or chemical effects [31]. Transport and dispersion models can be used to estimate immissions, which are based upon meteorological transmission models. The pollutants are released at the source and are then transported by prevailing air currents. During transport, they can be diluted, precipitated, or transformed by chemical reactions on their way to the immission location. However, immission calculations can be very uncertain, as many parameters are often not known exactly or vary strongly with time and location. Volatile metalloids such as Se, Hg, As, and Sb, can be transported both in gaseous form or enriched in particles. Other metals such as Cu, Pb, and Zn, are only transported as particles. Atmospheric deposition is an important source of metals in plants and soils. As for Pb, more than 90% of the total plant uptake can be attributed to atmospheric deposition [2] as well as for Cd, where atmospheric deposition has been reported to contribute to more than 50 % of the plant uptake of this heavy metal [32, 33].

Table 5
Estimated emissions (in 10^3 tons/year) of Cd, Pb, Cu, and Zn into the atmosphere

Element	Atmospheric Emissions	
	Natural	Anthropogenic
Cadmium	0.29	5.5
Lead	4	400
Copper	19	260
Zinc	36	840

Modified after Ref. 31.

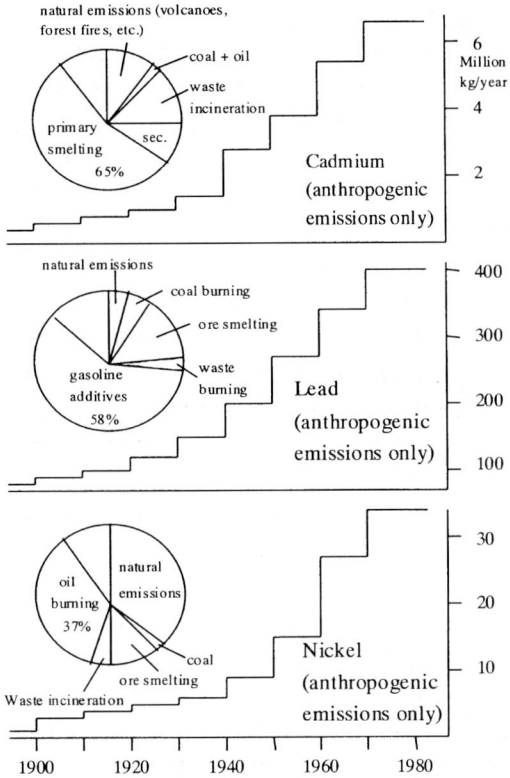

Fig. 6. Origin and chronological development of atmospheric emissions of cadmium, lead, and nickel (redrawn after Ref. 28).

5. ANTHROPOGENIC SOURCES OF HEAVY METALS

Heavy metals are released into the environment by many human activities. They are also used in a large variety of industrial products, which in the long term have to be deposited as waste. Heavy metal release into the environment occurs at the beginning of the production chain, whenever ores are mined, during the use of products containing them, and also at the end of the production chain. Table 6 gives an overview on the multiple uses and products, which contain heavy metals. The natural sources are dominated by parent rocks and metallic minerals, while the main anthropogenic sources are agricultural activities, where fertilizers, animal manures, and pesticides containing heavy metals are widely used, metallurgical activities, which include mining, smelting, metal finishing, and others, energy production and transportation, microelectronic products, and finally waste disposal. Heavy metals can be released into the environment in gaseous, particulate, aqueous, or solid form and emanate from both diffuse or point sources.

5.1. Agricultural Activities

The ever growing world population requires intensive land use for the production of food, which includes repeated and heavy input of fertilizers, pesticides, and soil amendments. Fertilizers are added to the soil in order to provide additional nutrients to crops or by changing soil conditions such as pH to make nutrients more bioavailable. Pesticides are used to protect crops. Soil amendments are often derived from sewage sludge, animal manure, and dredged sediments from harbours and rivers. Heavy metals from these sediments can be mobilized during dredging [34] as the reducing environment changes to oxidizing condition, thus remobilizing heavy metals. As soil, surface, and groundwaters are closely interconnected systems, metals introduced into soils can also affect aquifers or surface waters by infiltration. Also irrigation may trigger release of heavy metals.

An example of heavy metal release by the use of aquifer water for irrigation occurred in West Bengal, India and East Bangladesh. The local water level had been lowered due to increased water use by wells and strata containing pyrite (FeS_2) were exposed to oxidation. In pyrite, As substitutes in part for Fe, and this metal has been released into the aquifer by oxidation and decomposition of this mineral. In the aquifer, the toxic arsenite complex was formed, which heavily affects the local population [35-39].

5.1.1. Phosphatic Fertilizers

Phosphatic fertilizers contain various amounts of Zn, Cd, and other heavy metals depending from which parent rock the fertilizer has been produced. Those made from sedimentary rocks tend to have high levels of Cd, while those made from magmatic rocks have only small Cd concentrations [2]. The differences in heavy metal content are caused by impurities coprecipitated with the phosphates. Therefore, Cd input into agricultural soils varies considerably according to the Cd concentration of the fertilizer used [40, 41].

5.1.2. Pesticides

Pesticides are used for insect and disease control in high-production agriculture and can be applied as seed treatment, by spraying, dusting, or by soil application. Although metal-based pesticide are no longer in use, their former application lead to increased accumulation of heavy metals, especially of Hg from methyl mercurials, of As, and of Pb from lead arsenate into soils and groundwater. Table 7 lists some pesticide containing metals recommended in Ontario, Canada, from period 1892-1975 [42].

5.1.3. Sewage Effluents

Land application of waste water is widely used in industrialized countries for the last 50 to 100 years [43].

Table 6
Anthropogenic sources and uses of heavy metals, through which they can be introduced into the environment

As:	Additive to animal feed, wood preservative (copper chrome arsenate), special glasses, ceramics, pesticides, insecticides, herbicides, fungicides, rodenticides, algicides, sheep dip, electronic components (gaalium arsenate semiconductors, integrated circuits, diodes, infra-red detectors, laser technology), non-ferrous smelters, metallurgy, coal-fired and geothermal electrical generation, texile and tanning, pigments and anti-fouling paints, light filters, fireworks, veterinary medicine.
Be:	Alloy (with Cu), electrical insulators in power transistors, moderator or neutron deflectors in nuclear reactors
Cd:	Ni/Cd batteries, pigments, anti-corrosive metal coatings, plastic stabilizers, alloys, coal combustion, neutron absorbers in nuclear reactors
Co:	Metallurgy (in superalloys), ceramics, glasses, paints
Cr:	Manufacturing of ferro-alloys (special steels), plating, pigments, textiles and leather tanning, passivation of corrosion of cooling circuits, wood treatment, audio, video, and data storage
Cu:	good conductor of heat and electricity, water pipes, roofing, kitchenware, chemicals and pharmaceutical equipment, pigments, alloys
Fe:	Cast iron, wrought iron, steel, alloys, construction, transportation, machine-manufacturing
Hg:	Extracting of metals by amalgamation, mobile cathode in the chloralkali cell for the production of NaCl and Cl_2 from brine, electrical and measuring apparatus, fungicides, catalysts, pharmaceuticals, dental fillings, scientific instruments, rectifiers, oscillators, electrodes, mercury vapour lamps, X-Ray tubes, solders
Mn:	Production of ferromanganese steels, electrolytic manganese dioxide for Use in batteries, alloys, catalysts, fungicides, antiknock agents, pigments, Dryers, wood preservatives, coating welding rods

Table 6 (continued)
Anthropogenic sources and uses of heavy metals, through which they can be introduced into the environment

Mo:	Alloying element in steel, cast irons, non-ferrous metals, catalysts, dyes, lubricants, corrosion inhibitors, flame retardants, smoke repressants, electroplating
Ni:	As an alloy in the stell industry, electroplating, Ni/Cd batteries, arc-welding, rods, pigments for paints and ceramics, surgical and dental protheses, molds for ceramic and glass containers, computer components, catalysts
Pb:	Antiknock agents, tetramethyllead, lead-acid batteries, pigments, glassware, ceramics, plastic, in alloys, sheets, cable sheathings, solder, ordinance, pipes or tubing
Sb:	Type-metal alloy (with lead to prevent corrosion), in electrical applications, Britannia metal, pewter, Queen's metal, Sterline, in primers and tracer cells in munition manufacture, semiconductors, flameproof pigments and glass, medicines for parasitic diseases, as a nauseant, as an expectorant, combustion of fossil fuels
Se:	In the glass industry, semiconductors, thermoelements, photoelectric and photo cells, and xerographic materials, inorganic pigments, rubber production, stainless steel, lubricants, dandruff treatment
Sn:	Tin-plated steel, brasses, bronzes, pewter, dental amalgam, stabilizers, catalysts, pesticides
Ti:	For white pigments (TiO_2), as a UV-filtering agents (suncream), nucleation Agent for glass ceramics, as Ti alloy in aeronautics
Tl:	Used for alloys (with Pb, Ag, or Au) with special properties, in the electronics industry, for infrared optical systems, as a catalyst, deep temperature thermometers, low melting glasses, semiconductors, supraconductors
V:	Steel production, in alloys, catalyst
Zn:	Zinc alloys (bronze, brass), anti-corrosion coating, batteries, cans, PVC Stabilizers, precipitating Au from cyanide solution, in medicines and chemicals, rubber industry, paints, soldering and welding fluxes

Modified after Ref. 1.

There are some benefits of using the water and nutrients in sewage effluents such as recycling of nutrients, restoring of groundwater, preventing stream pollution, and cut down of commercial fertilizers. Nevertheless, several factors such as the transportation costs, the land use, the soil type, etc. have to be taken into account. Several reviews of hazards from heavy metal concentration in waste water have been conducted and phytotoxic symptoms when using waste water containing Cd, Zn, Cu, Ni, Pb, and especially B have been observed [44, 45]. Nevertheless, long-term observations in Germany indicate that even after use of sewage effluents since 1895, heavy metal concentrations in soil is still within tolerable limits [46].

Table 7
Pesticide-containing metals recommended in Ontario, Canada, from period 1892-1975

Chemical	Metal composition of product	Period of recommendation	Crops
	Insecticides		
Copper aceto-arsenite (Paris green)	2.3 % As, 39 % Cu	1895-1920 1895-1957	Apples, cherries Vegetables, small fruit
Calcium arsenate	0.8 – 26 % As	1910-1953	Fruit, vegetables
Lead arsenate	4.2 – 9.1 % As	1910-1975	Apples
	11-26 % Pb	1910-1971	Cherries
Mercuric chloride	6 % Hg	1932-1954	Cruciferous crops
Zinc sulfate	20-30 % Zn	1939-1955	Peaches
	Fungicides		
Copper sulfate-calcium salts	4-6 % Cu	1892-1975	Fruit and vegetables
Fixed copper salts	2-56 % Cu	1940-1975	Fruit and vegetables
Maneb	1-17 %	1947-1975	Fruit and vegetables
Mancozeb	16 % Mn, 2 % Zn	1966-1975	Fruit and vegetables
Methyl and phenyl mercuric salts	0.6-6 % Hg	1932-1972	Seed treatment
Phenyl mercuric acetate	6 % Hg	1954-1973	Apples
Zineb and ziram	1-18 % Zn	1957-1975	Vegetables
	Topkiller		
Calcium arsenite	30 % As	1930-1972	Vegetables
Sodium arsenite	26 % As	1920-1972	Vegetables

Modified after Ref. 2.

5.1.4. Biosolids

The term "biosolids" refers to sewage sludge, animal wastes, municipal solid waste, and some industrial wastes such as paper pulp sludge [2]. These biosolids are used for soil enhancement due to their increased content in nutrients and organic matter (OM), which is favourable for soil tilth, pore space, aeration, and water retention capacity. They contain nutrients, heavy metals, and pathogens such as *Escherichia coli*. The main heavy metals of concern in sewage sludge are Cd, Zn, Cu, Pb, Se, Mo, Hg, Cr, As, and Ni. Concentrations of these elements depend on the type and amount of discharges into the sewage treatment system and on the amount added in the conveyance and treatment system.

A comprehensive study on sludge application to croplands is presented in Ref. 47. The main results of this study are that metals stemming from sewage sludge generally remain in a narrow zone of application of 0-

15 cm depth. Mostly, sludge application exhibited a positive effect on plant growth, and phytotoxicity was only rarely observed. The most bioavailable sludge-borne metal was Zn, followed by Cd and Ni. Cr and Pb uptake by plants was observed to be insignificant. When harmful effects to certain plants (legumes) occurred, they were attributed to detrimental influence of heavy metals on microbial soil activity, e.g. in N_2 fixation.

Animal wastes and manure pose a major environmental concern due to their content of pathogens, public nuisance (flies, malodour, etc.), and their potential to contaminate surface water and groundwater. The concentration of heavy metals in animal wastes depends on a variety of factors such as class of animal (cattle, swine, poultry, etc.), age of the animals, type of ration, housing type, and waste management practice [2]. Heavy metals such as Cu, Co, and Zn originate from rations and dietary supplements fed to the animals. Although animal wastes are usually rather low in heavy metal content, input of excess N and salts as well as nutrient imbalance in plants poses a problem [48, 49].

5.2. Industrial Activities

The most important industrial activities, by which heavy metals are introduced into the environment, are mining, coal combustion, effluent streams, and waste disposal. In the past, only small attention has been paid to prevent introduction of these toxic and hazardous substances into the environment. In the meantime, loading of ecosystems with heavy metals has lessened considerably in many countries due to enhanced legislation concerning capture and treatment of pollutants. Unfortunately, this is not a global effect.

5.2.1. Mining

Most metals occurring in ore deposits have only low concentration. During the extraction process, large amounts of waste rock are produced, which still contain traces of heavy metals that have not been picked out of the ore-bearing rock. The waste rock is usually disposed of in mine tailings or rock spoils. In the case of pyrite, this mineral will weather in the tailing due to oxidizing environmental conditions and thus create acid mine drainage. The acid conditions also mobilize heavy metals form the waste rock. This mobilization can cause fatal environmental and health problems through respiration, drinking and cooking contaminated water, and eating food grown on soils influenced by irrigation. Numerous examples are known especially for the heavy metals As, Cd, Cu, Hg, and Pb.

A good example for the grave consequences of mining is the extraction of gold from placer deposits using an Hg amalgam. Hg is released into the river water by flushing and into the atmosphere by burning resulting in severe bioaccumulation in the food web. This has been observed especially in the Amazon where Hg release and disposal are uncontrolled [50, 51]. In Romania, cyanide from the heap leach extraction of Au contaminated the Tiza river and killed all life forms. About 400 km of the river were affected. If the polluted sediments were dredged, there is a grave risk of mobilizing even more heavy metals from the sediment into the water.

5.2.2. Coal and Petroleum Combustion

The combustion of fossil fuel contributes heavily to the release of heavy metals in the environment, especially into the atmosphere. Coal combustion is used to generate electric power in coal-burning power plants. Fly ash and flue gas desulfurization residues amount to more than 106 million tons in the US yearly [2]. Only one third of the fly ash is recycled by using it for cement making, concrete mixing, ceramics, and others [52]. The kind and concentration of heavy metals in coal residues depends on parameters such as composition of the parent coal, conditions during combustion, efficiency of emission control devices, storage and handling of the by-products, and climate. Notable heavy metals in coal residues are As, Cd, Mo, Se, and Zn, especially compared to their mobilization due to natural weathering. Table 8 lists the typical concentration of heavy metals in fly ash. Heavy metals in coal residues that are of special concern are As, Mo, and Se. Ref. 53 is a comprehensive review on coal residues and their characterization, potential for utilization, and potential hazards to plants and animals. As reported earlier, large amounts of Pb stemming from gasoline additives are released into the atmosphere from fuel burning for transportation. Also the use of fossil fuels and wood for home heating and cooking attributes to the increased introduction of heavy metals from combustion. In the light of a growing human population and an increased need for energy, release of heavy metals from these activities can be expected to increase in the future.

5.2.3. Indoor and Urban Environments

Mega cities and large urban areas are constantly growing all over the world. These areas are characterized by very high population densities, excess energy consumption, and extended industrial and transportation activities often combined with very low standards for housing, water and energy supply as well as waste collection and deposition. Extreme environmental pollution results from these factors.

Indoor air is mostly polluted with volatile organic compounds (VOCs) and gases from fabrics, floor coverings, dyes, pesticides, refrigerants, and heating and cooking fuels. Leaded pipes and paints containing lead add to elevated Pb levels in humans in urban areas [53]. The main input of Pb comes from house dust and potable water. The Pb enters the dust from automobile exhaust emissions and resuspension from floor coverings and carpets [2]. Automobile exhaust is responsible for the contamination of roadside soils and vegetation with heavy metals, which include Pb, Zn, Cd, Cu, and Ni [54] as well as platinum group elements (PGE) stemming from catalytic converters [55]. Contaminated zones may extend up to several hundred meters from the road depending on traffic intensity and location. Ref. 56 surveyed the concentration of As, Cu, Ni, Pb, Sn, and Zn in sediments from the East and Gulf of Mexico coasts of the United States. High levels of these metals are found at a variety of sites located near big cities. The use of waste incinerators for the thermal treatment of solid waste seems to be one of the main sources of atmospheric heavy metals such as Pb and Hg in many cities [57].

Table 8
Typical concentration of heavy metals in fly and bottom ash

Element	Fly ash		Bottom ash	
	Mean	Range	Mean	Range
Al	113,000	46,000-152,000	101,000	30,500-145,000
As	156.2	7.7-1385	7.6	<5-36.5
Ba	1880	241-10,850	1565	150-9360
Cd	11.7	6.4-16.9	<5	<5
Cr	247.3	37-651	585	<40-4710
Fe	76,000	25,000-177,000	105,000	20,200-201,000
Mn	357	44-1332	426	56-1940
Mo	43.6	7.1-236	14.4	2.8-443
Ni	141	22.8-353	216	<10-1067
Pb	170.6	21.1-2120	46.7	4.6-843
Sb	42.5	11-131	<10	<10
Se	14.0	5.5-46.9	4.1	<1.5-9.96
Sn	43.6	7.9-56.4	28.2	<9-90.2
Ti	6644	1310-10,100	5936	1540-11,300
V	271.5	<95-652	176	<50-275
Zn	449.2	27-2880	127	3.8-515

Modified after Ref. 2. The total numbers of observation were 39 for fly ash and 40 for bottom ash, respectively.

5.2.4. Solid Waste Disposal

Solid wastes are produced worldwide in immense amounts of thousands of millions of tons annually. The most important sources of heavy metals stem from wastes from industrial activities, especially energy generation, from mining, agricultural activities (animal manure), and domestic waste (e.g. batteries, tires, appliances, junked automobiles). These wastes are often disposed of without proper treatment at waste disposal sites, which do not meet the requirements necessary for a secure deposition. Over the past twenty years, some countries have emphasized legislation for those requirements. A site for safe waste deposition should meet some basis requirements. First, the site itself should be selected according to its geological and hydrogeological characteristics (low permeability underground such as clay, large distance to groundwater levels, no karst, no earthquakes or volcanic activities, no mass movements, etc.). Second, the site has to be equipped with barrier systems both on the base and the top of the deposit in order to prevent spreading of contaminants from the waste and the leachate into the environment. Third, leachate and gas collection systems should allow for collection and transfer of gas and leachate to treatment plants for heavy metal removal. Finally, the site should be constantly monitored by air sampling devices and wells sunk outside its periphery.

What is actually done in many countries often does not fulfil any of the abovementioned requirements. Wastes are deposited at sites where they come in contact with groundwater or are burned in open fires. Animal manure is often used as soil conditioner or fertilizer or for methane gas production. Once in the soil, the heavy metals bioaccumulate in the food web and pose a threat to consumers. The deposition of fly ash from fossil fuel combustion and the acid mine tailings from heavy metal bearing rocks mining leads to the mobilisation of heavy metals into surface waters, soils, and ground water [2]. Recycling activities for fly ash and treatment of acid mine drainage would be the ideal way of preventing heavy metal release into the environment, but such a protocol has not commonly been followed in the past.

REFERENCES

[1] F.R. Siegel, Environmental Geochemistry of Potentially Toxic Metals, Springer, Berlin, Heidelberg, 2002.
[2] D.C. Adriano, Trace Elements in Terrestrial Environments, Springer, Berlin, Heidelberg, 2001.

[3] F. Press and R. Sievers, Fundamentals of Geology, Freeman and Company, New York, 1994.
[4] R.L. Mitchell, in "Chemistry of the Soil", (F.E. Bear, ed.), p. 320-368, New York, 1964.
[5] B.J. Alloway, Heavy Metals in Soils, Blackie Academic, Glasgow, 1995.
[6] H. Jenny, The Factors of Soil Formation, McGraw-Hill, New York, 1941.
[7] J.I. Drever, The Geochemistry of Natural Waters, Prentice Hall, New Jersey, 1997.
[8] K.W. Kim and I. Thornton, Environ. Geochem. Health, 15 (1993) 119.
[9] H.J.M. Bowen, Environmental Chemistry of the Elements, Academic Press, London, 1979.
[10] P. Duchaufour, Pedology, Allen and Unwin, London, 1977.
[11] H.B. Bradl, in "Encyclopedia of Surface and Colloid Science" (A. Hubbard, ed.), p. 373-384, Marcel Dekker, New York, 2002.
[12] S.M. McLennan and R.W. Murray, in "Encyclopedia of Geochemistry", (C.P. Marshall and R.W. Fairbridge, eds.), p. 282-292, Kluwer Academic Publishers, Dordrecht, 1999.
[13] S.R. Taylor and S.M. McLennan, Rev. Geophys., 33 (1995) 241.
[14] K.K. Turekian and K.H. Wedepohl, Geol. Soc. Amer. Bull., 72 (1961) 175.
[15] A.P. Vinogradov, Geochemistry, (1962) 641.
[16] J.J. Connor and H.T. Shacklette, Background Geochemistry of some Soils, Plants, and Vegetation in the Conterminous United States, U.S. Geol. Survey Prof. Paper 574-F, Washington, D.C., 1975.
[17] H.T. Shacklette and J.G. Boerngen, Element Concentrations in Soils and other Surficial Materials of Conterminous United States: An Account of the Concentrations of 50 Chemical Elements in Samples of Soils and Other Regoliths, U.S. Geol. Survey Prof. Paper 1270, Washington, D.C., 1984.
[18] A.H. Brownlow, Geochemistry, Prentice Hall, New Jersey, 1996.
[19] R.E. Grim, Clay Mineralogy, McGraw-Hill, New York, 1953.
[20] K. Jasmund and G. Lagaly, Tonminerale und Tone - Struktur, Eigenschaften, Anwendung und Einsatz in Industrie und Umwelt, Steinkopff, Darmstadt, 1993.
[21] R.E. Grim, Am. Mineral., 22 (1937) 813.
[22] O. Altin, H.Ö. Özbelge and T. Dogu, J. Coll. Interf. Sci., 217 (1999) 19.
[23] F.R. Siegel, J. Geochem. Explor., 38 (1990) 265.
[24] J.W. Christensen, Global Science, Kendall/Hunt Publishers, Dubuque, Iowa, 1991.
[25] S. Lohman, Definitions of Selected Groundwater Terms – Revisions and Conceptual Refinements, U.S. Geol. Survey Water Supply Paper 1988, Washington, D.C., 1972.
[26] I. Colbeck, in "Airborne Particulate Matter" (T. Kouimtzis and C. Samara, eds.), p. 1-33, Springer, Berlin, Heidelberg, New York, 1995.
[27] H. Sievering, in "Encyclopedia of Environmental Science" (D.E. Alexander and R.W. Fairbridge, eds.), p. 9-10, Kluwer Academic Publishers, Dordrecht, 1999.
[28] J.O. Nriagu, Nature, 279 (1979) 409.
[29] D.H. Klein, Environ. Sci. Technol., 10 (1975) 973.
[30] U. Förstner, in „The Importance of Chemical Speciation in Environmental Processes" (M. Bernhard, F.E. Brinckmann and P.J. Sadler, eds.), p. 465-491, Springer, Berlin, Heidelberg, New York, 1986.

[31] U. Förstner, Integrated Pollution Control, Springer, Berlin, Heidelberg, New York, 1995.
[32] R.M. Harrison and M.B. Chirgawi, Sci. Total Environ., 83 (1989) 13.
[33] R.M. Harrison and M.B. Chirgawi, Sci. Total Environ., 83 (1989) 35.
[34] D.A. Darby, D.D. Adams and W.T. Nivens, in "Sediment and Water Interaction" (P.G. Sly, ed.), p. 343-351, Springer, Berlin, Heidelberg, New York, 1986.
[35] P. Bagla and J. Kaiser, Science, 274 (1996) 174.
[36] D. Dipankar, G. Samanta, B.K. Mandal, T.R. Chowdhury, C.R. Chanda, P.P. Chowdhury, G.K. Basu and D. Chakraborti, Environ. Geochem. Health, 18 (1996) 5.
[37] R.T. Nickson, J.M. McArthur, W.G. Burgess, K.M. Ahmed, P. Ravenscroft and K.M. Rahman, Nature, 395 (1998) 338.
[38] R.T. Nickson, J.M. McArthur, P. Ravenscroft, W.G. Burgess and K.M. Rahman, Appl. Geochem., 15 (2000) 403.
[39] M.R. Islam, R. Salminen and P.W. Lahermo, Environ. Geochem. Health, 22 (1996) 33.
[40] M. Hutton and C. Symon, Sci. Total Environ., 57 (1986) 129.
[41] H.P. Rothbaum, R.L. Goquel, A.E. Johnston and G.E.G. Mattingly, J. Soil Sci., 37 (1986) 99.
[42] R. Frank, K. Ishida and P. Suda, Can. J. Soil Sci., 56 (1976) 181.
[43]) C.W. Carlson and J.D. Menzies, BioScience, 21 (1971) 561.
[44] H. Bouwer and R.L. Chaney, Adv. Agron., 26 (1975) 133.
[45] D.G. Neary, G. Schneider and D.P. White, Soil Sci. Proc. Am. J., 39 (1975) 981.
[46] N. El-Bassam, C. Tietjen and J. Esser, Management and Control of Heavy Metals in the Environment, CEP Consultants, Edinburgh, 1979.
[47] D.C. Adriano, Biogeochemistry of Trace Metals, Lewis Publishers, Boca Raton, 1992.
[48] D.C. Adriano, P.F. Pratt and S.E. Bishop, Soil Sci. Soc. Am. Proc., 35 (1971) 759.
[49] D.C. Adriano, A.C. Chang, P.F. Pratt and R. Sharpless, J. Environ. Qual., 3 (1973) 396.
[50] D. Cleary, I. Thornton, N. Brown, G. Kazantis, T. Delves and S. Washinton, Nature 369 (1994) 613.
[51] D. Lacerda, O. Malm, J.R.D. Guimaraes, W. Salomons and R.-D. Wilken, in "Biogeodynamics of Pollutants in Soils and Sediments" (W. Salomons and W.M. Stigliani, eds.), p. 213-245, Springer, Berlin, 1995.
[52] C. Carlson and D.C. Adriano, J. Environ. Qual., 22 (1993) 227.
[53] P. Bullock and P.J. Gregory, Soils in the Urban Environment, Blackwell, London, 1991.
[54] J.V. Lagerwerff and A.W. Specht, Environ. Sci. Technol., 4 (1970) 583.
[55] K. Ravindra, L. Bencs and R. Van Grieken, Sci. Total Environ., 318 (2004) 1.
[56] K.O. Daskalakis and T.P. O´Connor, Environ. Sci. Technol., 29 (1995) 470.
[57] S.N. Chillrud, R.F. Bopp, J.M. Ross and A. Yarme, Environ. Toxicol. Chem., 33 (1999) 657.

Chapter 2

Interactions of Heavy Metals

H. Bradl[a], C. Kim[b], U. Kramar[c], and D. Stüben[c]

[a]Department of Environmental Engineering, University of Applied Sciences Trier, Umwelt-Campus Birkenfeld, P.O. Box 301380, D-55761 Birkenfeld, Germany
[b]Department of Physical Sciences, Chapman University, Orange, CA 92866, USA
[c]Institute for Mineralogy and Geochemistry, University of Karlsruhe, Kaiserstr. 12, D-76128 Karlsruhe, Germany

1. ANALYTICAL PROCEDURES FOR THE DETECTION OF HEAVY METALS (U. Kramar)

For environmental investigations, heavy metals are to be analyzed in a variety of natural and technogenic materials like aerosols, dust, water, stream and lake sediments, soils, vegetation or building materials like concrete, bricks, and slags.

The choice of which analytical method has to be applied depends on the material to be analyzed, the availability of the method and the goal of the investigation, respectively. This chapter gives an overview on the most important methods. Since most analytical methods require a solution, the analysis of environmental materials consists of two or three major steps: first, mechanical sample preparation (all materials), like grinding, sieving, drying, or homogenizing, second, sample dissolution or extraction (for solid samples, except for non-destructive methods like X-ray fluorescence), and third, the final instrumental determination.

1.1. Sample preparation

The analytical method chosen is influenced by many environmental geochemical considerations, analytical as well as administrative restrictions, and regulations by authority. The kind of information, which can be obtained from the analytical results (e.g., whether the bulk concentration of a heavy metal or the speciation can be determined), depends mainly on the sample preparation and sample digestion method. Sample preparation and digestion are main controlling factors influencing the quality of the analytical results (Table 1). Therefore, these preconditions are discussed first.

1.1.1. Soils and Sediments

Soils and sediments are best collected in high wet strength paper bags, in which they can be dried without removing the sample from the bag. The material is normally air dried at higher temperatures. Clay rich samples should not be dried above 65 °C to avoid baking of the samples. Drying at temperatures above 40 °C may cause losses of Hg. After drying, the material is disaggregated and sieved (soils to < 2mm). Only nylon sieves or stainless steel sieves with friction flange fitting should be used to avoid contamination by solder materials. Further sample treatment depends on the kind of the subsequent dissolution and analysis. Partial extraction is carried out directly using the sieved fraction. For total or near total digestions and instrumental (non-destructive) methods the material has to be ground finely. Since grinding of the sample also causes abrasion from the material of the grinding equipment, agate grinding vessels should be preferred to avoid contamination. For an overview on the sample preparation of soils and sediments, see Ref. 1.

1.1.2. Vegetation

Vegetation should not be collected in tightly closed plastic bags because this will cause rotting. Vegetation is best sampled in brown paper bags and air dried. The material then can be ground with a hammer mill with steel blades or better with ceramic blades.

1.1.3. Waters

The waters are stored in thoroughly cleaned bottles. Polyethylene or polypropylene bottles are preferred to glass. However, during production several heavy metals originating from the production process may be incorporated in the plastic and are often present in leachable form (Table 2). Therefore, heavy metal contents of waters may change between sampling and analysis due to adsorption effects on the container walls or by contamination due to extraction of heavy metals adsorbed on the walls contained in the material of the storage bottles. Ground and surface waters have to be filtered directly at the sampling site through 0.45 µm filters (cellulose acetate), as they may contain undissolved suspended matter. This dissolved matter can release or adsorb heavy metals.

Table 1
Some pathways for preparation, dissolution and analysis of environmental samples

Material	Sample Preparation	Dissolution	Analytical Method
Bricks	Dry	Partial extraction or (near) total digestion by strong acids	Atomic Absorption ICP-OES ICP-MS
Concrete	Crush and grind		XRF
Rocks	Thin sections		Microanalytical Methods
Soils and Sediments	Dry, disaggregate, and sieve	Partial extraction or (near) total digestion by strong acids	Atomic absorption ICP-OES ICP-MS
	Grind		XRF
	Thin sections		Microanalytical Methods
Vegetation	Dry, (ashing)	Acid digestion (Microwave)	Atomic Absorption ICP-OES ICP-MS XRF
	Microtome sections		Microanalytical Methods
Waters			Atomic Absorption ICP-OES, ICP-MS
	Solvent extraction or Ion exchange Precipitation		ICP-OES ICP-MS XRF

For an overview on the sample preparation of waters, see Ref. 2. To prevent these contaminations it is recommended to prepare the bottles by protracted soaking with 50% hydrochloric or nitric acid followed by rinsing with distilled water and conditioning with several aliquots of the water to be collected. To avoid loss of elements by adsorption on the wall of the storage bottles the samples have to be stabilized by acidification to pH 1.

1.2. Digestion methods

Digestion methods vary according to whether the sample, which has to be analyzed, is soil, sediment, building material like concrete, vegetation, or surface and groundwater.

1.2.1. Soils, sediments, and building materials

For most analytical methods solid sample material has to be transferred to a solution. The dissolution step introduces an additional step between sample preparation and final instrumental determination of the elements, however, it provides the possibility to determine either all or a particulate fraction of the trace element in the sample. For environmental investigations, aqua regia dissolution preferentially with reflux according to ISO standard 11466 or to German standard DIN 38414 is most commonly used as a strong digestion method. The sample is digested for several hours in a mixture of three parts of hydrochloric acid and one part of nitric acid. The remainder are bleached silicates. However, heavy metals bound in the silicate fraction are only partially released by aqua regia digestion [3].

To get access to the total amount of heavy metals bound to silicates, digestion with hydrofluoric acid and perchloric acid has to be used. Silicon is volatilized as SiF_4 and removed from the digested solution. Perchloric acid strongly absorbs water and therefore disables hydrolysis of SiF_6 and back reaction to SiO_2. Extreme care has to be taken working with this type of digestion. Hydrofluoric acid is one of the most hazardous mineral acids. HF appears to attack the calcium of the body and to react with the bone. In contact with organic material perchloric acids forms highly explosive perchlorates. Also, some inorganic perchlorates may explode. If organics are present in the sample, the material has to be oxidized before the digestion step, e.g. with H_2O_2 or nitric acid. Perchloric acids should only be used in special designed exhausts equipped with a facility washing down the perchlorates from the exhaust stream. For the characterisation of heavy metals in solid environmental materials like soils, sediments, sewage etc., sequential extraction schemes are often used. In general, these extraction schemes are designed to give information on the mobility of the respective heavy metal and also to give some ideas of its speciation.

Table 2
Possible heavy metal contamination by grinding and packaging materials

Packaging materials	Possible contaminations
Polyethylene	Zn, Cd
PVC	Zn, Cd
Rubber	Zn
Sieves	
Solder	Cu, Zn, Ag, Cd, Sn, Sb, Pb
Grinding equipment	
Steel and iron grinding plates	Co, Cr, Cu, Fe, Mo, Mn, Ni, V
Alumina ceramics	Co, Cu, Fe, Mn, Zn
Tungsten carbide	Co, W
Lubricants	Mo; V

Modified after Ref. 4.

The sequential extraction procedures consist of several subsequent steps. In a first step, the exchangeable fraction is determined followed by steps for the reducible fractions, carbonate bound, organic fractions, oxidizable fraction, and residual minerals. The three step extraction scheme, which is proposed by BCR (Community Bureau of Reference, see Ref. 5), is based on the method described by Ref. 6. It accounts for the water soluble, exchangeable, and carbonate bound fraction, the Fe-Mn oxide bound, the organic matter, and the sulfide bound fraction [7].

1.2.2. Vegetation

Vegetation is commonly digested by nitric acid in a microwave oven. If undigested residuals are observed, a small amount of hydrofluoric acid can be added.

1.3. Analytical methods

A variety of analytical methods is available for the determination of heavy metals in environmental media. The most common methods are optical spectroscopic methods on the one hand. These methods include UV-visible spectrophotometry, flame photometry, fluorimetry, atomic absorption spectrometry (AAS), and inductively coupled plasma atomic emission spectrometry (ICP-OES). On the other hand, microanalytical methods such as secondary ion mass spectrometry (SIMS), electron microprobe, proton induced X-ray fluorescence (PIXE), microsynchrotron X-ray fluorescence (μSYXRF), and micro X-ray absorption analysis are used to determine the spatial distribution of heavy metals in a sample.

1.3.1. Optical spectroscopic methods

Optical spectroscopic methods are based on the measurement of radiation intensity in the ultraviolet to infrared range of the spectrum emitted or absorbed by atoms or molecules (Fig. 1). Nowadays, the most commonly used spectroscopic methods are UV-visible spectrophotometry, flame photometry, fluorimetry, atomic absorption spectrometry (AAS), and inductively coupled plasma atomic emission spectrometry (ICP-OES). UV-visible spectrophotometry and flame photometry are less important for the determination of heavy metals at the trace element level. Therefore, only a short outline is added for completion. The atomic spectrometric methods AAS and ICP-OES, which are of higher importance for the determination of heavy metals, are described in more detail. With UV-visible spectrophotometry, light is transmitted through the sample solution containing a coloured complex, which absorbs specific wavelengths from the light spectra. The absorption is measured. The intensity of the coloured complex and consequently the specific absorption depends on the concentration of the analyte in the solution.

With flame photometry, a solution is aspirated in a flame and the emission lines of elements, which can be excited by the thermal energy of the flame, are observed at characteristic wavelength. The intensity of these lines is proportional to the concentration of the analyte in the solution and can be measured by simple flame photometers [3]. With fluorimetry, a solution or solid is irradiated by an intense (UV) - light source (e.g. laser) and fluorescence radiation of lower energy/ longer wavelength is emitted from the sample. Fluorimetry is mainly important for the determination of Uranium. Atomic absorption spectrometry (AAS) is a widely used analytical method [8-13]. If a valence electron in the atomic shell is hit by a photon of exactly the energy, which amounts to the difference between the ground state and an excited level, the photon is absorbed and the electron is moved to the excited level. This effect is used in atomic absorption analytics.

The heart of the atomic absorption instrument is the atomic cell, where the sample is introduced, desolvated, and atomized. Two basic atom cells are used: first, the flame and second, electrothermal heating of the sample cell. If the atomic cell is transmitted by a light beam of exactly the energy necessary to move the electron to the excited level, the analyte atoms formed in the atomic cell absorb this light of specific wavelength. The principle of AAS is shown in Fig. 2. As atomic absorption lines have a width of 0.001-0.003 nm, they are too narrow to be isolated by small monochromators. Therefore, the fundamental requirement of the light source is to provide a narrow line profile with little background. This is realized by light sources, which are emitting the line spectra of the analyte element only. Two kinds of light sources are used in atomic absorption spectrometry: hollow cathode lamps and electrode less discharge lamps. The latter are of interest only for some volatile elements like Hg, As and Se. More common is the use of hollow cathode lamps. The hollow cathode of these lamps is coated with the analyte metal. The cylinder of the lamp is filled with an inert gas which is ionised by an electric current. These ions are attracted by the cathode. Due to the bombardment by the inert gas ions the metal ions of the coating are excited and the characteristic line spectra of the analyte metal are emitted. The light consisting of the characteristic line spectra of the analyte is focussed into the atomic cell, were it is partially absorbed by exciting the analyte atoms.

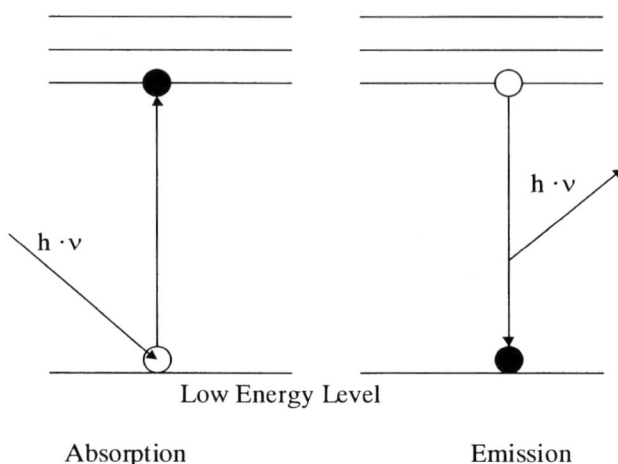

Fig. 1. Absorption and emission of light by electron orbital transition.

According to Beer's law the intensity ratio of the transmitted light beam (I_t) and the incident beam (I_0) of wave length λ depends on the concentration of the analyte atoms

$$I_t/I_0 = e^{(-\mu(\lambda)\,c\,l)} \quad \text{or} \quad \log(I_0/I_t) = K\,c \tag{1}$$

with $\mu(\lambda)$ = absorption coefficient for wavelength λ
  l = length of the absorbing path
  K = constant containing μ and l
  N = 1, 2, 3, 4, ...

The transmitting light of the specific wavelength, which is absorbed by the analyte atoms, is then separated by a monochromator from the other wavelengths.

With flame systems, the sample solution is dispersed into the flame by a nebulizer and spray chamber assembly. The analyte is atomized in the flame, most commonly an air acetylene burner, which provides a temperature of approximately 2300°C. Depending on the analyte a more oxidizing or reducing flame can be used, e.g. while magnesium should be analyzed in an oxidizing flame, chromium gives better sensitivity in a reducing flame. The air-acetylene flame is sufficient for most elements except those which form refractory oxides (e.g. Al, Mo, Si, and Ti). These can be analyzed using a nitrous oxide-acetylene flame, which is more reducing and provides a higher temperature (~2900°C).

With electrothermal atomisation (graphite furnace or ETAAS), some µl of the sample solution are introduced into a graphite tube. The sample is then dried just above the boiling point, ashed to remove as much of the matrix as possible, and atomized at high temperatures. Temperatures for ashing and atomisation depend on the analyte element. Since the atoms are released in the tube, they remain in the beam for a longer period than those atomized in the flame. Consequently, sensitivity of ETAAS is considerably higher then the sensitivity of flame AAS.

The light reaching the detection device from the atom cell contains various unspecific components (e.g. scattered radiation, "white" radiation from the flame, lines of emitted by other elements etc.) besides the specific radiation. To separate these components from the lines of the lamp, these unspecific components have to be removed by a line separation device. Most instruments are equipped with a monochromator, which enables the detection of only one wavelength at a time, followed by a photo-multiplier as detection system. These instruments are only capable for the determination of one element at a time. The rapid advance in technology and electronics enabled the design of multi-element instruments equipped with bank of 4-6 hollow cathode lamps or multielement hollow cathode lamps, polychromators and diode arrays as detection system, which overcome the limitations of single element instruments. With state of the art instruments, the registration and output device is based on a computer, which controls the instrument, stores the calibrations, and enables the monitoring of the output signal with the time. To solve specific analytic problems flame and electrothermal atomization, AAS can be combined with a variety of methods that allow the analyte to be introduced into the atom cell in an adequate form. Samples containing high dissolved or suspended solids can be analyzed using a flow injection system, which introduces discrete aliquots of a sample in a carrier stream. Hydride generation systems are commonly used for the determination of heavy metals, which form volatile hydrides (e.g. Bi, Pb, Sb) as well as other vapours like Hg or alkylated Cd. The vapours can be readily atomized in the flame, but commonly heated quartz tubes or T-pieces are used. For the determinations of elemental speciation AAS can be coupled with chromatography.

The detection limits offered by AAS for the determination of heavy metals are sufficient for most of the applications as long as only a limited number of components have to be analyzed. Flame atomic absorption is relatively easy to use and relatively robust against matrix interferences. Electrothermal AAS provides much lower detection limits but requires higher analytical skills. In contrast to atomic absorption analysis, which is based upon the absorption of a characteristic radiation, atomic emission spectroscopy uses the emission of the characteristic wavelength for the determination of the analyte element. Atomic emission spectroscopy belongs to the most useful and commonly used techniques for the analyses of heavy metals. Atomic emission spectroscopy provides rapid and sensitive results in a variety of sample matrices although the detection limits are higher than with AAS [14-18]. An atomic emission system consists of an excitation source, where a plasma is produced, in which the analyte is atomized and will be excited to emit its light of characteristic wavelength, a spectrometer, which separates the characteristic wavelength emitted by the analyte elements, a detector, where the intensities of the different analyte lines are measured, and a registration system for signal and data processing. When a sample solution is introduced into the emission source, it is vaporized, the compounds of the analytes will be dissociated, and ideally all lines of the analyte elements of interests will be excited.

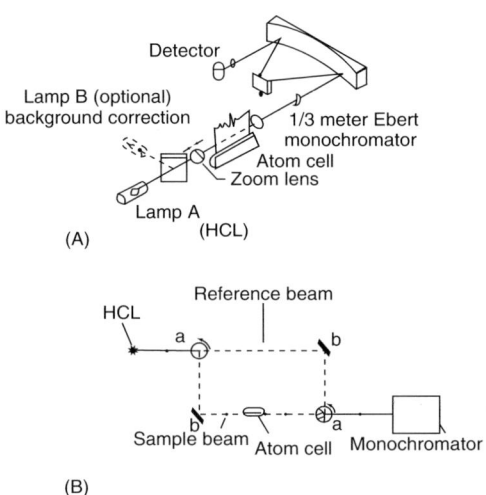

Fig. 2. Schematic principle of a flame atomic absorption (reprinted with permission from Ref. 14)

Plasma sources are able to provide all these presumptions, and consequently a variety of different plasma sources including inductively coupled plasma (ICP), microwave induced plasma (MIP), direct current plasma (DCP), and glow discharge, which allows analysing solids directly, are available. Nevertheless we will focus on ICP, since it is the plasma source most commonly used. The ICP is produced by a high frequency field (typically 27 MHz microwaves) in an inert gas (argon). The high frequency alternating current in a water cooled induction coil induces a changing magnetic and electric field (Fig. 3). Electrons of the gas are accelerated to high energies. By collision the gaseous atoms are ionised and excited. The resulting plasma is self-sustaining as long as RF and gas flow continue. The sample introduced in the plasma is desolvated, dissociated, atomized, and excited. Plasma temperature and residence time of the sample atoms in the plasma result in complete vaporization of the sample, formation to free atoms of the analyte, and excitation.

The intensity of the emitted light of characteristic frequency for the elements of interest is proportional to the concentration of the analyte element in the sample solution. The radiation emitted from the plasma is passed through an entrance slit into the spectrometer optics, where the different spectral components are separated.

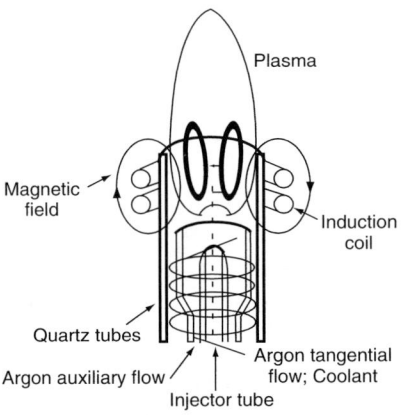

Fig. 3. Schematic sketch of a plasma torch (Reprinted with permission from Ref. 14).

The spectrometers are either equipped with a monochromator or a polychromator. With a monochromator, only one element at a time can be determined. Multielement capabilities are achieved by moving the secondary slit or the grating (Czerny-Turner configuration). Multielement determination machines using such sequential spectrometers are slower than machines equipped with polychromators, but the range of wavelength and detectable elements is very flexible.

Conventional polychromators are equipped with a set of secondary slits at fixed positions for the determination of certain individual wavelengths. Each slit is equipped with an own detector, e.g. photomultipliers, single charge-coupled devices (CCD), or charge injection devices (CID). These polychromators allow fast simultaneous determination of a fixed set of elements. Modern high end instruments are equipped with concave gratings, which focus the emission lines on the circumference of a Rowland circle (Fig. 4) and an array of CCD on the circumference, allowing simultaneous and sensitive determination of a great number of elements. With ICP, one of the most critical parts is the introduction of the sample into the plasma. In general, the sample is carried into the plasma by a carrier gas (argon) at the head of the plasma torch. Commonly, the torch consists of three concentric quartz tubes flown by argon streams. The aerosols of the sample and carrier argon flow in the central tube. The stream in the next outer tube adjusts the plasma position and the stream of the outer most tube acts as a cooling gas (Fig. 3).

Aerosols of liquid samples are generated by nebulizers and injected into the inner tube of the torch.

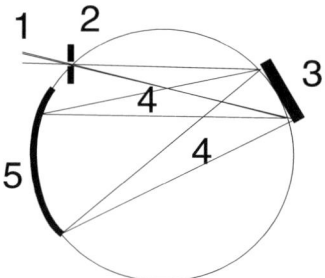

Fig. 4. Schematic sketch of a Rowland circle polychromator. 1: light emitted by the plasma, 2: entrance slit, 3: grating, 4: light dispersed by the grating to its different components, 5: exit slits or CCD (redrawn after Ref. 14).

As in atomic absorption spectroscopy, the sample introduction system can be coupled with a variety of sample preparation methods (e.g. hydride generation, slurry injection, and chromatography). Since the spectral lines of all elements contained in the sample are emitted from the plasma, an adequate line for the determination of an element of interest has to be selected with respect to possible overlapping interferences of lines from other elements contained in the sample, from plasma background, and of the intensity. The effect of partial overlaps can be reduced by high resolution instruments and/or adequate line correction methods. Such methods and assistance for line selections are included in the software of most modern instruments.

Due to its high analytical sensitivity in multi element determinations, ICP-MS is the high end instrument for simultaneous low level determinations of heavy metals. As with ICP-AES, the sample is introduced into plasma. Therefore the "front end" of ICP-MS is practically identical to ICP-AES. In contrast to ICP-AES, the spectral lines are not measured but the ions produced in the plasma are sampled and transferred for measurement into a mass spectrometer. In argon plasma the yield of ionisation is greater then 90% for most of the elements. Mass analysis includes the detection and separation of ions and, in the same process, isotopes of the elements under investigations. Since the plasma is burning at environmental atmospheric pressure and the mass spectrometer is operated at high vacuum ($\sim 10^{-5}$ Torr), a small portion from the central part of the plasma containing the sample must be extracted through a differential pumped interface. The interface consists of two water cooled cones, the sampler cone, and the skimmer cone. The high vacuum mass spectrometer is separated from the atmospheric pressure of the plasma by a pre-vacuum of ~1 Torr between sampler and skimmer cone (Fig. 5). Different kinds of mass spectrometers are used for mass separation and detection of the different ions. Most of the more inexpensive, low resolution instruments are equipped with quadrupole mass spectrometers, but also time of flight machines are available. The quadrupole machines allow fast scans of the entire isotope range and result in a quasi simultaneous determination, whereas time of flight machines provide real simultaneous determinations. With low resolution machines, a mass resolution of 1 amu (atomic mass units) and detections limits < 1 ng/L is achieved for most of the elements.

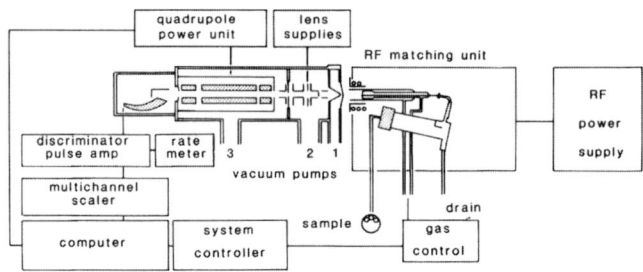

Fig. 5. Schematic sketch of an ICP-quadrupole mass spectrometer (reprinted with permission from Ref. 15).

The more expensive high resolution instruments of the latest generation contain double focusing mass spectrometers with electrostatic and magnetic analysers in series. They achieve mass resolutions of 1/300 – 1/10000 amu. Besides the high mass resolution, which eliminates most of the isotopic interferences in high resolution machines, the background signal is negligible compared to the low resolution machines. Therefore, detection limits are several magnitudes lower.

Compared to other analytical methods, ICP-MS has several advantageous capabilities: due to a large linear dynamic range, elements of major, minor, and trace levels can be determined at the same time (quasi-simultaneously). Detection limits are very low and more than 50 elements can be determined within 2 minutes out of less than 2 ml solution. If microconcentric nebulizers are used, the sample consumption can be reduced even more drastically.

Mass spectra are relative simple and easy to interpret. On the other hand, isobaric interferences from isotopes of other elements or even from "strange" compounds can occur. Most of the isotopic or compositional interferences can be resolved by high resolution ICP-MS machines but with a concurrent loss in sensitivity. Even with high resolution machines, the mass differences of some isobaric interferences (monoisotopic or isotope clusters) are too small to be resolved, e.g. the $^{40}Ar^+$ interference on $^{40}Ca^+$, the major isotope of Ca. One example of such a "strange" isobaric cluster is $^{40}Ar\,^{35}Cl$, which interferes on ^{75}As, but also other Ar containing polyatomic ions and oxides or hydrides of matrix elements may interfere.

Another complication, which may occur is caused by the sampling process. Since the plasma is introduced through a small orifice into the mass spectrometer, dissolved salts from the solution may clog the orifice during subsequent measurements. Nevertheless, a maximum of 0.2% dissolved salts (better 0.1%) can be tolerated. Another limitation depends on the sample injection into the plasma, also the matrix of the sample influences plasma forming conditions, which leads to matrix effects. Matrix effects and intensity loss due to clogging can be partly corrected by the use of internal standards.

In contrast to most other analytical methods, X-ray fluorescence analysis is neither based on reaction of the valence electrons of the analytes nor on the mass of the analyte compound, but on electron transfer between the inner electronic shells of the analyte atoms.

Nowadays, X-ray fluorescence is one of the most commonly used routine methods for the determination of major and trace element determinations in solids. The method is non-destructive and solids can often be analyzed directly in pressed powder pellets, bulk powder samples, or qualitatively even in the untreated sample.

Elements with atomic numbers from 4 (Beryllium) to 92 (Uranium) at concentrations from 0.x µg/g to the high percentage level can be analyzed using characteristic K_α- lines (KL_{III}) from 11.4 nm/0.1885 keV (Be K_α) to 0.0126 nm/ 98.4 keV (U K_α). If solid materials are analyzed by X-ray fluorescence without digestion and dilution, detection limits of 0.x µg/g are equivalent to those of 0.x µg/l. Nevertheless elements of higher atomic numbers (e.g. Cd $L_{III}M_V$ 0.3956 nm/3.133 keV; U $L_{III}M_V$: 0.09106 nm/ 13.61 keV) are often determined using their L- lines.

The characteristic X-ray lines of the elements can be determined either with sequential or simultaneous wavelength dispersive spectrometers by Bragg diffraction, using the wave phenomena of X-ray or in energy-dispersive systems using their energy characteristic. If a target is irradiated with photons or charged particles (electrons or ions), whose energies are exceeding the binding energy of the bound inner electrons, then an electron from inner orbitals of the target atoms can be ejected. The resulting atom is unstable and regains its ground state by transferring an electron from a high energy outer orbital to the vacancy in the inner electron shell (Fig. 6). The energy difference between the initial and final energy state of the transferred electron is released as a photon of characteristic energy and wavelength. With X-ray fluorescence spectroscopy, photons are used to excite the characteristic elemental X-rays from the sample. The X-ray fluorescence spectrum of a sample is a mixture of the different characteristic X-rays emitted from the atoms in the sample and of coherent (Rayleigh) and incoherent scattered components of the primary radiation source.

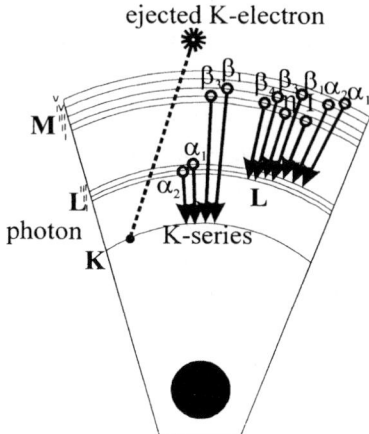

Fig. 6. Schematic sketch of the production of characteristic X-rays (K-series and L-series) after removing an electron from the inner orbitals by an incident photon.

The task of X-ray fluorescence spectrometers is to separate the different spectral components, to determine their intensities, and to calculate the elemental concentrations. Typically X-ray fluorescence spectrometers consist of a photon source for the excitation of the secondary X-rays, a sample support, an X-ray detection unit, and a data evaluation unit.

In most X-ray fluorescence spectrometers, an X-ray tube is used as photon source. Alternatively, radioactive isotopes can be applied for the excitation of the characteristic X-rays from the sample. X-ray tubes can be used in both wavelength dispersive and energy dispersive systems. Due to their lower beam intensities, application of radionuclides is restricted to energy dispersive systems. XRF spectrometers have two major objectives: first, to determine the spectral distribution of the X-rays emitted from the sample, and second, the measurement of the intensity of the selected spectral component. In wavelength dispersive XRF, the spectrum is dispersed into different wavelength's by Bragg diffraction at different crystals according to the Bragg equation

$$n\lambda = 2 d \sin \theta \tag{2}$$

with λ = wavelength of the radiation
 d = lattice distance of the analyzer crystal and
 θ = diffraction angle.
 $n = 1, 2, 3\ldots$

The wavelengths of characteristic lines used in X-ray spectrometry range from ~0.03 nm (Ba K) to ~10 nm (Be K). To cover the entire wavelength range, X-ray spectrometers are equipped with a set of analyzer crystals with adequate lattice distances. Intensities are measured by scintillation detectors for the detection of high photon energies and gas filled detectors (flow counter or proportional counter) for low photon energies. In most sequential XRF spectrometers, the flat crystal is mounted on the central axis of a rotating goniometer and a proportional counter and a scintillation counter mounted at the moving goniometer arm. The Bragg angle can be varied simply by rotating the crystal mount by θ and the detectors by 2θ (Fig. 7). Simultaneous spectrometers are designed with fixed crystal arrangements.

In modern simultaneous spectrometers up to 30 crystal - detector combinations are fixed at suitable angle positions for the peak- and background measurements of the application. In general, focusing crystal shapes are used. Simultaneous spectrometers are capable of simultaneous multielement determination, but elements and background positions are fixed and cannot be changed. With energy-dispersive systems, the separation of the different components in the X-ray spectrum is performed exclusively by the detector and the subsequent electronic components. Thus good energy resolution, low electronic noise, low temperature drift, and excellent linearity of the electronic components are required.

The required excellent energy resolution can only be provided by high resolution proportional counters or semiconductor detectors (Si(Li) or Ge). Due to the low efficiency of proportional counters at higher radiation energies, all energy dispersive spectrometers except some mobile equipments for low energy application are equipped with a semiconductor detector. In general, an energy dispersive spectrometer consists of a radiation source (X-ray tubes or radionuclides), sample support with the sample, detector, pre- and main amplifier, pile-up rejector, analogue digital converter, and multichannel analyser (Fig. 8). Energy dispersive (EDXRF) systems are available as stationary systems, which can be used for fast simultaneous multielement determinations in analytical laboratories, as compact equipments suitable for mobile field laboratories, and as hand held probes to be used directly on site for screening analyses. Some latest generation EDXRF instruments use polarized X-rays. Due to the polarization of the primary beam, background from scattered radiation is reduced and detection limits are improved [19-24].

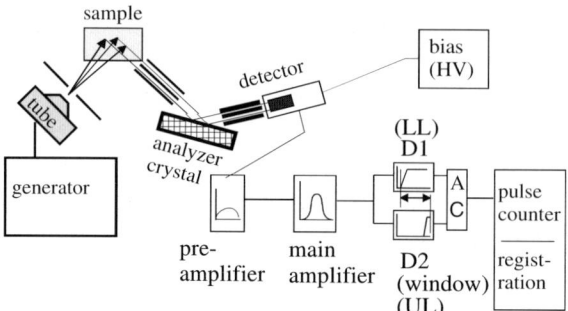

Fig. 7. Schematic sketch of a sequential wavelength dispersive X-ray fluorescence (redrawn after Ref. 14)

Solid as well as liquid samples can be analyzed by XRF. Liquid samples have to fill an adequate sample container with an X-ray transparent window (e.g. Mylar). Solid samples can be analyzed as pressed powder pellets, bulk powder samples, fused pellets, or as pills (diluted with a flux, e.g. Spectromelt). For quantitative analyses a flat and homogenous sample surface is needed. Solid and liquid samples can be prepared either as samples of "infinite thickness" or by thin film techniques. The characteristic radiation of the elements excited in the sample is partly absorbed on its way to the surface. A sample can be considered to be of infinitely thickness, if more then 95% of the radiation exited at the rear side is absorbed on it way to the surface. In this case no weighing of the sample is necessary.

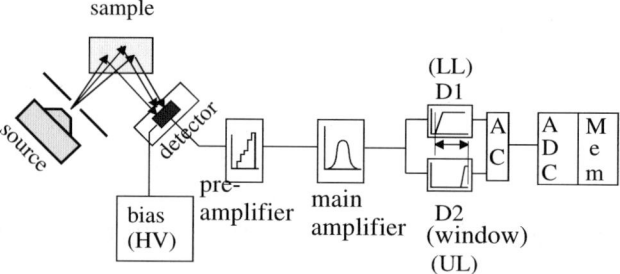

Fig. 8. Schematic sketch of an energy dispersive X-ray fluorescence (redrawn after Ref. 14).

When using thin film techniques, the amount of sample in the irradiated spot has to be known. This can be established by adding an internal standard. Thin specimens show a strictly linear dependency of the intensity of a characteristic line on the amount of the analyte. In thick specimens, primary radiation is absorbed along its path in the sample and secondary radiation on its way out. In addition, intense lines with energies above the absorption edge of the analyte may cause secondary excitation. These intensity variations (up to a factor of ~ 20) are strongly matrix dependent and have to be corrected. Different methods for mathematical correction of such matrix effects are implemented in the software of modern instruments [25-27]. The analysis of very small samples or residuals from evaporated waters by X-ray fluorescence is limited due to high background from radiation scattered at the sample carrier. Nevertheless, an X-ray beam dipping on a surface at very small incident angles ($< \alpha_{krit}$) is totally reflected at the surface of a specimen. In this case, the X-rays can penetrate into the substrate only for a few nm. This effect is used in total reflection XRF. The sample is prepared as a thin film on a well polished quartz block of total reflection. The characteristic radiation of the elements is excited by it, but the primary radiation does not penetrate the sample carrier. Background is drastically reduced and detection limits at the $\mu g\ g^{-1}$ level can be obtained from a few micrograms of solid sample and at the $ng\ g^{-1}$ level from a few micro litres of a liquid sample.

1.3.2. Microanalytical methods

To evaluate the effects of heavy metals on the environment the knowledge of their microdistribution and speciation has become increasingly important. The spatial distribution of heavy metals can be determined by microanalytical methods such as secondary ion mass spectrometry (SIMS), electron microprobe, proton induced X-ray fluorescence (PIXE), micro synchrotron X-ray fluorescence (μSYXRF), and micro X-ray absorption analysis [28-30]. When a solid sample is sputtered by primary ions of a few keV energy, a fraction of the particles emitted from the target is ionized. SIMS consists of analyzing these secondary ions with a mass spectrometer. By using secondary ion mass spectrometry, information about the elemental, isotopic, and molecular composition of the uppermost atomic layers of a sample specimen can be obtained. SIMS is the most sensitive elemental and isotopic surface analysis technique. The secondary ion yields will vary in large amounts according to the chemical environment causing serious matrix effects and on the sputtering conditions (ion, energy, and angle). With electron microprobe, the sample specimen is irradiated by a highly focused electron beam. The electrons excite the characteristic X-rays of the atoms at the surface of the sample. With routine electron microprobe analysis, detection limits of some hundred µg/g can be achieved. Therefore, electron microprobe is restricted to the determination of major and minor elements. With proton induced X-ray fluorescence (PIXE), the sample is irradiated by a focused beam of high energetic protons produced in an accelerator. These high energetic protons are very effective in removing electrons from the inner shells of the analyte atoms. With PIXE, detection limits in the range of some 0.1 µg/g are achieved at a spatial resolution of few micrometers. Since the protons are absorbed in the sample within a few nm, PIXE provides information about the surface of the sample. With micro synchrotron X-ray fluorescence (µSYXRF), an intense X-ray beam of a synchrotron is focused to a spot few µm in diameter. The sample is scanned with the microbeam and detection limits in the range of some 0.1 µg/g can be achieved with spatial resolutions of few micrometers. Since the X-rays are able to penetrate the sample, the nondestructive determination of the three-dimensional distribution of heavy metals is possible. With micro X-ray absorption analysis, the intense X-ray beam of a synchrotron is focused to a spot few µm in diameter, as is done with µSYXRF. Yet in contrast, the energy of the primary beam is scanned across the absorption edge of the element of interest and either the absorption of the X-rays or intensity variations of the characteristic X-rays of the analyte element are observed. The exact position of the absorption edge depends on the speciation of the respective element and oscillations near the absorption edge provide information on the neighborhood of the respective elements. Using µ-X-ray absorption analysis, the speciation even of some ten µg/g of an element can be determined nondestructively.

2. BIOGEOCHEMICAL PROCESSES REGULATING HEAVY METAL MOBILITY

Heavy metals are mobilized by various physical, chemical, and biological vectors. This chapter introduces the most important biogeochemical processes, which influence and regulate heavy metal mobility and immobility in environmental media. These mechanisms include sorption processes, redox reactions as well as weathering processes and speciation. These processes are influenced by chemical driving factors such as pH, redox potential, or chemical speciation of an element.

2.1. Sorption (C. Kim)

Sorption processes play an important role in reducing the transport of heavy metals in the environment. The precise mechanisms of sorption taking place in any given system are also critical in understanding and predicting the future mobility of sorbed metals. This section reviews different methods of heavy metal sorption and describes the utility of X-ray absorption spectroscopy (XAS) as a technique well suited for identifying the types of surface sorption complexes that can form between metals and mineral substrates.

Sorption is a general phrase that refers to the partitioning of a sorbate, in this case a metal cation, from the aqueous phase to a sorbent, in this case a mineral surface, and is used when the precise method of partitioning is not known or has not been conclusively identified. The extent, behaviour, and reversibility (or permanence) of sorption are all dependent to some degree on the actual mechanism of sorption onto the solid, of which several have been defined.

2.1.1. Introduction

Heavy metals represent a ubiquitous constituent of the near-surface environment, present in widely varying concentrations that typically have little impact on human behaviour and health. However, the mining of localized geological enrichments of metals used in industrial applications (e.g. copper, chromium, iron, lead, mercury, and zinc) as well as precious metals (e.g. gold, platinum, and silver) often results in highly elevated metal concentrations in the surrounding soils, sediments, water supplies, and atmosphere. Similarly, the use of these metals in industrial processes, when not properly regulated, has historically produced significant anthropogenic inputs of metals to both local and global environments. The detrimental human health effects associated with exposure to certain metals have long been established and remain a constant concern for both developing and fully industrialized nations. Therefore, the reduction or removal of metals from human exposure pathways is of considerable scientific and political interest.

Metal sorption to mineral surfaces was proposed as early as the 1930's by Goldschmidt [31, 32], who observed that concentrations of many trace elements in seawater were several orders of magnitude lower than would be predicted based on global inputs from weathering processes over geologic time. Ref. 33 later used solubility product calculations and laboratory experiments to study the behaviour of 13 trace metals in seawater (including

Ag, Bi, Cd, Co, Cr, Cu, Hg, Ni, Mo, Pb, V, W, and Zn), concluding that precipitation of the metals as chlorides, sulfates, carbonates, hydroxides, or sulfides was not a feasible explanation for their dramatically low concentrations as measured in seawater.

However, simple macroscopic uptake measurements showed that metal sorption onto hydrated iron and manganese oxides and organic matter is a significant process, by which metals such as Ag, Bi, Cd, Cu, Hg, Mo, Pb, and Zn might be scavenged from the aqueous phase and sequestered into the solid phase. This discovery was an important landmark in the development of a new field of study investigating sorption processes of metal contaminants onto mineral particles and the corresponding impacts of sorption on the release, transport, and potential bioavailability of these metals in the environment.

The purpose of this section is to provide an overview of sorption mechanisms at the mineral/water interface and the geochemical conditions that favor or inhibit sorption. This will include recent results of both macroscopic uptake experiments and X-ray spectroscopic studies. The rapid development and application of synchrotron-based X-ray techniques to metal sorption processes have greatly contributed to an understanding of metal surface complexation at the molecular level. These techniques, which include extended X-ray absorption fine structure (EXAFS), X-ray absorption near edge structure (XANES), and X-ray standing wave (XSW) spectroscopy, represent powerful new methods for probing the exact modes of metal sorption to solid surfaces. For a comprehensive overview of synchrotron-based X-ray techniques as applied to metal sorption on mineral surfaces, the author would refer to a review given in Ref. 34, which served as an integral reference and source of inspiration for this section.

While acknowledging the importance of natural organic matter, bacteria, and biofilms on metal sorption and precipitation in natural systems, this subchapter will instead focus on the fundamental inorganic processes involved in metal sorption. The reader is referred to the aforementioned review in Ref. 34 for a summary of metal uptake studies involving organic constituents.

2.1.2. Adsorption Mechanisms

Adsorption is a specific form of sorption that is characterized as an accumulation of sorbate at the interface between an aqueous solution and a solid adsorbent phase without the development of a three-dimensional molecular arrangement [35]; that is, there is insufficient sorption to induce the formation of a solid precipitate involving the sorbed species. Instead, adsorption implies the formation of some two-dimensional structural arrangement at the mineral/water interface. This can be further split into two types of adsorption. The first is alternately described as non-specific, physical, or outer-sphere adsorption, while the second is described as specific, chemical, or inner-sphere adsorption.

Outer-sphere adsorption of metals occurs when a (positively-charged) metal cation retains its outer hydration shell of water molecules and is attracted to a (negatively-charged) mineral surface through a combination of hydrogen bonding and electrostatic long-range Coulombic forces. Inner-sphere adsorption of metals, in contrast, occurs through the loss of these waters of hydration and the formation of direct chemical bonds with the mineral surface, typically with surface oxygen atoms. Inner-sphere adsorption can be additionally distinguished by the number of surface oxygen atoms to which the sorbate binds, where one, two, or more chemical bonds results in monodentate, bidentate, and (rarely) tridentate and tetradentate inner-sphere surface adsorption complexes [36]. Outer- and inner-sphere adsorption can be expressed by the following reactions, where $\underline{S}OH$ represents a surface hydroxyl site (and $\underline{S}$ is a surface metal atom of the sorbing mineral), typical of oxide surfaces in the presence of water [37], and M^{2+} represents a fully hydrated (waters omitted for clarity), divalent aqueous metal ion [35]:

$$\underline{S}OH + M^{2+}(aq) = \underline{S}OHM^{2+}(aq) \text{ outer-sphere adsorption} \qquad (3)$$

$$\underline{S}OH + M^{2+}(aq) = \underline{S}OM^{+} + H^{+} \text{ monodentate inner-sphere adsorption} \qquad (4)$$

$$2\underline{S}OH + M^{2+}(aq) = (\underline{S}O)_2M^{\circ} + 2H^{+} \text{ bidentate inner-sphere adsorption} \qquad (5)$$

The final products of these reactions are shown schematically in Fig. 9. Ref. 38 summarized several characteristics of outer- and inner-sphere adsorption that can be observed or determined macroscopically. Those most commonly reported in the literature involve issues of electrostatic repulsion and attraction as follows, with adaptation to the case of metal cations. In the case of outer-sphere adsorption, adsorption occurs largely on surfaces of opposite (negative) charge, less so near the point of zero charge (PZC, or pH_{pzc}), and very little below the pH_{pzc}, where the charge of the surface and sorbate are the same (both positive).

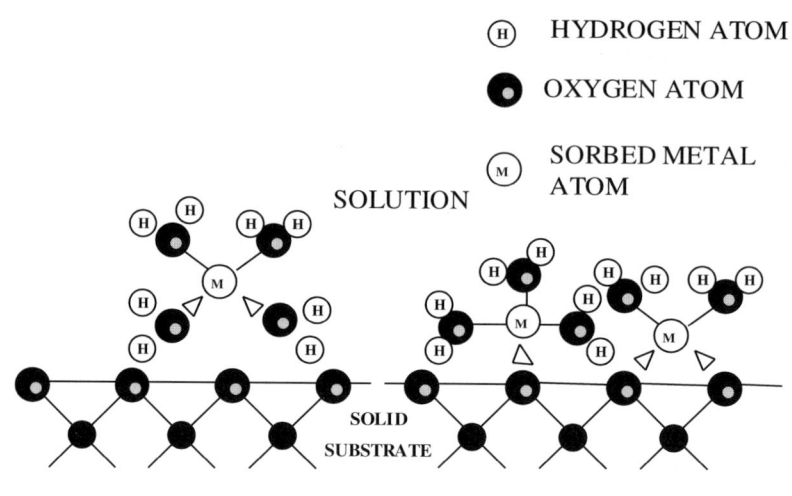

Fig. 9. Schematic sketch of possible sorption complexes at the mineral/water interface (represented by the horizontal line). The solid substrate is below this line and the solution is above the line. The circles labelled M represent sorbed metal atoms in various types of sorption configurations. The larger shaded spheres in the substrate and surrounding the metal in the solution phase are oxygen atoms (redrawn after Ref. 35).

Increasing the ionic strength, or the concentration of an indifferent background electrolyte, decreases metal adsorption densities measurably due to a neutralization of surface charge, thereby diminishing the attractive forces that encourage outer-sphere adsorption (Fig. 10a) In the case of inner-sphere adsorption, adsorption is generally insensitive to surface charge, with metal uptake capable of taking place to surfaces of zero charge or even like charge (both positive). Increasing the ionic strength of the solution has little effect on sorption densities (Fig. 10b). Generally, the effects of charge can be used to distinguish outer-sphere adsorption from inner-sphere adsorption, though it is also likely that in any system a combination of both types of surface adsorption complexes are present [39]. To summarize, outer-sphere adsorption is primarily an electrostatic-based attraction of metal cations to surfaces of opposite negative charge, while inner-sphere adsorption primarily involves the formation of chemical bonds with the surface and is less affected by surface charge. As one might imagine, the hydrogen bonding and long-range Coulombic forces of outer-sphere adsorption are generally weaker than the short-range electrostatic or covalent bonds that characterize inner-sphere adsorption.

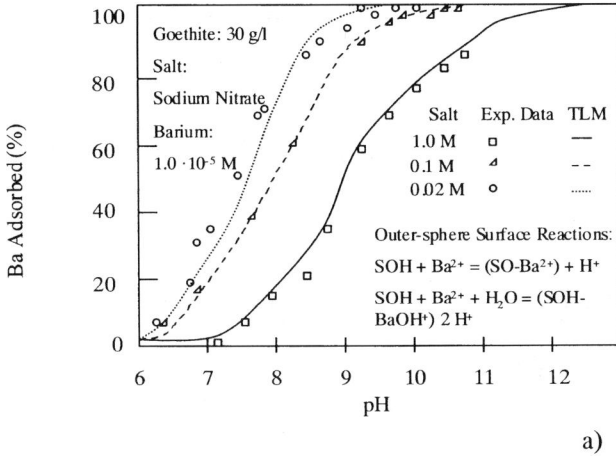

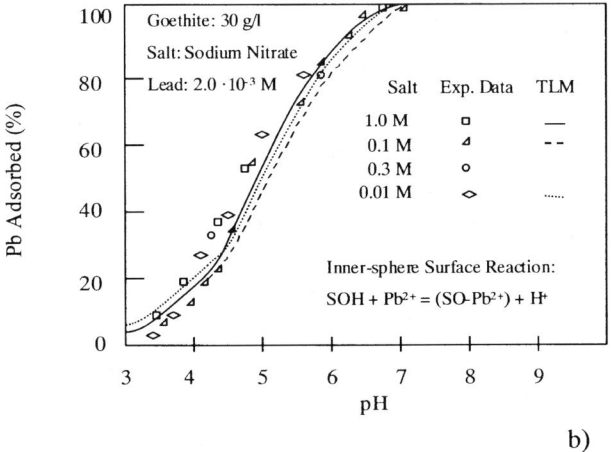

Fig. 10. Macroscopic uptake data of Ba^{2+} and Pb^{2+} on goethite as a function of pH and ionic strength (points=data, curves=triple layer model simulations). a) Ba^{2+} modeled as an outer-sphere complex, showing great dependency on ionic strength; b) Pb^{2+} modeled as an inner-sphere complex. (Redrawn after Ref. 40).

As a result, inner-sphere complexes are typically more tightly bound to the surface and are less easily removed back into solution via desorption. Adsorption as ternary surface complexes represents a sorption mechanism, in which the sorbing metal ion either serves as a bridge between the ligand L and the surface or is bound to the surface indirectly via an intermediate bridging ligand. These are defined as Type A and Type B ternary complexes (Fig. 11a, 11b), respectively, and can be represented thusly [40]:

$\underline{S}OH + M^{2+}(aq) + L^- = \underline{S}OML + H^+$ Type A ternary surface complex (6)

$\underline{S}OH + L^- + M^{2+}(aq) = \underline{S}LM + OH^-$ Type B ternary surface complex (7)

The final products of these reactions are shown schematically in Fig. 11. Examples of inorganic ligands prevalent in freshwater and seawater include bicarbonate, carbonate, chloride, nitrate, phosphate, sulfate, and sulfide, while organic ligands such as amino, carboxylic, fulvic, and humic acids are also common [41]. So the consideration of ligands and formation of ternary surface complexes must be included when addressing metal adsorption in natural systems.

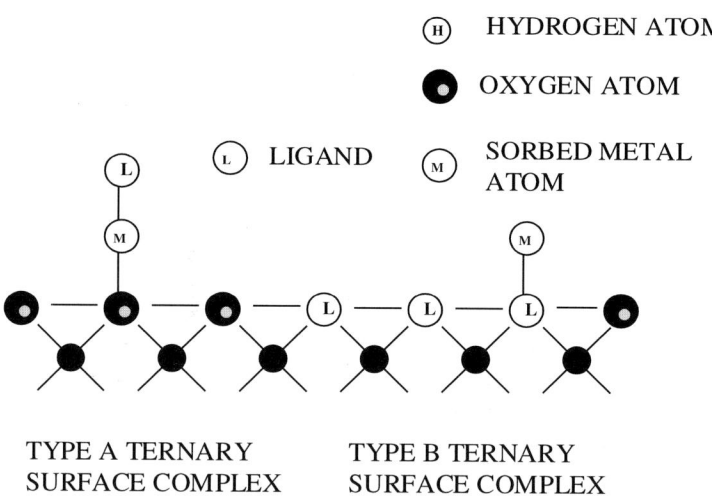

Fig. 11. Schematic sketch of a) type A ternary surface complex and b) type B ternary surface complex involving the metal, mineral surface, and ligand L, which may bond directly to the surface or primarily via the sorbing metal.

Ternary surface complexation can be promoted by the formation of a metal-ligand aqueous species, which may have a higher affinity for a mineral surface (e.g. by possessing a lower charge and therefore less potential electrostatic surface repulsion) than the uncomplexed metal or by initial adsorption of the ligand directly to the surface, similarly reducing surface repulsive forces and allowing adsorption of the metal to the surface via the ligand in a Type B ternary surface complex. However, the effects of ternary surface complexation on the extent of sorption are variable and may be associated with either an enhanced or a reduced degree of total metal uptake. In studies of Pb(II) sorbed to goethite in the presence of sulfate and carbonate, evidence of ternary surface complex formation accompanied increases in uptake by 30 and 18%, respectively [42, 43], while studies of Hg(II) sorbed to goethite in the presence of chloride featured both reduced total uptake (by as much as 80%) and indications of Type A ternary surface complexes [44, 45]. Similarly, dependent on the ligand, metal, and mineral surface being considered, a ternary surface complex may be more or less permanently bound than a direct inner-sphere adsorption species.

According to Ref. 46, ternary adsorption complexes should be statistically less stable than binary adsorption complexes, as the static Jahn-Teller deformation of a metal cation's coordination shell resulting from ligand complexation allows fewer available positions for metal binding than the uncomplexed species. However, examples exist where ternary surface complexes, usually involving organic ligands such as EDTA [47] or humic or fulvic acids [48-50], form strongly adsorbed species that limit the subsequent desorption from the surface that would have been more likely to occur in the absence of the ligands. As demonstrated in Fig. 9 and Eq. (3-5), inner- and outer-sphere adsorption complexes are typically envisioned as mononuclear species, i.e. as single metal atoms adsorbing to a surface.

However, multinuclear complexation is also possible in certain cases and may arise as a result of factors such as a propensity to form dimers, as with Cu(II) on TiO_2 [51], or high sorption densities of the metal on the solid leading to localized metal clusters and multinuclear complexes [52, 53]. Continuing with this line of thinking, sorption densities exceeding a monolayer of adsorbed metal atoms have a high likelihood of forming a solid surface layer of the metal (Fig. 12), usually as a hydroxide or oxide phase. Such three-dimensional solids no longer fulfill the definition of adsorption as described earlier and are instead referred to as surface precipitates. These precipitates represent one end in a continuum of sorption, first from mononuclear adsorbed ions to multinuclear adsorption complexes and finally to homogeneous surface precipitates, with the sorption species determined largely by the degree of metal loading onto the surface. It is worth noting that while published thermodynamic solubility information for solid crystalline phases is helpful when trying to either induce or avoid precipitation, the sorption-related surface precipitation described here may still occur well below calculated solubility limits (i.e. at metal concentrations below that expected to result in precipitation) due to localized accumulations of the metal at the mineral/water interface. In fact, multinuclear adsorption complexes and surface precipitates have been observed at metal surface loadings much less than one monolayer [53-59]. This highlights the point that the interfacial region is distinctly different from bulk water, both in terms of the ordering/structure of water molecules, charge distribution, and concentration of counterions, among other features. Natural roughness at the mineral surface can also present a multitude of steps, terraces, edges, corners, and defect sites that may prove more attractive to sorbing metals and more conducive to multinuclear complexes and local precipitation. Such conditions allow precipitates to form in undersaturated conditions and with characteristics (e.g. templates, thin films) that would not be found elsewhere. Besides homogeneous surface precipitation, other types of precipitation may occur which incorporate the aqueous metal species.

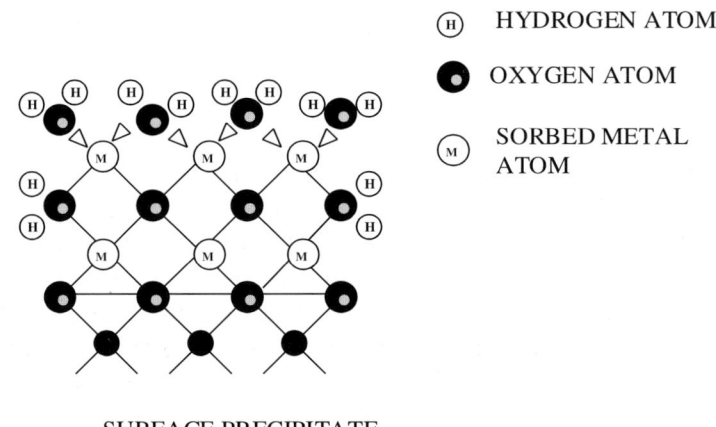

Fig. 12. Schematic sketch of surface precipitation at the mineral/water interface (represented by the horizontal line) following the same nomenclature as Fig. 9. Note the three-dimensional nature of the precipitate propagating away from the surface, distinguishing it from adsorbed species (redrawn after Ref. 35).

One type of precipitate occurs through dynamic dissolution of the solid surface and formation of a new heterogeneous solid phase that includes both material from the substrate and aqueous metal species. This coprecipitation phenomenon occurs more commonly among aluminum-based substrates; for example, Co(II), Ni(II), and Zn(II) ions exposed to aluminum oxide or certain phyllosilicates were shown to form layered double hydroxide (LDH) phases with solubilities lower than those of the initial sorbents [60-65]. This results in a potential for metal sequestration far greater than that expected if simple adsorption of the metals to an unreactive mineral surface were the primary sorption mechanism.

Yet another form of precipitation incorporating both metal from solution and the sorbing substrate occurs through the migration, diffusion, or encapsulation of the metal species into the crystalline structure of the solid, where it may substitute for other ions of the initial substrate. This method of metal incorporation is most common among clay minerals with large interlayer spaces. Pore spaces in poorly crystalline phases or aggregates of nanoscale particles are also attractive sites for adsorbed metals to become structurally assimilated, particularly during a phase transformation to a more crystalline mineral.

However, permanent structural incorporation is often dependent on several parameters including the metal's ion size if proper substitution is expected; for example, studies of various metals coprecipitated with ferrihydrite ($Fe(OH)_3$, am) and aged to facilitate phase transformation to goethite (α-FeOOH) and hematite (α-Fe_2O_3) showed that the extent of sorption increased and sorption reversibility decreased significantly for Mn(II) and Ni(II), consistent with structural incorporation into the crystalline phase [66]. In contrast, Pb(II) and Cd(II) displayed net desorption and constant high sorption reversibility with aging, likely due to decreasing surface areas over time and a lack of structural assimilation.

All metal (co-)precipitates that form through the various pathways described here accomplish the same result of reducing the amount of metals in solution, thereby affecting their mobility and limiting its exposure and bioavailability to organisms. For precipitates with lower solubilities than those of the original mineral sorbent, precipitation is a particularly effective method of permanent sequestration of metals into the solid phase, especially if it leads to sedimentation and burial. Long-range transport of such precipitates in the colloidal size fraction, however, can still present environmental risks if geochemical conditions change to redissolve these precipitates and release metals back into the aqueous phase.

2.1.3. Utility of X-Ray Absorption Spectroscopy in Determining Sorption Mechanisms

X-ray absorption spectroscopy (XAS) is uniquely suited for studying sorption reactions of heavy metal cations as it can directly determine the local molecular structure around a metal ion in a sorbed state. Some of the key features of XAS methods can be summarized as follows [35]. First, synchrotron-based XAS can be used to study most elements in solid, liquid, or gaseous states at concentrations ranging as low as tens of parts per million (ppm, or mg/kg). The high intensity of synchrotron radiation allows the study of very small or very dilute samples and experimental conditions of high or low temperature or pressure and controlled atmospheres, including the presence of fluids such as water. Second, XAS is an element-specific bulk method which provides information about the average local structural and compositional environment of the absorbing atom. Initially used to study model systems synthesized or prepared in the lab, it can also be used to study compositionally complex materials such as natural minerals and determine the speciation, or identity and proportion, of phases in a heterogeneous sample. Third, XAS is a local structural probe which "sees" only the two or three closest shells of neighbours around an absorbing atom (less than about 6 Å) due to the short electron mean free path in most materials. This allows one to obtain quantitative structural information about the local environments around elements in amorphous materials, such as silicate glasses and poorly-crystalline iron oxyhydroxides.

Fourth, for many systems, extended X-Ray absorption fine structure (EXAFS) analysis is capable of yielding average distances accurate to ±0.02 Å and average coordination numbers to ±10-20%, assuming that systematic errors have been minimized in the experiment and data analysis and that static and thermal disorder are small or can be well quantified.

Fifth, the time required for photon absorption is about 10^{-16} seconds, compared to about 10^{-12} to 10^{-14} seconds for interatomic vibrations. Thus XAS averages over all distances around an absorbing atom. XAS involves bombarding a sample with high energy X-rays, generating photoelectrons from a specific element in the sample when the incident beam possesses sufficient energy to excite or eject core electrons from that element. The generation of photoelectrons induces specific electronic scattering interactions between the central absorbing atom and neighbouring atoms [67]. The incident X-ray energy at that point corresponds to the absorption edge of the element for a given electron shell (K, L_{III}, etc.), while the electron scattering interactions give rise to the X-ray absorption near-edge structure

(XANES) and the extended X-ray absorption fine structure (EXAFS) regions (Fig. 13). The EXAFS and XANES regions can be analyzed in the case of samples featuring metal sorption to derive detailed molecular-scale information such as identity and number of nearest-neighbour and second-neighbour atoms around a central metal atom, interatomic distances between the sorbed metal and these neighbouring atoms, oxidation states, and degree of structural disorder, all of which are needed to characterize the mode(s) of metal sorption onto a specific surface. Additionally, EXAFS spectroscopy is non-destructive and samples can be run in situ (i.e., with water present) and require minimal preparation, allowing the analysis of samples under ambient conditions representative of complex natural environments.

Because of the molecular-level information which can be obtained, XAS is often capable of distinguishing between different sorption mechanisms that would be more difficult or impossible with any other analytical technique. For example, an EXAFS spectroscopy study of Hg(II) sorbed to goethite [68] showed first-neighbour oxygen and second-shell iron neighbours at distances of 2.02-2.05 (±0.01) Å and 3.19-3.28 (±0.02) Å, respectively (Fig. 14).

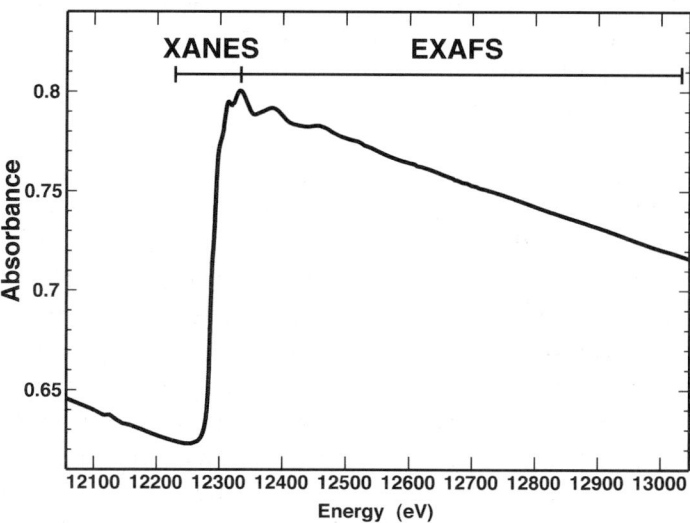

Fig. 13. X-ray absorption spectrum of cinnabar (HgS$_{hexagonal}$) indicating the X-ray absorption near edge structure (XANES) and extended X-ray absorption fine structure (EXAFS) regions.

Since the Hg-O interatomic distance is consistent with distances of Hg both in aqueous solution and in crystalline solids, it cannot be used to distinguish between Hg(II) in the aqueous, sorbed, and solid phases. However, the relatively short Hg-Fe distance determined from EXAFS analysis supports the formation of predominantly bidentate inner-sphere sorption complexes on the goethite surface (Fig. 15).

By comparison, outer-sphere complexation of Hg(II) would yield a much larger Hg-Fe distance of >6 Å, according to modelling results of such a complex using molecular modelling programs. Hg precipitation or multinuclear complex formation can also be ruled out due to the absence of Hg-Hg neighbour interactions in the EXAFS spectra (although such interactions are observed in other systems, such as Hg(II) sorbed to γ-alumina). In a companion study including the ligands chloride and sulfate, EXAFS studies show evidence for Hg(II)-Cl-goethite and Hg(II)-SO$_4$-goethite ternary surface complexation. These and other studies show that information gained from EXAFS spectroscopy can often be used effectively to differentiate between the many different mechanisms of sorption.

Table 3 is a representative summary of EXAFS studies of metal ion sorption complexes at mineral/solution interfaces in model systems in the absence of complexing ligands, and has been edited and updated from a much larger compilation of these results in Ref. 39. Table 3 shows the wide range of sorption mechanisms possible among the systems studied and demonstrates the variability of sorption complexes that can form with no other variables than the metal ion and the substrate. Table 4, also edited and updated from the same reference, lists EXAFS studies of selected metal cations on mineral surfaces in the presence of selected inorganic and organic ligands in model systems.

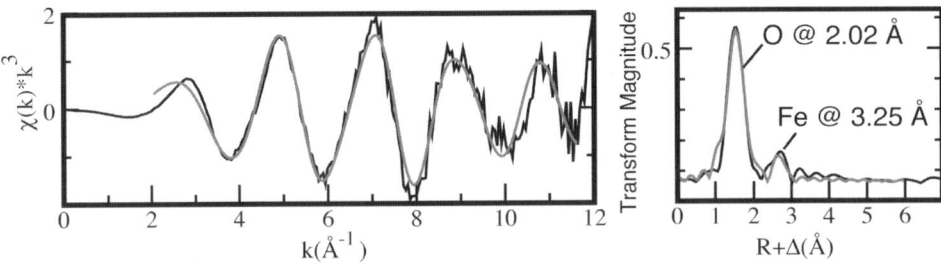

Fig. 14. Fits of the EXAFS spectrum (left) and corresponding Fourier transform (right) for a sample of Hg(II) sorbed to goethite at pH 6; black = raw data, gray = fit. Fitting results determine oxygen nearest neighbors at 2.02 Å and iron second-nearest neighbors at 3.25 Å, indicative of an inner-sphere adsorption surface complex [68].

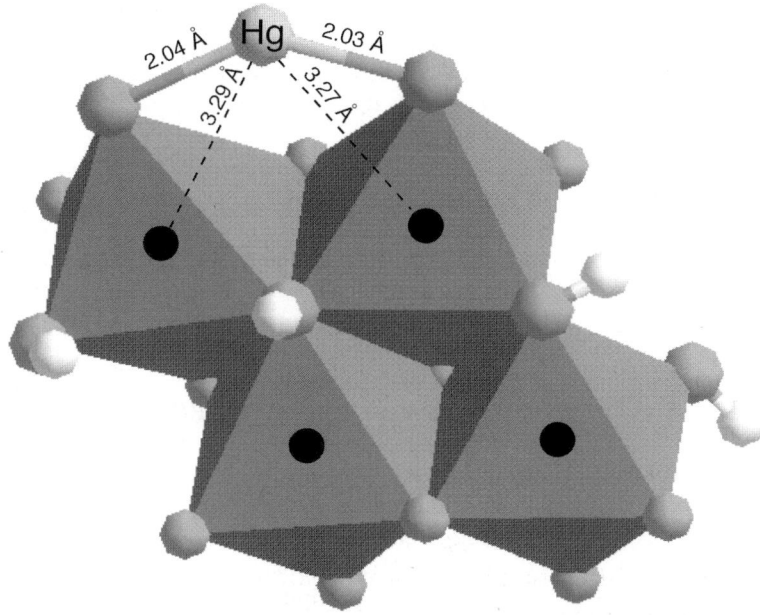

Fig. 15. Proposed Hg(II) bonding configuration on goethite based on EXAFS fitting results, with Hg(II) sorbing as a bidentate inner-sphere complex linked in a corner-sharing arrangement to oxygen atoms of adjacent Fe(O,OH)$_6$ octahedra [68].

In several cases, ternary surface complexes play a role in the effects of these ligands on the sorption of the metals to the solid surfaces.

2.1.4. Model approaches for heavy metal adsorption (H.B. Bradl)

To predict the fate and transport of heavy metals in various environmental media such as soil, surface water, and groundwater, is of great importance. As surface waters and shallow aquifers are especially vulnerable to heavy metal contamination from land application of sewage effluent and sludge, discharge from mining, recharge of contaminated surface waters, and landfilling, the prediction of heavy metal mobility in aquifers and soils is of special interest. For this purpose, both conceptual and quantitative model approaches have been developed.

First theoretical models for adsorption of metal ions on oxides surfaces appeared approximately 30 years ago connected with experimental studies of oxide surfaces such as titration. Theoretical models have been increasingly applied to adsorption data.

Table 3
Summary of EXAFS spectroscopy studies of metal ion sorption complexes at mineral/solution interfaces in model systems in the absence of complexing ligands, updated and edited from a more comprehensive version in Ref. 69. Individual references should be consulted for specific experimental variables.

Sorbate	Sorbate	Dominant Sorbate Geometry or Phase	Reference
Cr(VI)	Magnetite	Cr(III)OOH precipitate	[70]
	Goethite	IS-mono and bidentate	[71]
	am-FeS	(Cr(III),Fe)(OH)$_3$ precipitate	[72]
Cr(III)	am-SiO$_2$	IS-monodentate	[56]
	γ-Al$_2$O$_3$	IS-multinuclear	[73]
	HFO	(Cr,Fe)OOH precipitate	[55]
	Goethite	(Cr,Fe)OOH precipitate	[74]
	HMO	IS-multidentate (initial) OS (CRO$_4^{2-}$) (seconds later)	[75]
Fe(III)	Quartz (1011)*	IS-bidentate	[76]
Co(II)	α-SiO$_2$	IS-multinuclear	[77]
	γ-Al$_2$O$_3$	IS-multinuclear	[52, 53]
	γ-Al$_2$O$_3$	Co(II)-Al LDH precipitate	[50]
	α-Al$_2$O$_3$	Co(II)-Al LDH precipitate	[51, 78]
	TiO$_2$ (rutile)	IS-multinuclear	[53]
	Calcite	lattice substitution	[79]
	Kaolinite	Co(II)-Al LDH precipitate	[64]
	Hectorite	IS (edge sites)	[80, 81]
	Smectite	OS (charge sites-low pH) Co(II)-Al LDH precipitate (high pH)	[82]
Ni(II)	am-SiO$_2$	a-Ni(OH)$_2$ precipitate	[83]
	γ-Al$_2$O$_3$	Ni(II)-Al LDH precipitate	[80]
	Gibbsite	Ni(II)-Al LDH precipitate	[83]
	Ferrihydrite	Ni-coprecipitate	[83]
	HMO	OS-mononuclear	[84]
	Kaolinite	Ni(II)-Al LDH precipitate	[62]
	Pyrophyllite	Ni(II)-Al LDH precipitate	[63]
Cu(II)	am-SiO$_2$	IS-monodentate (mono+ dimers) + Cu(OH)$_2$ precipitate	[85]
	γ-Al$_2$O$_3$	IS-mono or bidentate	[85]
	Goethite	multidentate polymer (pH 8)	[86]
	Ferrihydrite	IS-bidentate (monomers + dimers)	[87]
	TiO$_2$ (anatase)	IS-dimers	[51]
	Birnessite	IS-tridentate, interlayer	[88]
	Montmorillonite	OS + edge sites(mono+dimers)	[89]
Zn(II)	γ-Al$_2$O$_3$	IS bidentate $^{[IV]}$Zn (low Γ) Zn(II)-Al precipitate (high Γ)	[65]
	Ferrihydrite	IS-bidentate$^{[IV]}$Zn) (pH 6)	[90, 91]
	Goethite	IS-bidentate$^{[VI]}$Zn	[92]

Table 3 (continued)

Sorbate	Sorbate	Dominant Sorbate Geometry or Phase	Reference
Zn(II)	HMO	OS-mononuclear	[84]
	Goethite	IS-bidentate$^{[VI]}$Zn	[92]
	HMO	OS-mononuclear	[84]
	Pyrophyllite	Zn(II)-Al LDH precipitate	[93]
	Silica	IS	[94]
	Gibbsite	IS (high SA), Zn(II)-Al LDH precipitate (low SA)	[94]
	Calcite	Lattice substitution	[79]
Cd(II)	HFO	IS (edge sites)	[95]
	Goethite	IS-mononuclear	[96]
	Lepidocrocite	IS-bidentate	[96]
	Akaganeite	IS	[97]
	Schwertmannite	IS	[97]
	γ-MnOOH	IS-bidentate (mononuclear)	[98]
Au(III)	Goethite	IS-bidentate	[99]
Hg(II)	Bayerite	IS-mono- and bidentate	[68]
	γ-Al$_2$O$_3$	IS-dimers	[68]
	Goethite	IS-bidentate	[100, 68]
Pb(II)	γ-Al$_2$O$_3$	IS-monodentate	[101, 54]
	Boehmite	IS-multinuclear	[102]
	Goethite	IS-bidentate-mononuclear	[103]
	Ferrihydrite	IS-bidentate	[83]
	Hematite	IS-bidentate-mononuclear	[103]
	Birnessite	IS-bidentate-mononuclear	[75]
	Calcite	Lattice substitution	[104]

Abbreviations: am = amorphous, HFO = Hydrous Ferric Oxide, HMO = Hydrous Manganese Oxide. Modified after Ref. 39.

An overview on these model approaches is given in Ref. 105-107. Since the 1990s, experimental confirmation of surface stoichiometries is possible by using surface spectroscopic techniques such as EXAFS (extended X-ray adsorption fine structure) or XANES (X-ray adsorption near edge structure). These techniques provide a deeper inside into the nature and the environment of the adsorbed species and lead to a sharper description of the surfaces involved. Thus, the fit of theoretical models to experimental data is improved [45, 59, 74, 108-110]. There are two different approaches to adsorption modelling of heavy metal adsorption. The empirical model approach aims at empiric description of experimental adsorption data, while the semi-empirical or mechanistic model approach tries to give comprehension and description of basic mechanisms.

Table 4
EXAFS spectroscopy studies of selected metal cations on mineral substrates in the presence of selected inorganic and organic ligands in model systems (edited and updated after Ref. 39)

Ternary System	References
Co(II)	
Co(II)-EDTA-pyrolusite	[111]
Ni(II)	
Ni(II)-EDTA-am-SiO_2	[112]
Cu(II)	
Cu(II)-humate-goethite	[113]
Cu(II)-bipyridine-am-SiO_2	[114]
Cu(II)-glutamate-α-Al_2O_3	[115]
Cu(II)-humate-illite	[116]
Cu(II)-xanthate-ZnS	[117]
Cu(II)-phosphate-goethite	[118]
Cu(II)-chloride-palygorskite	[119]
Cd(II)	
Cd(II)-sulfate-goethite	[120]
Cd(II)-phosphate-goethite	[120]
Cd(II)-humate-goethite	[120]
Cd(II)-citrate-goethite	[120]
Cd(II)-oxalate-goethite	[120]
Hg(II)	
Hg(II)-chloride-goethite	[121]
Hg(II)-chloride-goethite	[45]
Hg(II)-sulfate-goethite	[45]
Hg(II)-chloride-bayerite	[45]
Hg(II)-sulfate-bayerite	[45]
Hg(II)-chloride-γ-Al_2O_3	[45]
Hg(II)-sulfate-γ-Al_2O_3	[45]
Pb(II)	
(Pb)II-sulfate-goethite	[102]
Pb(II)-sulfate-boehmite	[102]
Pb(II)-phosphate-goethite	[102]
Pb(II)-phosphate-boehmite	[102]
Pb(II)-chloride-goethite	[122]
Pb(II)-EDTA-goethite	[123]
Pb(II)-carbonate-goethite	[43]
Pb(II)-sulfate-goethite	[42]
Pb(II)-sulfate-goethite	[124]
Pb(II)-malonate-hematite	[125]

Modified after Ref. 39.

Ref. 126 and 127 give a general overview on those model approaches. In the empirical model, the model form is chosen a posteriori form the observed adsorption data. To enable a satisfying fitting of experimental data the mathematical form is chosen to be as simple as possible and the number of adjustable parameters is kept as low as possible. Parameters are adjusted according to only a limited number of variables such as equilibrium metal concentration in the liquid phase and are therefore of only limited value. Nevertheless, empirical models can be very useful if one only aims at the empirical description of experimental data. In the mechanistic or semi-empirical model, the mathematical form is chosen a priori by setting up equilibrium reactions linked by mass balances of the different components and surface charge effects. As the number of adjustable parameters is higher the mathematical form of mechanistic models is more complex than that of empirical models. Due to the variety of components taken into account a higher number of experimental variables are required, which makes mechanistic models in general more valid than empirical models. Yet the difference between empirical and mechanistic models is often not very distinct. Simple empirical models may be extended by considering additional mechanisms such as competition for sorption sites or heterogeneity of solid phase. One of the main differences between the two model approaches is that mechanistic models include electrostatic terms, whereas empirical models do not.

Empirical models are usually based upon simple mathematical relationships between concentration of the heavy metal in the liquid phase and the solid phase at equilibrium and at constant temperature. This equilibrium can be defined by the equality of the chemical potentials of the two phases [128]. These relationships are called isotherms. Monolayer adsorption phenomena of gases on homogeneous planar surfaces were first explained mathematically and physically by Langmuir in 1916 [129]. Langmuir`s theory was based upon the idea that, at equilibrium, the number of adsorbed and desorbed molecules in unit time on unit surface are equal. The lateral interactions and horizontal mobility of the adsorbed ions were neglected. Later, statistical thermodynamics were incorporated and new isotherms for homogeneous surfaces were derived [130].
The classical thermodynamic interpretation of adsorption is given by Gibbs [131] who introduced the idea of a dividing surface (the so called Gibbs surface). He also proved that, in any case of adsorption, the excess adsorbed amount is the solely applicable and acceptable definition which should be considered in every calculation and measurement. An isotherm of multilayer gas-solid adsorption has been developed by Brunauer, Emmett, and Teller [132], the so called BET equation. The isotherms most

commonly used for empirical description of heavy metal adsorption on soils are referred to as general purpose adsorption isotherms or GPAI. One well-known isotherm is the Langmuir isotherm, which has been originally derived for adsorption of gases on plane surfaces such as glass, mica, and platinum [133]. It is applied for adsorption of heavy metal ions onto soils and soil components in the form

$$q_i = b\left(\frac{Kc_i}{1+Kc_i}\right) \tag{8}$$

where the quantity q_i of an adsorbate i adsorbed is related to the equilibrium solution concentration of the adsorbate c_i by the parameters K and b. The steepness of the isotherm is determined by K. K can be looked upon as a measure of the affinity of the adsorbate for the surface. The value of b is the upper limit for q_i and represents the maximum adsorption of i determined by the number of reactive surface adsorption sites. The parameters b and K can be calculated from adsorption data by converting equation (8) into the linear form:

$$\frac{q_i}{c_i} = bK - Kq_i \tag{9}$$

Then the ratio q_i/c_i (the socalled distribution coefficient K_d) can be plotted against q_i. If the Langmuir equation can be applied, the measured data should fall on a straight line with slope of –K and x intercept of bK. The Freundlich equation has the form

$$q_i = ac^n_i \tag{10}$$

where a and n are adjustable positive valued parameters with n ranging only between 0 and 1. For n = 1 the linear C-type isotherm would be produced. The parameters are estimated by plotting log q_i against log c_i with the resulting straight line having a y intercept of log a and a slope of n. The Freundlich equation will fit data generated from the Langmuir equation. Converting the Freundlich equation (10) to the logarithmic form, the equation becomes

$$\log q_i = \log a + n \log c_i \tag{11}$$

Considering the adsorption of heavy metals, q_i is equated to the total adsorbed metal concentration (M_T in mg kg^{-1}) and c_i is equated to the dissolved metal concentration (M_S in mg l^{-1}) in the batch solution at equilibrium with the solid. Defining log a as a constant, the equation becomes

$$\text{Log } M_T = C + n \log M_S \tag{12}$$

This form of the equation can be used to relate the amount of heavy metal adsorbed on specific media to the dissolved concentration of free metal ions. A generalized Langmuir-Freundlich isotherm can also be used as a model base for the interpretation of competitive adsorption isotherms.

The Langmuir equation for adsorption of heavy metal ions in soils and clays has been derived and applied by many authors [134-139]. Also deviations between experimental data and calculated behaviour have been observed, which has been explained by the presence of competition of different adsorbates for the adsorption sites on the surface. Consequently, the original Langmuir equation (8) had to be modified to include competitive effects and can be expressed as the so called competitive Langmuir equation:

$$q_1 = \frac{bK_1c_1}{1+K_1c_1+K_2c_2} \tag{13}$$

A well known situation for competitive behaviour is the influence of pH on heavy metal adsorption. As it can be shown in Figure 16, pH and ionic strength effects on As(III) adsorption on a Wyoming montmorillonite can be interpreted as a competition between protons and heavy metal for the adsorption sites [140]. Another source of deviations observed between experimental data and calculated behaviour according to single-site isotherms is the heterogeneity of adsorption sites, which means that the interaction between metal and surface site cannot be described by a single affinity parameter. This phenomenon is frequently encountered when dealing with clays due to imperfections in the crystal lattice and the different nature and position of charges on the surface. There are two different ways, by which heterogeneity effects can be included into modified single-site Langmuir-type isotherms. First, a discrete number of different types of sites, which is characterized by different concentration and affinity for the adsorbate can be taken into account.

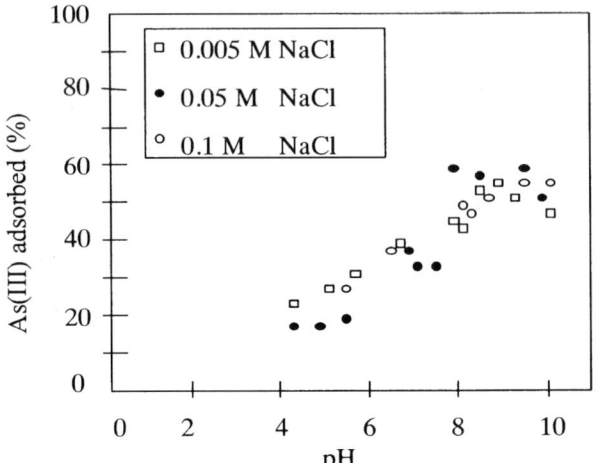

Fig. 16. Adsorption of As(III) on Wyoming bentonite as a function of pH and ionic strength. Reaction conditions: 25 g/l clay, [As(III)]₀ = 0,4 µM, reaction time = 16 h (redrawn after Ref. 140).

Adsorption is expressed as the sum of the adsorption on Z types of sites, each one following the Langmuir isotherm [114, 140] resulting in the multisite Langmuir isotherm

$$q_1 = \sum_{j=1}^{Z} \frac{b_1 K_1 c}{1 + K_1 c} \tag{14}$$

with 2Z adjustable parameters and j referring to each adsorption site. Second, a single type of site with a continuous distribution of the affinity parameter can be considered. To do this, it is assumed that the affinity parameter in the single-site isotherm is continuously distributed according to a site affinity distribution function (SADF). An overall isotherm can then be derived by integrating the single-site or local isotherm along SADF. If Φ_t (C) is the overall isotherm and $\Psi(K, C)$ the local isotherm, the overall isotherm can be built according to

$$\Phi_t(c) = \int \Psi(K, c) f(k) \, dk \tag{15}$$

where f(k) is the SADF and f(k)dk is the fraction of sites with K comprised among k and k+dk. By taking Eq. [1], which is the single-site Langmuir as the local isotherm, analytical solutions of Eq. (14) have been calculated for three types of distribution function f(K), which are of the forms [141]

Langmuir-Freundlich: $$\Phi_{(c)} = \frac{(Kc)^\beta}{1+(Kc)^\beta} \qquad (16)$$

Generalized-Freundlich: $$\Phi_{(c)} = \left(\frac{Kc}{1+c}\right)^\beta \qquad (17)$$

Toth: $$\Phi_{(c)} = \frac{Kc}{\left[1+(Kc)^\beta\right]^{1/\beta}} \qquad (18)$$

These equations are characterized by the three adjustable parameters b, K, and β. β is a heterogeneity index ranging from 0 to 1 (corresponding to very flat to very sharp distribution). For β=1 all composite isotherms will revert to the single-site Langmuir isotherm. While modifications considering influence of competition and surface heterogeneity have extended the original Langmuir isotherm on the one hand, the number of adjustable parameters has been increased. Often, this model is too flexible in respect to experimental data. This is also of importance when discussing mechanistic models.

General purpose adsorption isotherms do not take into account the electrostatic interactions between ions in solution and a charged solid surface as it is the case in most surfaces encountered when dealing with soils such as clay minerals, metal (hydr)oxides, and others. Adsorption as a function of pH and ionic strength is described as a competition for adsorption sites only. The effects of modifying the electric properties of the surface due to the adsorption of charged ions and its effect on affinity parameters cannot be taken into account in using GPAI.

The term "mechanistic models" therefore refers to all models, which describe adsorption by accounting for the description of reactions occurring between ions in solution and the charged surface. Models available may vary in the description of the nature of surface charge, the number and position of potential planes, and the position of the adsorbed species.

The two main reactions occurring are ion exchange, which is mainly of electrostatic nature, and surface complexation, which is mainly of chemical nature. Surface complexation models allow the description of macroscopic adsorption behaviour of solutes at mineral-aqueous solution interfaces [142]. Combined with an electric double layer model, this is a powerful approach to predict ion adsorption on charged surfaces predominant in soils such as clays and metal (hydr)oxides [143].

There are different electrostatic models available, which can be distinguished by the way the double layer at the solid/solution interface is described. The three models used most are the constant capacitance model, the diffuse layer model and the triple layer model, which describe the double layer by two, three and four potential adsorption planes [144].

The Constant Capacitance Model was developed by Stumm, Schindler and others [145-147] and considers the double layer as consisting of two parallel planes (Fig. 17). The surface charge σ_0 is associated to the one plane and the counter charge σ_1 is associated to the other plane.

The model contains the following assumptions: first, all surface complexes are inner-sphere complexes formed through specific adsorption; second, the constant ionic medium reference state determines the activity coefficients of the aqueous species in the equilibrium constants and no surface complexes are formed with ions from the background electrolyte; third, surface complexes exist in a chargeless environment in the standard state; and fourth, surface charge drops linearly with distance x from the surface and is proportional to the surface potential Ψ through a constant capacitance G:

$$\sigma_0 = G \Psi \tag{19}$$

The surface charge σ_0 is simply calculated by summation of all specifically adsorbed ions while all non-specifically adsorbed ions are excluded from plane 0. In this simple model, the only adjustable parameter is the capacitance G, which has to be optimized by regression of the experimental adsorption data. As for the application of the constant capacitance model (CCM) to adsorption of heavy metal ions onto clays and metal (hydr)oxides a combined ion exchange-surface complexation model with two kinds of binding sites was proposed [148]. One kind of site consists of a weakly acidic site ($\equiv$XH), which can undergo ion exchange with both Me^{2+} and Na^+ ions, while the other kind of site is formed by amphoteric surface hydroxyl groups ($\equiv$SOH), which form surface complexes $\equiv SOMe^{2+}$ and $(\equiv SO)_2Me$ and bind Na^+ as outer sphere complexes.

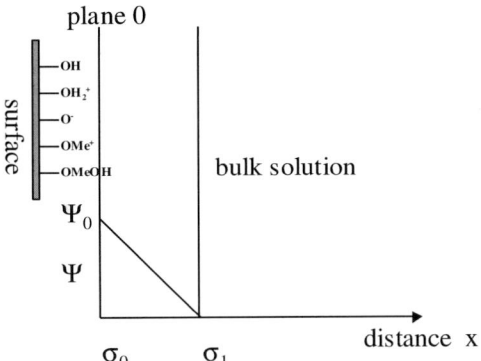

Fig. 17. Schematic illustration of the interface according to the Constant Capacitance Model (CCM) (redrawn after Ref. 126).

The CCM is looked upon as a limiting case of the basic Stern model [149] for high ionic strengths where $I \geq 0.1$ mol L^{-1}, although it is more often applied to lower ionic strengths in the literature [126]. The CCM is the simplest of the surface complexation models with the least number of adjustable parameters. It can only be used for the description of specifically adsorbed ions and is unable to describe changes in adsorption occurring with changes in solution ionic strength. The generalized diffuse layer model was introduced by Ref. 150 and developed further by Ref. 151. The model contains the following assumptions: first, all surface complexes are inner-sphere complexes formed through specific adsorption; second, no surface complexes are formed with ions from the background electrolyte; the infinite dilution reference state is used for the solution and a reference state of zero charge and potential is used for the surface. Three different planes are introduced (Fig. 18). First there is the surface plane 0 where ions are adsorbed as inner sphere complexes, second the plane d, which represents the distance of closest approach of the counter ions, and third a plane, after which surface potential is considered to drop to zero. The surface charge σ_0 is determined as the sum of all specifically adsorbed ions like it is calculated in the CCM. Yet the capacitance G is calculated by the Gouy-

Chapman theory and the ionic strength is taken into account. For a z:z electrolyte the relation $\sigma_0 = f(\Psi)$ can be calculated as:

$$\sigma_0 = -\sigma_d = \sqrt{8\varepsilon\varepsilon_0 RTI10^3} \sinh\left(\frac{zF\Psi_0}{2RT}\right) \tag{20}$$

where ε is the dielectric constant, ε_0 the permittivity of free space, and I the medium ionic strength. The DLM has been presented as a limiting case of the Stern model for low ionic strength $I \leq 0.1$ mol L^{-1}. The advantage of the DLM is that it is able to describe adsorption as a function of changing solution ionic strength and has only a small number of adjustable parameters. The CCM and the DLM have both been developed as limiting cases for high and low ionic strength. The triple layer model (TLM), however, can be applied to the whole range of ionic strengths and is a version of the extended Stern model [152, 153]. This model comprises four planes (Fig. 19), and electrolyte and metal ions can be adsorbed as inner or outer-sphere complexes depending on where the different ions are located. The adsorption of ions on the additional plane β creates a charge σ_β and electroneutrality can be expressed as:

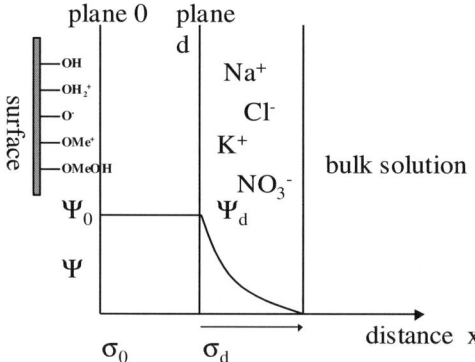

Fig. 18. Schematic illustration of the interface according to the Diffuse Layer Model (DLM) (redrawn after Ref. 126).

$$\sigma_0 + \sigma_\beta + \sigma_d = 0 \qquad (21)$$

Considered that the regions between planes 0 and β and between β and d are plane condensers with capacitance G_1 and G_2, respectively, the relation between charge and potential is given by:

$$\Psi_0 - \Psi_\beta = \frac{\sigma_0}{G_1} \qquad (22)$$

and

$$\Psi_\beta - \Psi_d = \frac{\sigma_0 + \sigma_d}{G_1} = -\frac{\sigma_d}{G_2} \qquad (23)$$

The relation between charge and potential on the diffuse plane d can be calculated by the Gouy-Chapman theory as follows:

$$\sigma_d = \sqrt{8\varepsilon\varepsilon_0 RTI 10^3} \sinh\left(\frac{zF\Psi_d}{2RT}\right) \qquad (24)$$

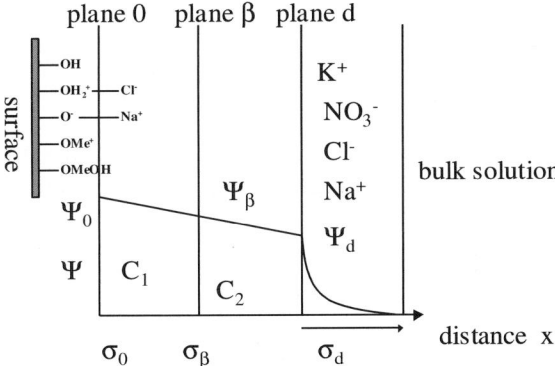

Fig. 19. Schematic illustration of the interface according to the Triple Layer Model (TLM) (redrawn after Ref. 126).

In a more general approach, the adsorption of metal ions can occur either at the 0 plane or the β plane [154]. If the TLM is to be applied, the determination of the two capacitances G_1 and G_2 is necessary. The TLM is more complex and contains more adjustable parameters than the other models described above. It offers the advantage of being more realistic because both inner- and outer-sphere surface complexation reactions can be taken into account.

There are other model approaches such as the ONE-pK model and the TWO-pK model [155-157]. These models are special cases of a more generalized model called the MUltiSIte Complexation model (MUSIC) which considers equilibrium constants for the various types of surface groups on the various crystal planes of oxide minerals [158, 159]. These models are very complex and involve a large number of adjustable parameters.

Once the set of equilibrium reactions and the related material balances have been defined the model can be fit to the experimental data by adjustment of unknown parameters such as site concentration and species formation constants. There are two critical points when defining the model structure. First, often the set of equilibrium reactions is more or less hypothesized, and second, the model has too many adjustable parameters with respect to experimental constraints, i.e. the model structure becomes too flexible. Defining the model structure follows in fact a trial-and-error approach where the model definition is also a part of the overall fitting procedure to the experimental data. As a result, the mechanistic model approach is reduced to a semi-empirical one as it was discussed earlier. If the model is too flexible different sets of adjustable parameters may result in similar description of experimental data [160, 161].

Also the mathematical form of the model and the quality of the experimental data may cause poor parameter identifiability. Therefore, it is often difficult to choose from different models and little information can be derived about the physical reality. In order to overcome these difficulties it is best to introduce as many constraints as possible for both model form and parameter values and to determine as many variables experimentally as possible [126]. For example, concentration or adsorption of all species in chemical equilibria as well as surface charges and potentials should be calculated and initial and final concentrations of all soluble components should be measured in order to obtain the numerical solution of the model. Often, only a simplified approach is used, i.e. the acid-base properties of the absorbent in absence of the heavy metal of interest are determined by titration. Then, heavy metal adsorption is determined as a function of pH or ionic strength [162].

Alternatively, it is possible to use all experimental variables available simultaneously [163]. In this modelling approach, three dependent variables (heavy metal adsorption, acid-base titration, and surface charge) were expressed as a function of three independent variables (pH, ionic strength, and heavy metal concentration in the solution at equilibrium) by using a multivariate non linear least squares procedure for fitting. It was shown that all models used were able to successfully simulate heavy metal adsorption on clays as a function of pH and heavy metal concentration at equilibrium. However, most adjustable parameters (e.g. the formation constants) are estimated with large uncertainty. The best way to overcome the problem of poor identifiability is the further increase of calculated variables, which can be determined experimentally.

As for surface potentials, good agreement between the measured zeta potential and the calculated diffuse layer potential in a TLM for the sphalerite/water interface has been reported [164], but for other oxide/water and clay/water interfaces such correspondences have not been observed [165-167]. As for the determination of adsorbed species at the interface, several spectroscopic methods can be used for the determination of surface reactions and species, which are important for the adsorption process [124, 168, 169].

2.1.5. Geochemical parameters influencing adsorption

Adsorption of heavy metal ions is influenced by a variety of parameters, the most important ones being pH, type and speciation of metal ion involved, heavy metal competition, and in case of soil, soil composition and aging [170]. For soils, soil pH is the most important parameter influencing metal-solution and soil-surface chemistry. The dependence of heavy metal adsorption on e.g. clays on solution pH has been noticed early [171]. The number of negatively charged surface sites increases with pH. In general, heavy metal adsorption is small at low pH values. Adsorption then increases at intermediate pH from near zero to near complete adsorption over a relatively small pH range; this pH range is referred to as the pH-adsorption edge. At high pH values, the metal ions are completely removed. Fig. 20 shows the pH dependence of Cd, Cu, and Zn adsorption onto a sediment composite, which consists basically of Al-, Fe-, and Si-oxides. 50% of the copper is adsorbed at pH 4.1, and the slope of the Cu adsorption curve is steeper than the Cd or Zn slopes.

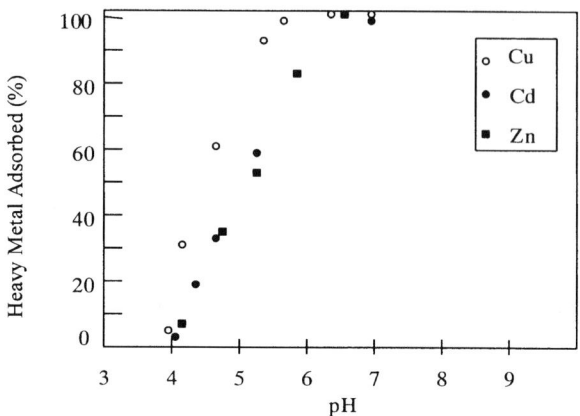

Fig. 20. Cd, Cu, and Zn adsorption onto sediment composite in 10^{-3} M NaNO$_3$ (redrawn after Ref. 172).

Fig. 21 shows the adsorption of different heavy metals onto soil humic acid [170]. 50 % of the Cd or Zn is adsorbed between pH 4.8 –4.9. In general, adsorption of heavy metals onto oxide and humic constituents of soil follows the basic trend of metal-like adsorption, which is characterized by increased adsorption with pH [173, 174]. The pH is a primary variable, which determines cation and anion adsorption onto oxide minerals.

Universally consistent rules of metal selectivity cannot be given as it depends on a number of factors such as the chemical nature of the reactive surface groups, the level of adsorption (i.e. adsorbate/adsorbent ratio), the pH at which adsorption is measured, the ionic strength of the solution, in which adsorption is measured, which determines the intensity of competition by other cations for the bonding sites, and the presence of soluble ligands that could complex the free metal. All these variables may change the metal adsorption isotherms. Competition from monovalent metal in background electrolytes has relatively little effect on adsorption of heavy metals, although presence of Ca ions does suppress adsorption on Fe oxide [175]. Preference or affinity is measured by a selectivity or distribution coefficient K_d [176].

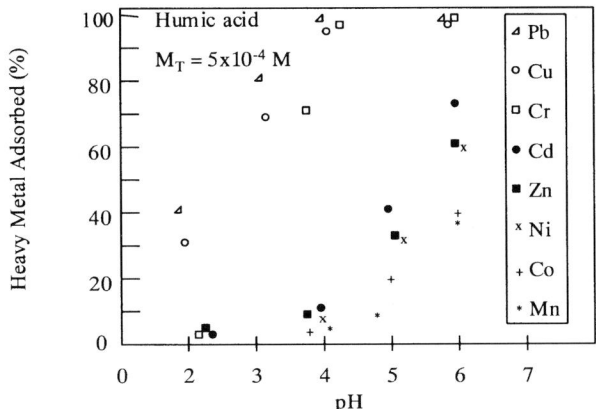

Fig. 21. Adsorption of Pb, Cu, Cr, Cd, Zn, Ni, Co, and Mn onto humic acid as a function of pH (redrawn after Ref. 170).

The reduction of this selectivity with increased adsorption is observed for metal adsorption on both clays as soil components and pure minerals [177, 178].

As for soils, the soil type and composition plays an important role for heavy metal retention. In general, coarse-grained soils exhibit lower tendency for heavy metal adsorption than fine-grained soils. The fine-grained soil fraction contents soil particles with large surface reactivities and large surface areas such as clay minerals, iron and manganese oxyhydroxides, humic acids, and others and display enhanced adsorption properties. Clays are known for their ability to effectively remove heavy metals by specific adsorption and cation exchange as well as metal oxyhydroxides [179]. Soil organic matter exhibits a large number and variety of functional groups and high CEC values which results in enhanced heavy metal retention ability mostly by surface complexation, ion exchange, and surface precipitation [180, 181]. X–ray absorption spectroscopy and ESR studies suggest that Pb, Cu, and Zn form inner-sphere complexes with soil humic acid [182]. Also ageing may play an important role for heavy metal retention as stable surface coatings are formed as a function of time and heavy metal retention onto aged soils acquires a more irreversible character [172].

2.2. Redox Reactions

The oxidation state of heavy metal ions depends on the redox conditions encountered in the environment. Changes in redox conditions may trigger changes in the oxidation states of the individual metal. Especially Hg, As, Se, Cr, Mn, and Fe are sensitive to these changes. For redox transformation of individual metals, the reader is referred to the individual chapters in the behaviour of selected heavy metals. Redox potentials are mainly influenced by pH and temperature. In the case of soils, sediments, and rivers, these conditions may change according to season and amount of rainfall. In the case of agricultural activities, draining and flooding of e.g. rice paddies will strongly influence metal behaviour [183]. Table 5 shows the redox sequence for elements O, N, Mn, Fe, S, and C in wetlands and other submerged soils. Sulfide-bound metals can easily be solubilized upon aeration of sediments, which can be an important heavy metal source when polluted harbour sediments are spread on the ground [183]. Fig. 22 shows some trends in heavy metal solubility in relation to pH and Eh. Under reducing conditions, heavy metals are removed from solution as sulfide minerals. In slightly reducing or slightly oxidizing environments and for medium to acidic pH range, the surface reactive Fe and Mn oxyhydroxides are solubilized when Fe(III) and Mn(IV) are reduced to soluble species [184].

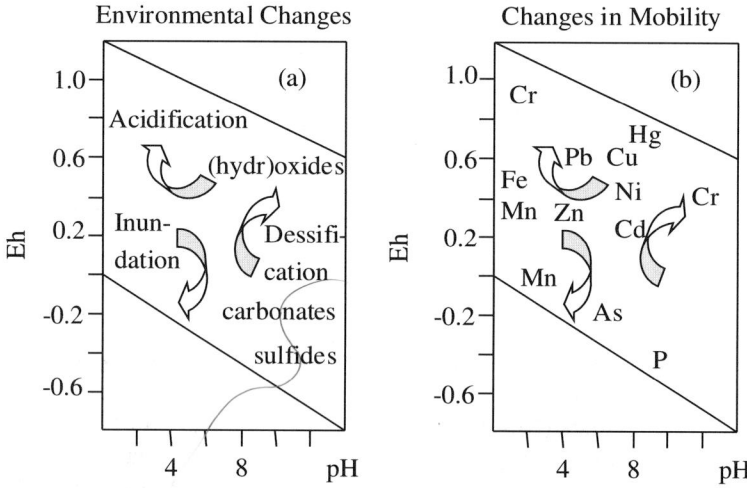

Fig. 22. Trends in heavy metal solubility in relation to pH and Eh (in the absence of dissolved and organic matter). a) main minerals controlling the solubility of heavy metals; b) trends of increasing solubility (redrawn after Ref. 184).

2.3. Weathering

As it has been stated before, weathering refers to the disintegration and alteration of rocks and minerals and results in soil formation by physico-biogeochemical processes. The most important factor for rock weathering is climate. Chemical weathering requires water, which is lacking in arid climates, so physical weathering will predominate. On the contrary, in temperate and warm climates, biogeochemical weathering is enhanced resulting in more clay formation. The most important biotic processes are decomposition of organic matter, respiration of living biomass, which leads to CO_2 and H_2CO_3 release, and release of humic and other organic acids.

2.4. Driving Factors

The most important geochemical factors are oxidation-reduction reactions, hydrolysis, carbonation, precipitation-dissolution reactions, and acid-base reactions. The biogeochemical processes discussed above are mainly, yet not exclusively, influenced by the four master variables pH, redox potential, cation exchange capacity, and chemical speciation of the individual element. These factors are discussed below.

2.4.1. pH and Redox Potential

pH and redox potential are crucial for heavy metal mobility in the environment. Fig. 23 summarizes the solubility patterns of heavy metals and their mobility in the absence of significant metal-organic interactions. Thus, mobility of heavy metals can in general be increased by three major factors: first, lowering the pH, which can dissolve and/or desorb metals from the solid phases, second, altering the redox conditions to induce moderate to high redox potential, and third, increasing salt concentration.

Table 5
Oxidation-reduction reactions of primary importance in wetland soils, rice paddies, and sediments

Element	Redox couple	
	Oxidized species	Reduced species
Oxygen	Oxygen	H_2O
Nitrogen	Nitrate (NO_3^-)	NH_4^+, N_2O, N_2
Manganese	Mn^{4+} (manganic: MnO_2)	Mn^{2+} (manganous: MnS)
Iron	Fe^{3+} (ferric: $Fe(OH)_3$)	Fe^{2+} (ferrous: FeS, $Fe(OH)_2$)
Sulfur	SO_4^{2-} (sulfate)	S^{2-} (sulfide: H_2S, FeS)
Carbon	CO_2 (carbon dioxide)	CH_4 (methane)

Table 5 (continued)
Oxidation-reduction reactions of primary importance in wetland soils, rice paddies, and sediments

Element	Reaction	Redox potential for reactions (mV)[a]
Oxygen	$0.5\, O_2 + 2\, e^- + 2\, H^+ = H_2O$	700 - 400
Nitrogen	$NO_3^- + 2\, e^- + 2\, H^+ = NO_2^- + H_2O$	220
Manganese	$MnO_2 + 2\, e^- + 4\, H^+ = Mn^{2+} + 2\, H_2O$	200
Iron	$FeOOH + e^- + 3\, H^+ = Fe^{2+} + 2\, H_2O$	120
Sulfur	$SO_4^{2-} + 8\, H^+ + 7\, e^- = 0.5\, S_2^{2-} + 4\, H_2O$	-75 to -150
Carbon	$CO_2 + 8\, H^+ + 8\, e^- = CH_4 + 2\, H_2O$	-250 to -350

[a]Redox potentials are approximate values and will vary with soil pH and temperature. Reprinted with permission from: D.C.Adriano, Trace Elements in Terrestrial Environments, Springer, New York, 2001, p. 47.

The knowledge of these factors is essential for evaluating the potential for heavy metal removal, e.g. for decontamination and remediation purposes. pH and Eh also strongly influence reaction of heavy metals with natural and synthetic complexing agents.

2.4.2. Complexing Agents

Heavy metals are usually complexed with natural ligands such as humic or fulvic acids or anthropogenic complexing agents such as EDTA or NTA. Complexation will alter metal reactivity, affecting properties such as catalytic activity, toxicity, and mobility [185]. The adsorption of a heavy metal onto the surface of a hydrous oxide is also represented as the formation of a metal complex. As hydrous oxide surfaces display amphoteric properties, they are able to coordinate with ligands as well. These three components -metal, ligand, and reactive surface- afford the formation of a ternary complex. This ternary complex can be exceedingly stable and may possess properties, which are very different from those of the individual component species. The formation of a ternary surface complex can be explained by two different mechanisms. First, bonding of the complex occurs through the metal to the surface:

$$\underline{S}\text{-OH} + Me^{n+} + H_m Lig \leftrightarrow \underline{S}\text{-OMe-Lig}^{(n-m-1)+} + (m+1)\, H^+ \qquad (25)$$

where Lig represents the ligand and $\underline{S}$-OH represents a hydroxyl functional group on the oxide surface.

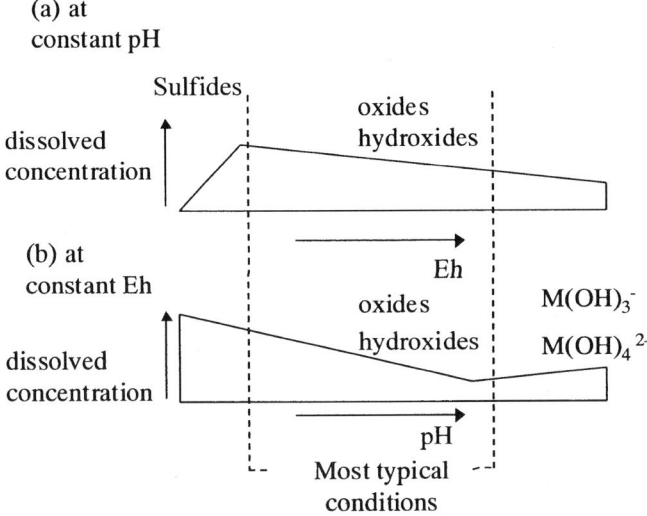

Fig. 23. Effect of Eh (a) and pH (b) on the solubility of heavy metals (redrawn after Ref. 183).

The surface complex is designated as "metal-like" or "type A" [186, 187]. This mechanism is usually characterized by increasing adsorption with increasing pH (Figure 24a). Second, the ligand may form a bridge between the surface and the metal, which is only possible when it is multidentate so it can coordinate with both species:

$$\underline{S}\text{-OH} + Me^{n+} + H_m Lig \leftrightarrow \underline{S}\text{-Lig-Me}^{(n-m-1)+} + (m+1)\, H^+ + H_2O \qquad (26)$$

Adsorption via a ligand bridge is classified as "ligand-like" or "type B" and occurs preferably at low pH (Figure 24b). A variety of studies has been conducted on metal complex adsorption. Only a few studies have examined the adsorption of metal-inorganic complexes. The majority of studies on ternary complexes has focused on the adsorption of metals complexed with EDTA and related chelates. The presence of SO_4^{2-} has been reported to increase Cd(II) adsorption onto goethite over that in the presence of the more inert co-ion NO_3^- [188]. This behaviour was explained by metal-like ternary surface complex formation:

$$\underline{S}\text{-OH} + Cd^{2+} + SO_4^{2-} \leftrightarrow \underline{S}\text{-OCd}^+ \text{-} SO_4^{2-} + H^+ \qquad (27)$$

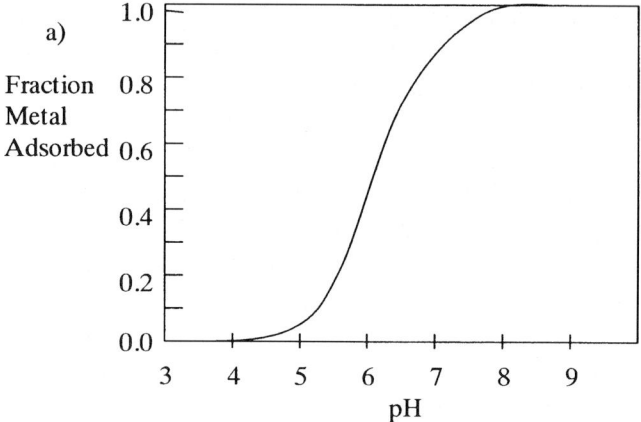

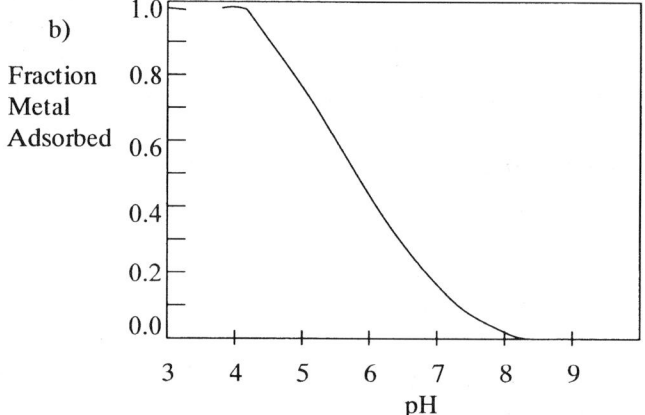

Fig. 24. Schematic representation of metal-like (a) and ligand-like adsorption (b).

Similar reactions have been suggested the formation of 1:2 Cu : P_2O_7 surface complexes on iron oxyhydroxide [189] and Ag^+: $S_2O_3^{2-}$ complexes on amorphous iron oxide [190]. This mechanism has been doubted by the results of some spectroscopic examinations [191]. EXAFS has been used to evaluate several ligands that have shown enhancement of Cd(II) adsorption onto oxides on goethite. No local coordination between S and Cd and between P and Cd could be found. It was suggested that Cd sorption enhancement due to sulphate and phosphate resulted from the reduction of oxide surface charge caused by anion adsorption and could not be attributed to the formation of ternary complexes.

Ternary complex formation can both enhance and diminish heavy metal adsorption by soils depending on pH conditions and complexing agents involved. As for humic acid, it is known that under acidic to neutral pH conditions, significant amounts can be adsorbed to positively charged soil mineral surfaces (such as Fe- and Al-oxides and oxyhydroxides), which may lead to charge reversal [192]. Humic-coated mineral surfaces strongly adsorb heavy metal ions, which will lead to diminished heavy metal mobility in groundwater [193, 194]. At higher pH values, the relative abundances of anionic forms of humic acid increase in aqueous solution. Aqueous complexation between these ligands and metals can significantly enhance heavy metal mobility [194, 195]. Stable anionic complexes (e.g. those with EDTA) are not as strongly adsorbed as the sole metal ions at higher pH, as the negatively charged surface repulses such complexes [196].

Various studies have been conducted on metal-EDTA complex adsorption as EDTA has strong complexing abilities and is widespread in the environment due to its numerous commercial and industrial uses. The adsorption of metals on various oxides of iron, aluminium, titanium, and silicon has been studied and has always been found to be ligandlike, as described in Fig. 24a with significant adsorption occurring at low pH decreasing to almost zero at pH near neutral. At very low pH (2 to 3) the complex becomes unstable so divergence of metal ad EDTA adsorption occurs.

Only very little difference occurs between adsorption of different divalent metal types -EDTA complexes onto the same surface [197-199]. Studies of adsorption of Co(II)-, Cu-, Ni-, Pb-, and Zn-EDTA onto goethite showed overlapping adsorption (Fig. 25). The only exception was Pd-EDTA, which has a much larger aqueous stability constant. The formation of adsorbed Cd-EDTA has been implicated in inhibiting the desorption of Cd(II) from goethite [200]. Co(II)-EDTA adsorption onto goethite [201] and a poorly-crystalline iron-oxide coated sand [202] exhibited ligand-like behaviour. The adsorption of Co(II)-EDTA onto several subsurface sediments was similar to that onto common Fe and Al oxides [201].

The adsorption of metal-EDTA complexes onto several hydrous oxides was modelled using a surface complexation reaction [123]:

$$\underline{S}\text{-OH} + \text{Me-EDTA}^{2-} + H^+ \leftrightarrow \underline{S}\text{-EDTA-Me}^{2-} + H_2O \quad (28)$$

A constant capacitance electrical double layer expression was employed. The surface stability constants for this reaction are provided in Table 6. The surface complexation constants were found to be similar for all metals for each oxide (except for Pd). All these metals form quinqedentate complexes with EDTA. For trivalent metals such as Co(III) and Cr(III), hexadentate complexes are formed [198]. Although the modelling studies assume a direct, inner-sphere bonding where the interactions with the surface are dominated by the chelating abilities of EDTA, FTIR spectroscopy and EXAFS showed no indications of inner-sphere complexation between Pb-EDTA and goethite [203]. Spectra confirmed hexadentate coordination between the EDTA and Pb but exhibited no evidence of EDTA-Fe-specific interactions. It was suggested that the mechanism of Pb-EDTA adsorption was through hydrogen bonding between the complex and goethite surface sites, which might explain the very similar behaviour of metal-EDTA for Cu, Zn, Pb, Ni, Cd, etc.

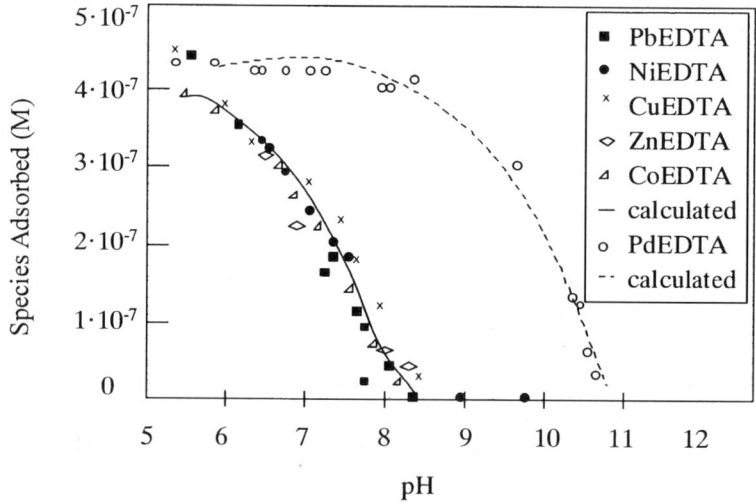

Fig. 25. Adsorption of Co(II)-, Cu-, Ni-, Pb-, and Zn-EDTA onto goethite (redrawn after Ref. 198).

Table 6
Surface complexation constants for adsorption of metal-EDTA onto oxides using constant capacitance model, 0 ionic strength
($\underline{S}$-OH + Me-EDTA^{2-} + H$^+$ ↔ $\underline{S}$-EDTA-Me^{2-} + H$_2$O)

Element	Goethite	HFO	δ-Al$_2$O$_3$	γ-Al$_2$O$_3$
Ca	12.26	-	11.09	-
Cd	-	-	11.54	-
Co(II)	11.05	-	-	11.97
Cu	11.44	-	11.08	-
Ni	11.26	9.36	11.55	11.44
Pb	11.18	-	11.03	-
Pd	15.26	11.32	-	-
Zn	10.85	-	11.54	-

NTA is a triprotic acid with four possible coordination sites, which forms strong complexes with metals, but not as strong as EDTA. Therefore, adsorption characteristics of metal-NTA complexes are different as compared with EDTA. Studies of adsorption of Co-NTA onto gibbsite [204] and Pb-NTA onto TiO$_2$ [205] showed that chelation of the metal had only small effects on the adsorption of the metal onto the surface. Obviously, the oxide surface competes for the individual metal and the ligand, respectively, and the Co(II)-NTA complex is broken in favour of individual ion adsorption. Spectroscopic evidence suggested the formation of weak mono- and binuclear metal-like outer-sphere complexes.

2.4.3. Type and Chemical Speciation of Metal

Type and speciation of a given metal is another important factor in assessing its environmental behaviour. The term speciation refers to its distribution amongst its chemical forms or species [184]. In general, solid species tend to be less mobile than colloidal or dissolved forms. Also complexation plays an important role as it has been discussed above. Certain physicochemical properties such as electronegativity and ionic potential (i.e. the charge/radius ratio) influence the order, in which metals are sorbed onto solid components. The following elements display multiple oxidation states in soils and aquatic systems: As (-III, 0, III, V), Cr (III, VI), Cu (I, II), Fe (II, III), Hg (0, II), Mn (II, III), and Se (-II, II, IV, VI). For example, hexavalent Cr (VI) is much more mobile and toxic than the trivalent Cr (III). Also, monomethyl Hg (CH$_3$Hg$^+$) is the most toxic of all Hg aquatic species. Table 7 lists the general effects of some geochemical factors on heavy metal mobility and bioavailability.

Table 7
Effect of soil/sediment factors on heavy metal mobility and bioavailability

Soil factor	Causal process	Effect
Low pH	Decreasing sorption of cations onto oxides of Fe and Mn	Increase
	Increasing sorption of anions onto oxides of Fe and Mn	Decrease
High pH	Increasing precipitation of cations as Carbonates and hydroxides	Decrease
	Increasing sorption of cations onto oxides of Fe and Mn	Decrease
	Increasing complexation of certain cations by dissolved ligands	Increase
	Increasing sorption of cations onto (solid) humus material	Decrease
	Decreasing sorption of anions	Increase
High clay Content	Increasing ion exchange for trace cations (at all pHs)	Decrease
High OM (solid)	Increasing sorption of cations onto humus material	Decrease
High (soluble) humus content	Increasing complexation for most trace cations	Decrease/ Increase?
Competing ions	Increasing competition for sorption sites	Increase
Dissolved inorganic ligands	Increasing trace metal solubility	Increase
Dissolved organic ligands	Increasing trace metal solubility	Increase
Fe and Mn oxides	Increasing sorption of trace cations with increasing pH	Decrease
	Increasing sorption of trace anions with decreasing pH	Decrease
Low redox	Decreasing solubility at low redox potential as metal sulfides	Decrease
	Decreasing solution complexation with Lower redox potential	Decrease/ Increase

Reprinted with permission from: D.C.Adriano, Trace Elements in Terrestrial Environments, Springer, New York, 2001, p. 56.

As it has been discussed above, type and speciation of an individual heavy metal determines its mobility between environmental media and its availability to organisms. In the next chapter, the effect of heavy metals on living organisms, their bioavailability, their accumulation in the food chain, and their impacts on ecosystems as well as on human health is discussed in detail.

3. ECOTOXICOLOGICAL EFFECTS OF HEAVY METALS

Ecotoxicology developed from the traditional fields of toxicology and environmental chemistry and can be defined as "the study and effect of toxic agents in ecosystems" [206]. Ecotoxicology is an interdisciplinary science, which deals with the interactions among organisms, toxic agents (here: heavy metals), and the environment, and integrates disciplines such as environmental biogeochemistry, toxicology, and ecology. The vast majority of ecotoxicological studies have been performed at the individual level, yet investigations including studies of toxic effects at the cellular, individual, and population level have been performed [207]. In the following chapter, the pathways of heavy metals into ecosystems and living organisms, their bioavailability, bioaccumulation, and general health effects are discussed. As for the health effects of individual heavy metals, the reader is referred to subchapter 4 "Individual behaviour of selected heavy metals".

3.1. Pathways of Heavy Metal Access

In order to cause any effect in a living organism, heavy metals have to come into contact with this organism. There are three principal ways, through which this might happen. The first pathway is through the atmosphere or through atmospheric deposition to water and soil, the second is through drinking contaminated water or using it for cooking and crop irrigation, and the third is through accumulation in the food web.

3.1.1. Respiration

Heavy metals can enter organisms by respiration of natural and anthropogenic emissions. The sources of these emissions have been discussed in the first chapter of this book. Heavy metals can be volatile (mostly Hg) or particulate. These substances are released into the atmosphere in the order of magnitude of several thousands of tons annually [208]. These numbers can be expected to grow due to the increasing world population and industrial activities in countries striving to increase their industrial bases (e.g., China and India).

Respiration of metal pollutants through dust is one of the most serious threads to humans working in industrial workplaces. Health problems such as "black lung" disease, silicosis, and radiation sickness have been recognized early [209]. Table 8 lists some health problems arising from the respiration and contact of heavy metals with humans in the workplace. They may cause a variety of health damages including cancer, liver, and kidney diseases, abortions, neurological and visual damages, negative effects on the immune system, allergies, cardiovascular toxicity, and anaemia.

3.1.2. Water

The second pathway of entering organisms is through drinking water contaminated with heavy metals, which can be ingested directly by drinking, or indirectly by using this water for cooking and irrigation. These contaminations can be both of natural and of anthropogenic origins. One should be aware that to date more than one third of the world population have no access to clean water for cooking, drinking, personal hygiene, and sanitation, which is threatening especially for infants and children. Contaminated drinking water is one of the major hazards in West India and Bangladesh, where more than 20 million of the 120 million people living in this country are affected by arsenicosis (see also subchapter 4.1.3. Ecotoxicological Effects of Arsenic).

3.1.3. Food

The third direct pathway is through foods with high natural or bioaccumulated contents of heavy metals. One of the main routes of entering the food chain is plant uptake. If soils contain high natural metal contents, are amended with metal-bearing sludges, or are irrigated with water contaminated with metals, then some plants will hyperaccumulate those metals, which results in polluted food crops and animal forage. Then the heavy metals are transferred through higher trophic levels to humans. The access of an individual heavy metal to the food web is determined by how the metal is bound to a soil, the soil phase it is bound to, and its chemical form. Pollutants can be present in soils as particulates, liquid films, absorbed ions, adsorbed ions, and liquid phases in pores [210]. Fig. 26 shows an example of the distribution of some heavy metals in solid phases of some Polish soils. Five solid phases have been identified, from which heavy metals (i.e. Mo, Zn, Cd, Cu, Pb, Ni, and Cr) are able to enter the food web [211, 212]. These are readily soluble phases, exchangeable sites, Fe and Mn oxy/hydroxides, organic matter, and residual phases. While metals from the first four phases are easily released, this is not the fact for metals bound to the residual phase for most environmental conditions encountered. Each of the metal studied is unique in the way it is bound to the different phases. For example, chromium can be found mainly in the residual phase with only 2% in an easily soluble phase. More than 20% of cadmium, on the other hand, is available from easily soluble and exchangeable phases.

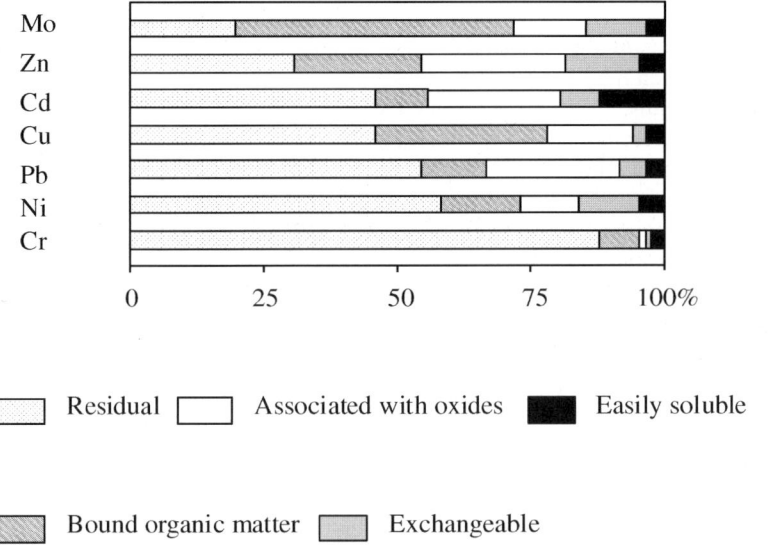

Fig. 26. Distribution of some heavy metals in solid phases of soils in Poland (redrawn after Ref. 209).

3.2. Bioavailability and Bioaccumulation

In order to be assimilated, heavy metals will have to be mobile, to be transported to the organism, and be bioavailable to it. Contrary to organic contaminants, heavy metals can not be degraded. The term "bioavailability" has different meanings when different disciplines are concerned, and must therefore be defined carefully.

3.2.1. Definition

In ecotoxicology, bioavailability can be defined as "the portion of a chemical in the environment that is available for biological action, such as uptake by an organism" [213]. In sediment-associated contaminant, bioavailability can be defined as "the fraction of the total contaminant in the interstitial water and on the sediment particles that is available for bioaccumulation", whereas bioaccumulation is "the accumulation of contaminant via all routes available to the organism" [214]. When dealing with pharmacology, bioavailability refers to "the fraction of an orally administered dose that reaches the blood of an animal" [215], a definition, which is most commonly used by mammalian toxicologists [216].

Table 8
Health effects of some heavy metals

Carcinogens which may occur in the workplace
Human carcinogens: As, Cd, Cr
Possible human carcinogens: Co, Pb, Ni
Lung carcinogens include: As, Be, Cd, Cr

Some reported cancers caused by or associated with certain occupations and industries
As: lung, skin: pesticides, others
Cr: lung: metals, welders
Ni: lung, nose: metals, smelters, engineering

Substances linked to occupations liver disease
As: Cirrhosis, angiosarcoma, hepatocellular carcinoma: pesticides, wood, vinters, smelters
Be: granulomatous disease: ceramics

Substances reported to have damaged the kidney in the workplace
Cd: nephrotoxicity: welding, engineering
Pb: nephrotoxicity: chemicals, paint, batteries
Hg (inorganic): nephrotoxicity: chemicals, paints

Possible factors influencing reproductivity outcomes based on experimental data
As: fetotoxic, teratogen, transplacental carcinogen: agriculture, wood preserving
Cd: spontaneous abortions, impaired implantation, teratogen
    Male and female damage: engineering, chemicals, batteries, paints, smelting
Cr: teratogen: chemicals, engineering
Pb: decreased fertility, fetotoxic, impaired implantation, teratogen, sperm damage, hormonal alterations: various
Mn: decreased fertility, impaired implantation: various
Hg: fetotoxic, teratogen, menstrual disorders: chemical, pesticides
Se: fetotoxic
Tl: : fetotoxic
Triethyl Pb: spontaneous abortions

Substances found in the workplace reported to have caused neurological damage
As: peripheral neuropathy: metal production, pesticides
Pb: encephalopathy and peripheral neuropathy: general
Mn: encephalopathy, ataxia, later Parkinson disease-like symptoms occur, acute psychosis: engineering, aircraft industry, steel, aluminium, magnesium, and cast iron production
Hg: tremor, weakness, peripheral neuropathy is uncommon, chronic exposure leads to ataxia, mental impairment: chemicals, pharmaceuticals, dentistry, plastic, paper, various
Ni: headache: engineering
Tl: encephalopathy, ataxia (high doses)
Sn (organic): encephalopathy

Table 8 (continued)
Health effects of some heavy metals

Substances known or associated with visual damage in the workplace
Pb: optical neuropathy: foundry industry
Hg: cranial nerve palsies: chemicals

Substances reported to have caused immune system effects in the workplace
Ni: hypersensivity: metals engineering

Reported respiratory effects of certain workplace substances:
Metals: especially Pt, Ni, Cr, Co, and V: occupational asthma, metal and engineering workers

Substances known to be absorbed through or damaging to the skin in the workplace
As: skin cancer: agriculture, lead workers, dyers, copper smelters, brass makers, chemicals, textiles, painters, pesticide users
Cr: allergic contact dermatitis: percutaneous absorption: engineering and chemicals
Ni: allergic contact dermatitis: metals, engineering, jewellery

Sustances linked with cardiovascular toxicity:
Arsine: cardiac arrhythmia
As: myocardial injury
Sb: hypertension
Cd: hypertension
Co: myocardial injury
Pb: myocardial injury, hypertension

Reported adverse effects of chemicals on the blood
As: aplastic anaemia: glass, paints, enamels, pesticides, tanning agents
Cu: red blood cells: engineering
Pb: red blood cells, porphyria: general

Modified after Ref. 209.

The biological response and risk is a function of the dose of the heavy metal. For metals, which are essential for metabolism, there can be three ranges: first, the deficiency range, where biological activities can be increased by increasing the dose, second, the buffering or normal range, where biological functions are optimal, and third, the toxicity range, where further increases in concentration inhibit metabolism and may be even lethal to the organism. The range of concentration depends on the physical and chemical nature of the individual metal, the sensivity or tolerance of the receptor organism, and the nature and properties of the environmental medium concerned, e.g., soil or aquatic systems [183].

3.2.2. Bioavailability in the Soil-Plant System

The uptake and bioaccumulation of heavy metals by plants is of importance because of impact of soils by anthropogenic emissions and its consequences for human uptake. Fig. 27 shows the main three categories, into which plants can be grouped: excluders, indicators, and accumulators. Excluders include members of the grass family such as sudangrass, bromegrass, and others. These plants are insensitive to heavy metals over a wide concentration range. Indicators include grain and cereal crops such as corn, soybean, wheat, oats, etc., and accumulators include the mustard and *Compositae* families such as lettuce, spinach, etc. and tobacco. Extreme accumulators are known as hyperaccumulators, which can be found on heavily contaminated soils and near ore deposits. These plants have developed a tolerance mechanism and can be used for the phytoremediation, i.e. the use of plants for the removal of heavy metals in soils, sediments, sludges and waters. Phytoremediation and its basic physiological plant mechanisms are discussed in detail in the last chapter of this book. In contrast to hyperaccumulators, excluders have developed avoidance or exclusion mechanisms. Indicators are plant species that respond to soil metal concentration displaying linear curves, while accumulators and excluders display logarithmic curves [217].

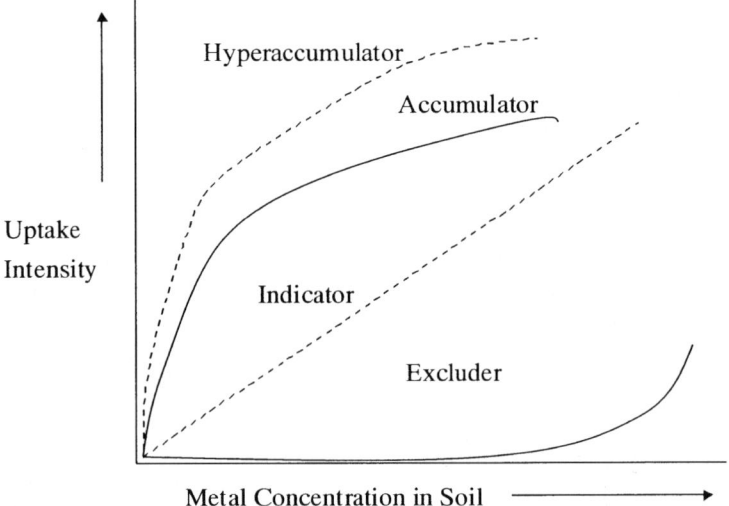

Fig. 27. Relative uptake and bioaccumulation potential among plant species (redrawn after Ref. 183).

In order to assess bioavailability of individual metals for certain plant species, chemical extraction techniques are used most commonly. In general, the readily soluble plus the weakly adsorbed (i.e. exchangeable) metal fraction is considered to be bioavailable. Organisms such as worms can also be used to assess bioavailability. The main categories used as extractants are dilute acids such as HCl or H_2SO_4, and chelating agents such as EDTA and DTPA. The method most widely used is described in Ref. 6. This method and variations of it include in general the selection of chemical reagents from the least to the most aggressive in sequential fashion and from the least to the greatest extremes in temperature and stirring. The various fractions can be described as follows: water soluble (metal exists in soil solution either in free ionic or complexed form), exchangeable (metal is sorbed by electrostatic attraction to negatively charged exchange sites), sorbed (adsorption onto specific exchange sites on colloidal surfaces), organic (complexed with soil organic matter), oxide-crystalline or amorphous Fe/Mn oxides (specific adsorption onto Fe/Mn oxides), carbonate (precipitated and/or occluded in soils high in free $CaCO_3$), sulfide (highly insoluble compounds of metal sulfides in poorly aerated soils), and residual (fixed within the crystalline lattices of alumosilicate particles).

3.2.3. Bioavailability in Aquatic Systems

Similar to heavy metal uptake by plants, aquatic organisms vary in their metal uptake. They can be grouped into two categories: regulators (excluders) and accumulators (non-excluders) [218]. Fig. 28 shows some examples for aquatic organisms and their different strategies for uptake, accumulation, and excretion of heavy metals [219]. Regulators are characterized by their low metal uptake, while accumulators are characterized by their high metal uptake. While regulators are able to control metal accumulations and keep their intracellular metal concentrations within a narrow range over a broad external concentration range of heavy metals, accumulators are capable of adopting a detoxification system with an elevated metal body level even in noncontaminated environments.

The most important factors influencing bioaccumulation in aquatic systems are compound characteristics (e.g. solubility), sediment characteristics (e.g. CEC, pH), water quality (e.g. temperature), biological characteristics (e.g. organism behaviour, modes of feeding), source of water, and age and size length of the individual organism [218, 220]. In general, bioavailability and toxicity of heavy metals correlates directly to concentrations of the free metal ion rather than to total or complexed metal concentrations.

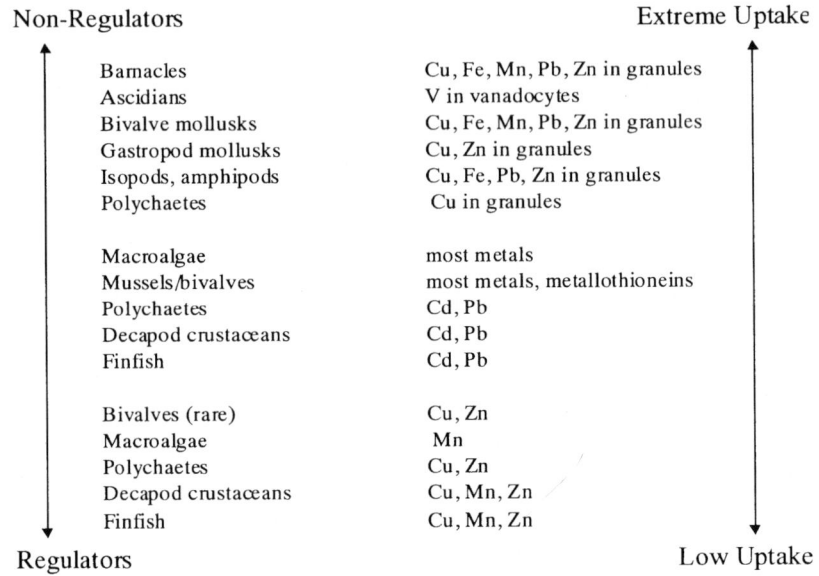

Fig. 28. Aquatic organisms employing different strategies for the uptake, accumulation, and excretion of heavy metals (redrawn after Ref. 219).

The most common mechanisms to limit uptake of heavy metals are altering the chemical speciation in the surrounding environment to reduce bioavailability, complexing the metal at the organism surface, decreasing the permeability of epithelial surfaces by introducing extracellular barriers, reducing transport into the cell, and undertaking behavioural avoidance activity [183]. As bioavailability is influenced mainly by solubility and mobility of the individual metal, all factors influencing these properties will of course be of importance for bioavailability as well. As for soils, the master variables are pH, soil organic matter, redox potential, presence of Fe/Mn oxyhydroxides, etc. As for aquatic systems, such as fresh and salt water, the most important variables include pH, dissolved organic matter, suspended particulate matter, ionic strength, alkalinity, and salinity [218]. Once heavy metals are absorbed, taken up or assimilated by an organism, they may unfold both adverse and positive effects depending on the kind and concentration of metal. Some metals are essential for organisms, while others are not. Among essential elements are Fe, Zn, Cu, Mn, Se, Cr, Co, and Mo [212]. Toxicological effects to humans are well known, especially those of Cd, As, Hg, and Pb. As for the toxicological effects of individual heavy metals, the reader is referred to the following chapter.

4. INDIVIDUAL BEHAVIOUR OF SELECTED HEAVY METALS

The following chapter deals with the individual behaviour of selected heavy metals, viz. their chemical and physical characteristics, their sources and applications, and their ecotoxicological effects.

4.1. Arsenic
4.1.1. Chemical and Physical Character of Arsenic

Arsenic is a steel-gray, brittle, crystalline metalloid with three allotropic forms of yellow, black, and gray colour and has atom number 33. It belongs to group V-A of the periodic table and resembles phosphorus chemically. Its atom weight is 74.92 and upon contact with air, it tarnishes and rapidly oxidizes when heated to arsenous trioxide (As_2O_3). The density of the ordinary stable form (gray As) is 5.73 g cm^{-3}, it melts at 817°C, and sublimes at 613°C. Common oxidation states are –III, 0, III, and V. Elemental As is formed by reduction of arsenic oxides. Arsenic covalently bonds with most nonmetals and metals and forms stable components with its trivalent and pentavalent states. The most important as compounds are white As (As_2O_3), the sulfide, Paris green (Cu acetarsenite [$3Cu(AsO_2)_2 \cdot Cu(CH_3COO)_2$]), Ca arsenate, and Pb arsenate. Paris green, Ca arsenate, and Pb arsenate are widely used as pesticides and poisons.

There are two forms of As which are needed by end-product manufacturers: white arsenic (arsenic trioxide As_2O_3) and the metal form, which is used in some lead and copper alloys. The primary sources of As are copper and lead ores. Arsenic compounds are mainly used for their toxicity as pesticides, herbicides, and silvicides. Until the 1940s, inorganic As solutions were used in the treatment of various diseases, such as syphilis and psoriasis. Inorganic As is still used as an antiparasitic agent in veterinary medicine and in homeopathic remedies. Organic arsenicals have largely replaced inorganic As compounds as herbicides. As is also used as a feed additive, e.g. in poultry feeds to control coccidiosis and to promote growths. Mobility and bioavailability of As is primarily influenced by the chemical species of As present, soil or groundwater pH, presence of oxides of manganese or iron and clay minerals, redox potential, and competing ions (Fig. 29). The primary species of As in soils and natural water are arsenate(V) and arsenite(III). As(III) is the most mobile and soluble and therefore the most toxic species. As solubility is controlled by the formation of $Mn_3(AsO_4)_2$, $FeAsO_4$, and $Ca(AsO_4)$. In water, As(V) is the most common species.

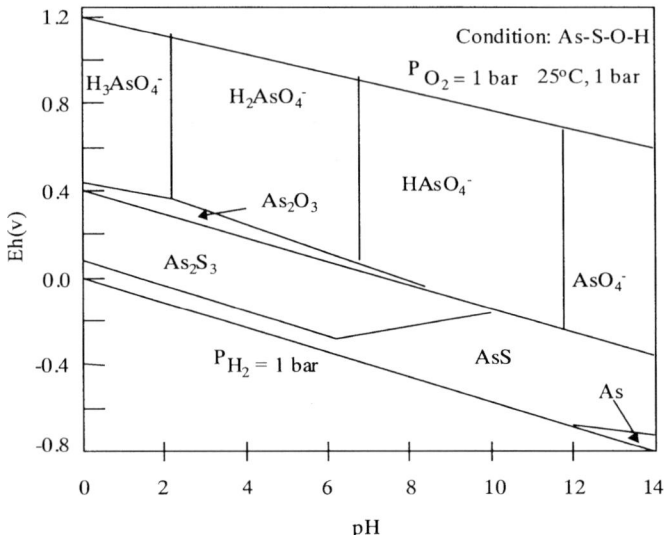

Fig. 29. Predicted Eh-pH stability field for arsenic under the system As-S-O-H, with assumed activities of As = 10^{-6}, S = 10^{-3} (redrawn after Ref. 183).

It is formed preferably under conditions of high dissolved oxygen, alkaline pH, high redox potential, and reduced content in organic matter while the formation of As(III) species is favored by opposite conditions.

4.1.2. Sources and Applications of Arsenic

Arsenic is ubiquitous in nature and can be found in soil, water, groundwater and other media in detectable concentrations (Table 9). Arsenic ranks 52nd in crustal abundance, ahead of Mo and is found in high concentrations in sulfide deposits such as arsenides, sulfides, and sulfosalts. As concentration in soils depends on the parent rock from which the soil is derived. Soils overlying sulfide ore deposits may contain several hundred ppm of As, while background soil concentration ranges from 0.2 to 40 ppm [221]. In general, sedimentary rocks contain much higher levels of As (0.3 to 500 ppm) than igneous rocks (1.5 to 3.0 ppm). Coal and its by-product fly ash also contain significant quantities of As and combustion of coal and disposal of fly ash is an important anthropogenic source of As in the environment.

Table 9
Commonly observed arsenic concentrations (ppm) in various environmental media

Material	Average Concentration	Range
Igneous rocks	1.5 – 3	0.06 - 113
Limestone	1.7	0.1 - 20
Sandstone	4.1	0.6 - 120
Shale	14.5	0.3 - 500
Phosphate	22.6	0.4 - 188
Petroleum	0.18	< 0.003 – 1.11
Coal	13	trace - 2000
Coal Ash		
Fly ash	156	8 - 1385
Bottom ash	8	< 5 - 36
FGD sludge	25	< 5 - 53
Oil ash	112	75 - 174
Sewage sludge	14.3	3 - 30
Soils (world, normally)	7.2	0.1 - 55
Soils (USA, noncontaminated)	7.4	
Forest soils (Norway)	-	0.59 – 5.70
Common crops	-	0.03 – 3.50
Drinking water (USA, µg/L)	2.4	0.5 – 2.4
River water (µg/L)		≥ 5
Lake Superior water (µg/L)		0.1 – 1.6
Japan, several lakes (µg/L)		0.2 – 1.9
Groundwater (USA, µg/L)		≥ 10
Great Lakes sediments	-	0.50 – 14.00
Ocean sediments	33.7	< 0.40 - 455

Reprinted with permission from: D.C.Adriano, Trace Elements in Terrestrial Environments, Springer, New York, 2001, p. 223.

The three main anthropogenic sources of As are pesticide manufacture and use, mining and smelting, and combustion of coal and its by-products. Inorganic arsenicals as Pb, Ca, Mg, and Zn arsenate, Zn arsenite, or Paris green have been widely used as pesticides in orchards but have been banned in the USA and Europe. Paris green (copper acetoarsenite) has been used to control the Colorado potato beetle (*Leptinotarsa decemlineata*), while calcium arsenate has been applied to cotton and tobacco fields as an insecticide for boll weevil (*Anthonomus grandis*), beetle, and other insect control. Lead arsenate was a successful insecticide in the control of fruit trees, especially for codling moths (*Cydia pomonella*) in apple trees and hornworm (larvae of *Manduca quinquemaculata*) on tobacco. Arsenic trioxide has been used as a soil sterilant, sodium arsenite has been used for aquatic weed control. Inorganic As compounds have been replaced by organic arsenicals as herbicides. The use of organoarsenical herbicides such as MSMA (monosodium methanearsonate), DSMA (disodium methanearsonate), and cacodylic acid (CA) has grown rapidly (Table 10). More than 90% of the As used in agriculture is MSMA and DSMA [222]. In the last years, MSMA and DSMA is replaced by glyphosate [N-(phosphonomethyl)glycine] due to economical and environmental considerations.

4.1.3. Ecotoxicological Effects of Arsenic

Arsenic is well known for its suicidal and homicidal effects since the Middle Ages. As for plants, As phytotoxicity is influenced by the chemical form of As itself, soil properties, and environmental conditions. The trivalent form As(III) is considered to be much more toxic than the pentavalent form As(V). As(III) is predominant in reducing conditions and is more mobile in sediments and groundwater [223]. Arsenic is more mobile in coarse-grained soils than in soils having higher content in fines such as clay minerals as clay are characterized by higher contents of oxide minerals, which are known to be effective sorbers of As [224]. As both As(III) and As(V) show high affinities for proteins, lipids, and other cellular components, they readily accumulate in living tissues [225]. As for ecotoxicological effects of As in aquatic invertebrates, As uptake by mussels, crustaceans, and mollusks has shown to be directly proportional to substrate As concentrations [225-228]. In contrast, As seems not to bioaccumulate in fish [229]. Fish are able to retain more than 99% of the ingested As in an organic form.

In general, As does not biomagnify in the food chain and bioconcentration factors in aquatic organisms are relatively low, except for algae [230]. The toxicity of arsenicals in humans has been found to conform to the following order, from highest to lowest toxicity: arsines > inorganic arsenites > organic trivalent compounds arsenoxides) > inorganic arsenates > organic pentavalent compounds > elemental arsenic [230, 231]. Solubility in water and body fluids seems to be directly related to toxicity. As(0) is of low toxicity because of its virtual insolubility, while the highly toxic arsenic trioxide is fairly soluble. The two most common exposure routes of As to humans are by ingestion and inhalation. The drinking of groundwater contaminated with As may well be the main risk for humans. There are some regions and areas where consumption of As-contaminated drinking water stemming from wells drilled into As-rich strata or from water contaminated by toxic wastes has led to As poisoning of the population.

Table 10
Names and uses of some important arsenical pesticides

Pesticide	Application rates and method	Commercial uses
Arsenic acid	0.0035 m³/ha of the 75% concentrate	Cotton dessicant
CA (cacodylic acid)	3.4 – 11.2 kg/ha	Lawn renovation; general weed controll
DSMA (Disodium methanearsonate)	2.24 – 4.26 kg/ha	Cotton and non-crop areas
MSMA (monosodium Methanearsonate)	2.24 – 4.26 kg/ha	Cotton and non-crop areas
Calcium arsenate	2.24 – 28 kg/ha	Cotton insecticide Fruits, vegetables, Potatoes
Lead arsenate	3.3 – 67.3 kg/ha	Fruits, vegetables, Nuts, turf
Paris green	1.12 – 17.9 kg/ha	Bait and mosquito larvicide
Sodium arsenic	1.12 – 22.4 kg/ha	Baits, nonselectice Herbicide, rodenticide, dessicant

Modified after Ref. 183.

Such incidents have been reported from a variety of countries, but the most extensive metal poisoning has been reported for West Bengal (India) and Bangladesh [232]. In Bangladesh alone, more than 20 million people are estimated to be exposed to As-contaminated drinking water. In India and Bangladesh, the source of As is due to the geological condition of the underground strata.

Two different mechanisms may cause elevated As levels in groundwater. First, Quaternary sediments covering almost the entire alluvial region of the river Ganges contain As-rich pyrite. Due to the high extraction rates of groundwater from these strata, air enters the aquifer and the pyrite is oxidized and mobilized from the vadose zone. Second, it is possible that organic matter commingled with As-rich Fe oxyhydroxides, which were deposited within the Ganges basin, has been reduced to the soluble state by the anoxic nature of the geologic strata and leaches upon exposure to aquifer water [233, 234]. Clinical symptoms of As poisoning manifest in three stages, which include dermatitis, keratosis, conjunctivitis, bronchitis, and gastroenteritis in the initial stage, followed by peripheral neuropathy, hepatopathy, melanosis, depigmentation, and hyperkeratosis in the second stage, and gangrene in the limbs and malignant neoplasms in the third stage.

Toxicity of As in this region is exacerbated by malnutrition, illiteracy, bad food habits, and long-term intake. As doses of 1 to 3 mg kg^{-1} per day are usually fatal. Chronic oral exposure to inorganic As at doses of 0.05 to 0.1 mg kg^{-1} is associated with neurological or haematological signs of toxicity [183], and recent studies in Taiwan have showed significant dose-response relationships between long-term exposure to inorganic As in drinking water and risk of malignant neoplasms of the liver, nasal cavity, lung, skin, bladder, kidney, and prostate [235, 236].

4.2. Cadmium
4.2.1. Chemical and Physical Character of Cadmium

Cadmium is a soft, ductile, silver-white, lustrous metal with an atom weight of 112.4, density of 8.64 g cm^{-3}, and a melting point of 321°C. There are eight stable isotopes: ^{106}Cd, ^{108}Cd, ^{110}Cd, ^{111}Cd, ^{112}Cd, ^{113}Cd, ^{114}Cd, and ^{116}Cd with the following percentages of abundance, respectively: 1.22%, 0.88%, 12.39%, 12.75%, 24.07%, 12.26%, and 7.58%. Cadmium is a transition metal in group II-B of the periodic table (like Hg and Zn). Like Zn, Cadmium is almost always divalent in all stable compounds, and it forms hydroxides and complex ions with ammonia and cyanide, and also a variety of complex organic amines, sulfur complexes, chlorocomplexes, and chelates. Cadmium forms precipitates with carbonates, arsenates, phosphates, oxalates, and ferrocyanides. It is readily soluble in nitric acid and is widely used in low-melting alloys. The mobility and bioavailability of Cd depends mainly on its chemical species.

Cd in soils and sediments appears mainly in the exchangeable fraction, followed by the Fe-Mn oxides and the residual fractions. The rest was detected in the carbonate fraction while the organic fraction contributed only an insignificant amount of Cd [237]. Several studies implicated that Cd in soils contaminated by anthropogenic activities such as mining and smelting seems to be more bioavailable than Cd from unimpacted soils [238-241]. In the soil solution, most of the Cd is present as free Cd^{2+} and CdHCO$_3^+$, while most of the Cd added to calcareous soils was rapidly adsorbed or precipitated in the solid phase [242] and organo-Cd complexes were minimal (Fig. 30a). If the soil contains high concentrations of Cl$^-$ and SO$_4^{2-}$, Cd will occur primarily as chloro- and sulfato-complexes [243-246]. While other metals such as Cu, Pb, Hg, or Zn are greatly influenced by the presence of organic ligands, a number of studies have shown that either free Cd^{2+} ions or Cd complexed by inorganic ligands are the dominant Cd species in most sludged and arable soils [242, 247, 248]. The distribution of Cd aqueous species can be predicted based on a generalized river water composition [249, 250] and is shown in Fig. 30b. Calculations were made with MINTEQ A2 using a concentration of 1 µg/L total dissolved Cd. In waters with a pH < 6, all of the dissolved Cd will exist as the free Cd^{2+}, while at pH between 6 and 8.2, carbonate species such as $CdHCO_3^+$ and $CdCO_3^0$ predominate. At pH between 8.2 and 10, essentially all of the Cd will exist as a neutral carbonate complex $CdCO_3^0$. As complexation of Cd with organic matter is weak because of the competition for binding sites with Ca, only small amounts of Cd occur complexed with organic ligands [251]. Adsorption is the main operating mechanism of the reaction of Cd at low concentrations with soils. Most studies conducted found that adsorption behaviour of Cd in soils can be described by either the Langmuir or the Freundlich isotherm.

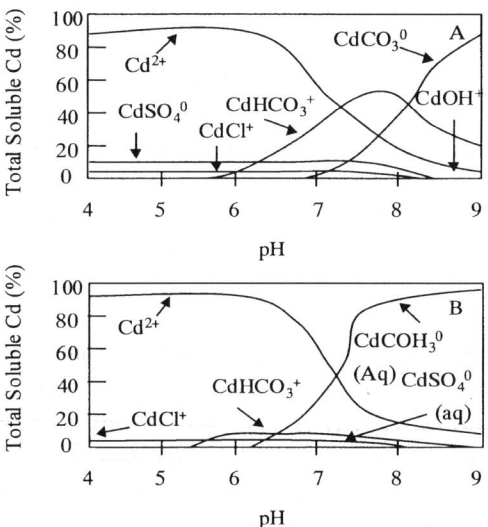

Fig. 30. (a) Calculated distribution of cadmium species between pH 4 and 9 in the solution of a typical calcareous soil (soil at P_{CO2} = 0.01 atm); (b) calculated distribution of cadmium aqueous species as a function of pH based on mean composition of river water of the world (redrawn after Ref. 242 and 250).

Adsorption of Cd by hydrous iron oxide was found to conform to the Langmuir isotherm [252]. Cd adsorption was demonstrated to be a fast process where >95% of the adsorption took place within the first 10 minutes and equilibrium was attained within an hour [253]. Fig. 31 shows Cd adsorption isotherms for two soils, a loamy sand and a sandy loam, as a function of pH. The sorption capacity of the soil increases approximately three times per unit increase in pH. In addition to adsorption, precipitation can play an important role in controlling Cd levels in soils. In general, Cd solubility in soils decreased as pH increased [254] with the lowest values for calcareous soils (pH 8.4). The precipitation of $CdCO_3$ occurs in sandy soils with low CEC, low content in organic matter, and alkaline pH and controls Cd solubility at high Cd concentrations [255]. Precipitation occurs in general at higher Cd^{2+} activities while ion exchange predominates at lower Cd^{2+} activities. Studies of behaviour of Cd^{2+} in the presence of $CaCO_3$ showed that initial chemisorption of Cd^{2+} on $CaCO_3$ was very rapid, while $CdCO_3$ precipitation at higher Cd concentrations was slow [243]. Chemisorption may regulate Cd^{2+} activity in calcareous soils by producing much lower solubilities than predicted by the solubility product for $CdCO_3$. Cd adsorption is influenced by variable parameters, the most important being pH, ionic strength, and exchangeable cations [256].

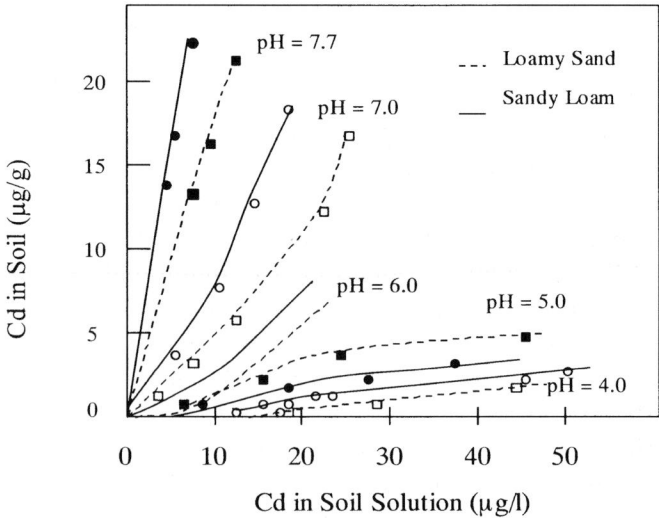

Fig. 31. Cadmium adsorption isotherms for two soils as influenced by soil texture and pH (redrawn after Ref. 252).

In the presence of Cl⁻, uncharged ($CdCl_2^0$) and negatively charged complexes of Cd with Cl⁻ ligands (e.g. $CdCl_3^-$, $CdCl_4^{2-}$ etc.) will form. The chloro-species of Cd are less strongly adsorbed than the Cd^{2+}. Cd Adsorption is also influenced by the presence of organic ligands such as EDTA, NTA, or others [248]. The presence of dissolved organic C or chelates could prevent metal coprecipitation with $CdCO_3$ or minimize adsorption of metals onto solid phases [252]. Cd adsorption is also strongly influenced by the presence of competing cations such as divalent Ca and Zn. These cations compete with Cd for sorption sites in soils or are able to desorb Cd from the soils [253, 257]. Experiments with pure clays showed that Cd^{2+} competes with Ca^{2+} for clay adsorption sites while with field soils, Cd^{2+} was preferably adsorbed over Ca^{2+} [258]. Obviously, soil colloids carry various specific adsorption sites with higher bonding energy for Cd than pure clays. Nevertheless, at typical environmental concentrations, the presence of alkaline-earth elements has only small effect on the adsorption of Cd on amorphous iron oxyhydroxides [259].

4.2.2. Sources and Applications of Cadmium

Cadmium is a by-product of the Zn industry, which is recovered from the smelting and refining of Zn concentrates. It is mainly used in alloys, in electroplating, in pigments, as stabilizers for polyvinyl plastics, in batteries, and for protecting iron and steel against corrosion. The use of Cd has been restricted worldwide due to environmental considerations. Nevertheless, Cd can be found in many consumer goods and is also used as a fungicide.

In nature, Cd exists in the II oxidation state. Cd is ranked 64th in crustal abundance [260] with an average concentration of 0.15 to 0.20 ppm. Cd is closely related to Zn and is therefore found mainly in Z, Pb-Zn, and Pb-Cu-Zn ores. Table 11 shows the normal Cd concentration found in various environmental media. The main anthropogenic sources of Cd are the use of phosphate fertilizers, land application of municipal sewage sludge, atmospheric deposition, and mining and smelting activities. As ores containing Cd are used for he production of phosphate fertilizers, these products may contain Cd concentrations of up to 340 ppm [245]. Nevertheless, long-term studies showed no significant increase in Cd uptake by plants growing on soils treated with these fertilizers [261, 262]. Municipal sewage sludge is widely used as a soil structure builder, N and P source, and its land application is currently increasing. Yet this application may increase the probability of contaminating the food chain with Cd [183].

Table 11
Normal cadmium concentrations (ppm) in various environmental media

Material	Average Concentration	Range
Igneous rocks	0.082	0.001 – 0.60
Metamorphic rocks	0.06	0.005 – 0.87
Sedimentary rocks	3.42	0.05 - 500
Recent sediments	0.53	0.02 – 6.2
Crude oil	0.008	0.0003 – 0.027
Coal	0.10	0.07 – 0.18
Fly ash	11.7	6.5 - 17
Phosphate rocks	25	0.2 - 340
Phosphated fertilizers	4.3	1.5 – 9.7
Sewage sludges	74	2 - 1100
Soils (world, nonpolluted)	0.35	0.001 – 2.0
Fruits (USA, $n = 3202$)	0.005	0.0043 – 0.012
Crop grains (field USA, n = 1302)	0.0047	0.014 – 0.21
River sediments (polluted)	-	30 - > 800
Freshwater (µg/L)	0.10	0.01 - 3
Seawater (µg/L)	0.11	<0.01 – 9.4

Modified after Ref.183.

Both US EPA and the EU have established limits on annual and cumulative Cd amounts in soils in the form of sewage sludge. Cd is also emitted into the environment by atmospheric deposition from smelter, incineration of plastics and pigments containing Cd, burning fossil fuel, coking, emissions from steel mills, and metallurgical processes. Cd contamination of environmental media in the vicinity of such activities has been reported from the United States [263], Japan [264] and England [265]. Near roads, automobile emissions are sources of Cd and other metals [266].

Mining activities producing wastewater containing significant concentrations of Cd can affect large areas. One of the most famous cases is the Jinzu River basin in Japan, where inhabitants had been exposed to Cd over a 30 year period. River water containing Cd had been used for the irrigation of rice fields resulting in the itai-itai disease [267]. The pollution source of the paddy soils in the Jinzu River basin is the Kamioka mine located about 40 km upstream of the Jinzu River. Cd concentrations in the paddy surface soils were reported to be in the range of 1.35 to 6.88 mg kg^{-1} and in unpolished rice 0.37 mg kg^{-1}. Remediation was accomplished by covering the surface soil with 25 cm of uncontaminated soil and by applying several soil amendments. After remediation, the Cd concentrations in unpolished rice dropped below 0.1 mg kg^{-1}, which is the usual Cd concentration value for rice produced in unpolluted paddy fields.

4.2.3. Ecotoxicological Effects of Cadmium

Cd is known to be toxic for plants as well as for invertebrates and vertebrates to a much higher extent and in smaller concentrations than e.g. Zn, Pb, or Cu. Phytotoxicity depends mainly on plant species and Cd concentration in the medium. The typical symptoms of Cd toxicity in plants resemble Fe chlorosis accompanied by necrosis, wilting, red-orange leaf coloration, and general growth reduction, which has been described for different species such as rice and radish [268, 269].

Cd interferes with plant metabolic processes, which cause root growth retardation, suberization, damage to internal and external root structures, decreased root hydraulic water conductivity, interference with nutrient absorption and translocation leading to nutrient imbalance, reduction of chlorophyll content, interference with enzymatic activities related to photosynthesis, and decrease in stomatal opening and conductance [270-274]. Tree species have been reported to be sensitive to soil Cd as well, but at much higher concentrations than for agronomic and horticultural crops.

As for invertebrates, comparisons between Cd contaminated and uncontaminated woodlands showed differences in the taxa especially of millipedes, earthworms, and wood lice, which were severely reduced in the contaminated area [275]. Cd hyperaccumulator invertebrates include earthworms (Annelida – *Lumbricus terrestris, Aporrectodea longa*), wood lice (Isopoda – *Oniscus asellus*), and snails and slugs (Mollusca – *Cepaea nemoralis, Discus rotundatus*). Cd accumulates in general in freshwater organisms, but does not biomagnify in the food chain as does Hg.

In humans, Cd interferes with the metabolism of Ca, vitamin D, collagen, and causes bone degenerations such as osteomalachia (or osteoporosis). Ca loss through urinary excretion is promoted. Long-term inhalation and oral exposure to Cd affects the kidneys and lungs, leading to proteinurea, decreased glomerular filtration rate, and emphysema. Cigarette tobacco contains about 1 ppm Cd. Smoking a pack of cigarettes gives a total intake of about 3 μg Cd [276]. In general, transfer of Cd to food crops is a major problem as the consumption of those crops is the most critical exposure pathway for Cd for the general population.

4.3. Chromium

4.3.1. Chemical and Physical Character of Chromium

Chromium has atom number 24 and is a member of group VI-B of the periodic table with an atom weight of 52.0, density of 7.2 g cm^{-3}, and a melting point of 1857°C. There are four stable isotopes: ^{50}Cr, ^{52}Cr, ^{53}Cr, and ^{54}Cr with the following percentages of abundance, respectively: 4.31%, 83.76, 9.55, and 2.38%. It is a silvery, lustrous, malleable metal, which dissolves easily in nonoxidizing mineral acids but not in cold aqua regia or in HNO$_3$, and is therefore used in corrosion-resistant alloys. Cr is mostly found in the oxidation state III, which is the most stable, but also occurs in the oxidation states 0 and VI. In soils and sediments, two trivalent forms (viz. the Cr^{3+} cation and the anion CrO$_2^-$) and two hexavalent anions (Cr$_2$O$_7^{2-}$ and CrO$_4^{2-}$) occur. The trivalent forms coordinate readily with ligands containing oxygen and nitrogen. Compounds with oxidation states below II are reducing, while those with oxidation states greater than III are oxidizing. The hexavalent form is relatively toxic compared to the trivalent form. Fig. 32 shows the distribution of Cr(III) species as a function of pH. The oxidation state of Cr is very important for its mobility. Fig. 33 presents the predicted Eh-pH stability field for chromium species in aqueous systems.

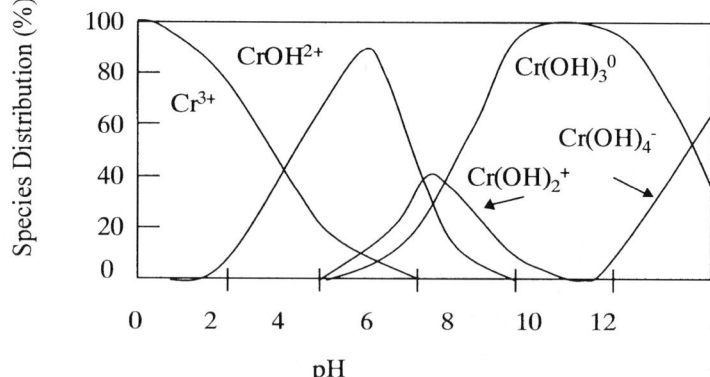

Fig. 32. Distribution of Cr(III) species as a function of pH where the solution is in equilibrium with Cr(OH)$_3$(s) (redrawn after Ref. 183).

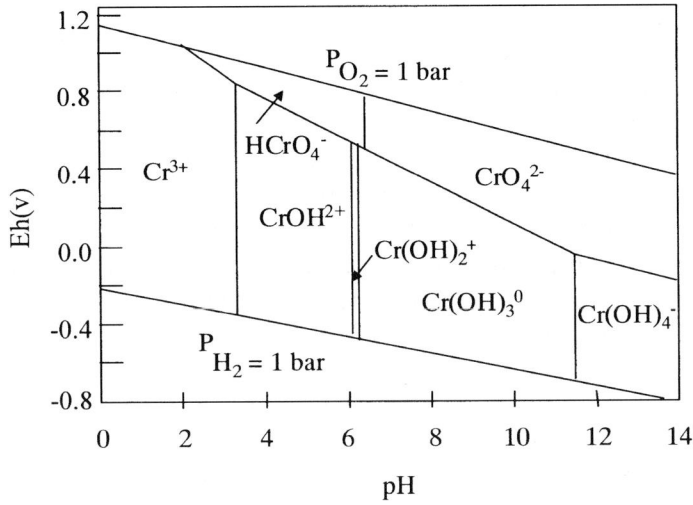

Fig. 33. Predicted Eh-pH-stability field for chromium species in aqueous systems (redrawn after Ref. 183).

The hydrolytic forms of Cr(III) are prone to precipitate starting at about pH 4.5 [277]. Adsorption and precipitation behaviour of Cr in soils is controlled by a variety of factors such as redox potential, oxidation state, pH, soil minerals, competing ions, complexing agents, and others. These factors control most of the partitioning processes of Cr between the solid and the aqueous media in soils. The most important among these are the hydrolysis of Cr(III) and Cr(VI), redox reactions of Cr(III) and Cr(VI), and adsorption/desorption and precipitation of Cr(VI). Hexavalent Cr species are adsorbed by a variety of soil phases with hydroxyl groups on their surfaces such as Fe, Mn, and Al oxides, kaolinite and montmorillonite [278-282]. Figure 34 shows the adsorption of hexavalent Cr onto various adsorbents as a function of pH [282]. The adsorption increases with decreasing pH due to the protonation of the hydroxyl groups. Obviously, Cr(VI) adsorption is favoured if the surfaces are positively charged and display high pH_{pzc} values at low to neutral pH. This reaction can be described as a surface complexation reaction between the Cr(VI) species and the surface hydroxyl sites. Fe oxides exhibit the strongest affinity for Cr(VI) followed by Al_2O_3, kaolinite and montmorillonite. Cr(VI) adsorption was found to be greatest in lower pH materials enriched with kaolinite and crystalline Fe oxides [282]. Cr(III) is rapidly and specifically adsorbed by Fe and Mn oxides and clay minerals, with about 90% of added being adsorbed within 24 hrs. Adsorption increases with increasing pH and content of soil organic matter, while it decreases in the presence of competing cations or dissolved organic ligands in the solution. Both Freundlich and Langmuir isotherms can be used to describe adsorption behaviour of Cr(III) on solid phases [283-285]. Trivalent Cr is known to be extensively hydrolyzed in acid solutions to species such as $Cr(OH)^{2+}$, $Cr_2(OH)_4^{2+}$, or $Cr_6(OH_{12})^{6+}$. The increased adsorption of Cr(III) with increasing pH is caused by cation exchange reactions of the hydrolyzed species. Cr(III) is preferably adsorbed by clay minerals to Cr(VI) to an extent of 30 to 300 times. The high affinity of Cr for Fe oxides was confirmed by experiments where Cr(III) was added to soil and a large fraction of the added Cr was extracted with the Fe oxides [286].

4.3.2. Sources and Applications of Chromium

The most important Cr ore is chromite [(Fe, Mg)O(Cr, Al, Fe)$_2$O$_3$], with a world production in the order of more than 9 million tons. Cr is mostly used in the manufacture of stainless steel, for refractory purposes due to its high melting point and chemical inertness, in the making of mortars and castables.

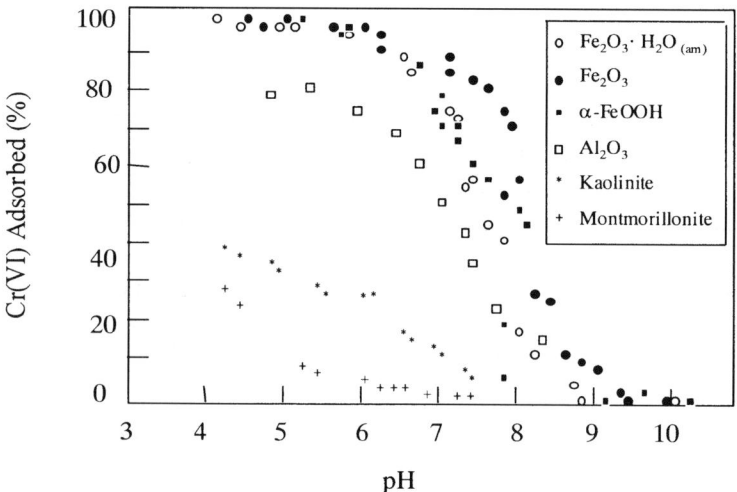

Fig. 34. Sorption of Cr(VI) by various absorbents (redrawn after Ref. 282).

It is also necessary for the making of Cr chemicals, which are used in leather tanning, catalysts, pigments, drilling muds, textiles, and others. Cr is abundant in the earth's crust and ranks 21st among the elements in crustal abundance. Table 12 shows commonly observed chromium concentrations (ppm) in various environmental media. Cr is used in a wide variety of applications, e.g. in the paper industry, chemical industry, in fertilizers, metal works and foundries, leather tanning and finishing, and power plants. The main anthropogenic sources of Cr are atmospheric depositions from electric furnaces, steel production, and coal-fired power plants [287]. Various industrial processes may release wastewater contaminated with Cr such as electroplating, leather tanning, metal processing, textile and fur dying, and animal glue manufacture. Fertilizers and sewage sludges may contain several hundred to thousand ppm of Cr [288, 289]. Although sewage sludges are increasingly used in arable lands, Cr stemming from those sludges showed to be relatively nonbioavailable to plants, although high application rates can enrich the soil with metals.

4.3.3. Ecotoxicological Effects of Chromium

Cr is an essential element in animal and human nutrition [290, 291]. As for plants, there is no consistent evidence for the stimulatory effects of Cr on plant growth, although such effects have been reported [292], while its nonessentially has been demonstrated conclusively [293]. Some plant species such as *Sutera fodina*, *Dicoma niccolifera*, and *Leptospermum scoparium* are able to accumulate appreciable amounts of Cr [183].

Table 12
Commonly observed chromium concentrations (ppm) in various environmental media

Material	Average Concentration	Range
Continental crust	125	80 – 200
Ultramafic igneous	1800	1000 – 3400
Basaltic igenous	200	40 - 600
Granitic igneous	20	2 – 90
Limestone	10	< 1 - 120
Coal	20	10 – 1000
Fly ash	247	37 - 651
Phosphate fertilizers	-	30 – 3000
Sewage sludges	74	2 - 1100
Soils	40	10 – 150
Freshwater (µg/L)	1	0.1 - 6
Seawater (µg/L)	0.3	0.2 – 50

Modified after Ref.183.

Accumulations of up to 48,000 ppm Cr have been reported. Cr(III) is required to maintain normal glucose metabolism, which was first demonstrated in rats [294]. Cr deficiency in humans causes impaired glucose tolerance, glycosuria, and elevations in serum insulin, cholesterol, and total triglycerides [295]. In animals, symptoms such as impaired growth, altered immune function, disturbances in aortic plaque and size, corneal lesion formation, and decrease in reproductive functions have been observed additionally. In contrast to Cr(III), Cr(VI) is toxic. Both Cr(III) and Cr(VI) are potent human carcinogens. The major target organ for Cr(III) and Cr(VI) is the respiratory tract. Classical symptoms are perforations and ulcerations of the septum, bronchitis, decreased pulmonary function, and pneumonia. Cr is the second most skin allergen after Ni causing allergic contact dermatitis [296]. If Cr(VI) is ingested, it is rapidly converted into Cr(III) due to the action of stomach acids.

4.4. Copper
4.4.1. Chemical and Physical Character of Copper

Copper has atom number 29 and is a reddish, malleable, ductile metal with very good heat and electricity conductivity. It belongs to group I-B of the periodic table, has an atom weight of 63.55, a melting point of 1083°C, and a specific gravity of 8.96 g cm^{-3}. There are two natural isotopes, ^{63}Cu and ^{65}Cu, with relative abundances of 69.1% and 30.9%, respectively. The radioactive isotope ^{64}Cu has a short half-life of only 12.8 hours and is therefore used as a tracer. Cu occurs in the I and II oxidation states. In the II state, it is isomorphous with Zn^{2+}, Mg^{2+}, and Fe^{2+} ions. Cu forms a variety of sulfides, sulfates, carbonates, and occurs native as well. The most abundant Cu mineral is chalcopyrite ($CuFeS_2$), which contains 34% of Cu. It is ranked 26th in crustal abundance behind Zn [260] with average crustal concentrations of 24 to 55 ppm [297, 298].

Copper in soils may occur in several forms that are partitioned between the solution and the solid phases. Distribution of Cu between different soil constituents is mostly influenced by the presence of soil organic matter, and Mn and Fe oxides. Cu shows a strong affinity for soil organic matter so that the organic-fraction Cu is high compared to the that for other metals even though the absolute amounts are low [299]. The most important sinks for Cu in soils are Fe and Mn oxides, soil organic matter, sulfides, and carbonates while clay minerals and phosphates are of lesser importance [300]. Adsorption maxima among soil constituents decrease in the order Mn oxide > organic matter > Fe oxide > clay mineral. Specific adsorption seems to play a more important role than non-specific adsorption (i.e. cation exchange). Sorption isotherms indicate preferential adsorption of Cu onto soil organic matter associated with the clay fraction of the soil [301]. Fig. 35 shows the adsorption of Cu onto various soil constituents [302]. Mn oxide and soil organic matter are the most likely to bind Cu in a nonexchangeable form. Sorption of Cu has been shown to follow either the Langmuir or the Freundlich isotherms [303, 304]. Cu in soil solution exists primarily in a form complexed with soluble organics [305]. Complexation by organic matter in the form of humic and fulvic acids is an effective mechanism of Cu retention in soils. It has been shown that Cu is most extensively complexed by humic materials [306] in comparison to other metals. The following preference series for divalent ions for humic acids and peat is indicated as follows: Cu > Pb > Fe > Ni = Co = Zn > Mn = Ca [183]. Synthetic chelating agents such as ETDA, DTPA, and others combine with heavy metals to increase their levels in soil solution. The stability of metal-synthetic chelating agents is a function of soil pH. CuDTPA is unstable in acidic soils, moderately stable in slightly acidic soils, and stable in alkaline and calcareous soils while CuEDTA is most stable in slightly acidic to neutral soils (pH 6.1 to 7.3). In acidic soils with pH below 5.7 CuEDTA becomes unstable since Fe displaces Cu.

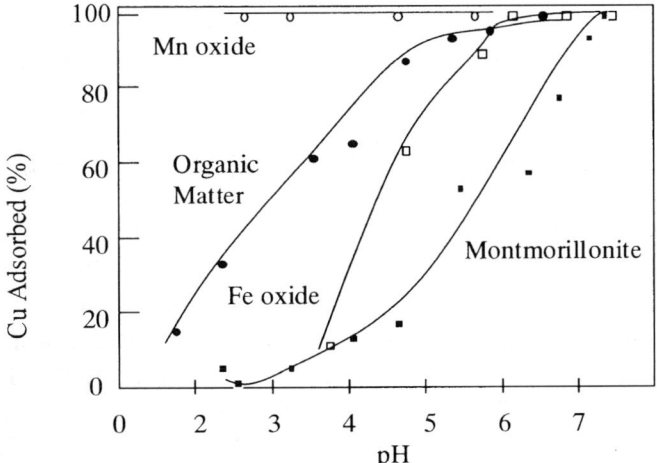

Fig. 35. Adsorption of Cu by different soil constituents as a function of pH (redrawn after Ref. 183).

4.4.2. Sources and Applications of Copper

Copper is widely used for wire production and in the electrical industry. Its main alloys are brass (with zinc) and bronze (with tin). Other applications are kitchenware, water delivery systems, fertilizers, bactericides and fungicides, feed additives and growth promoters, and as an agent for disease control in livestock and poultry production. Table 13 shows commonly observed copper concentrations in various environmental media. The main sources of copper are copper fertilizers, which are widely used in agriculture. Examples are $CuSO_4 \cdot 5H_2O$, CuO, Cu_2O, and $CuSO_4 \cdot 3 Cu(OH)_2$. Synthetic Cu chelates such as $Na_2CuEDTA$ and $NaCuHEDTA$ and natural lignosulfates and polyflavonoid have been used as well. Co compounds such as $CuSO_4$ and $Cu(OH)_2$ are used as fungicides and bactericides [307, 308] in a variety of agricultural crops such as pome, stone, and citrus fruits as well as on grapevines, hops, vegetables, coffee, cocoa [309], banana [310], and tea. The application of those fungicides and bactericides can lead to Cu accumulation in soil at phytotoxic levels.

$CuSO_4$ is also used as a feed additive in swine and poultry to stimulate growth and prevent dysentery [311-313]. Dietary levels of up to 250 ppm are reported [314]. Most of this Cu is excreted in the manure. If the manure is applied to soils, this may lead to potential accumulation and toxic effects, e.g. to sheep [315].

Cu is also emitted by metallurgical processing for Cu, iron, and steel production, and coal combustion. Heavy metal contamination caused by industrial emissions is well documented. The deposition rate of heavy metals from smelters is a function of distance [316].

Table 13
Commonly observed copper concentrations (ppm) in various environmental media

Material	Average Concentration	Range
Igneous rocks	125	80 – 200
Sandstone	30	6 – 46
Limestone	6	0.6 - 13
Shale and clay	35	23 - 67
Coal	17	1 – 49
Fly ash	185	45 - 1452
Sewage sludges	690	100 - 1000
Soils	30	2 – 250
Freshwater (µg/L)	3	0.2 - 30
Seawater (µg/L)	0.25	0.05 – 12

Modified after Ref.183.

4.4.3. Ecotoxicological Effects of Copper

Copper is one of the seven well known micronutrients (Zn, Cu, Mn, Fe, B, Mo, and Cl), which are essential for plant nutrition [317, 318], although it is only needed in small amounts of 5 to 20 ppm. Concentrations of <4 ppm are considered deficient, and concentrations >20 ppm are considered toxic [319]. Cu is a constituent of a number of plant enzymes, which trigger a variety of physiological processes in plants such as photosynthesis, respiration, cell wall metabolisms, seed production, and others.

As for aquatic systems, the free Cu^{2+} ion is considered to be the most toxic form of Cu to aquatic life rather than the complexed forms [320, 321]. Although Cu is an essential micronutrient for animals, it is toxic to aquatic life at concentrations approximately 10 to 50 times higher than normal. In fish, Cu interferes with branchial ion plasma ion concentration, hematological parameters, and enzyme activities [322]. A review of acute toxicity data is given in Ref. 323.

In humans, Cu is needed for synthesis of a variety of copper-catalyzed enzymes such as ferrooxidase I and II, which catalyzes the oxidation of Fe^{2+} to Fe^{3+}, cytochrome C oxidase, which is the terminal oxidase in the respiratory chain, lysyl oxidase, and many others. Cu deficiency in humans causes anemia, bone and cardiovascular disorders, mental and nervous system deterioration, and defective keratinization of hair as well as reduction in levels of neurotransmitters, dopamine, and norephedrine, and in defective myelination in the brain stem and spinal cord. A daily dietary intake of <2 mg Cu per day for adults may suffice. Nevertheless, Cu toxicity in humans is very rare and is usually associated with long-term intake of cow's milk or with severe malnutrition in infants and young children.

4.5. Lead
4.5.1. Chemical and Physical Character of Lead

Lead has atom number 82 and is a bluish-gray metal of bright luster. It is soft, malleable, ductile, a poor conductor of electricity, and very resistant to corrosion. Pb belongs to group IV-A of the periodic table, has an atomic weight of 207.2, a melting point of 328 °C, and a specific gravity of 11.4 gcm^{-3}. It occurs in two oxidation states, II and IV. In most inorganic compounds, it is in the II oxidation state. There are four stable isotopes, ^{204}Pb, ^{206}Pb, ^{207}Pb and ^{208}Pb with relative abundances of 1.48, 23.6, 22.6 and 52.3%. Two radioactive isotopes are used as tracers (^{210}Pb, $t_{1/2}$ = 22 a, and ^{212}Pb, $t_{1/2}$ = 10 h). Chloride and bromide salts are slightly soluble in water, whereas carbonate and hydroxide salts are almost insoluble.

The chemistry of Pb in soils is affected by three main factors: first, specific adsorption to various solid phases, precipitation of sparingly soluble or highly stable compounds, and third, formation of relatively stable complexes or chelates that result from interaction with soil organic matter. Fig. 36 shows predicted aqueous monomeric chemical speciation of lead as a function of pH, while Fig. 37 displays the predicted Eh-pH-stability field for Pb. Pb undergoes hydrolysis at low pH values and displays multiple hydrolysis reactions. Above pH 9, the formation of $Pb(OH)_2$ is important, while $Pb(OH)^+$ is predominant between pH 6 and pH 10. Adsorption of Pb onto soils and clay minerals has been found to conform to either the Langmuir or the Freundlich isotherm over a wide range of concentrations [138, 324]. Carbonate content in soils plays an important role in controlling Pb behaviour. In noncalcareous soils, Pb solubility is controlled by different Pb hydroxides and phosphates such as $Pb(OH)_2$, $Pb_3(PO_4)_2$, $Pb_4O(PO_4)_2$, or $Pb_5(PO_4)_3OH$, depending on pH. With increasing pH, the formation of Pb orthophosphate, Pb hydroxypyromorphite, and tetraplumbite phosphate is possible as well as formation of $PbCO_3$ in calcareous soils [325]. The presence of Mn and Fe oxides may exert a predominant role on Pb adsorption in soils. It was found that Pb adsorption onto synthetic Mn oxide was up to 40 times greater than that to Fe oxide, and that Pb was adsorbed more strongly than any other metal studied (Co, Cu, Mn, Ni, and Zn) [326].

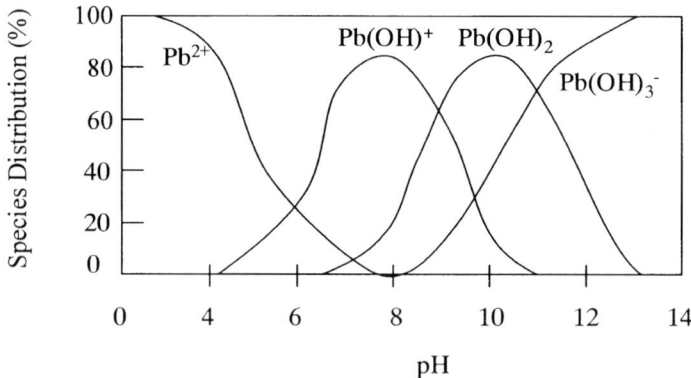

Fig. 36. Predicted aqueous monomeric chemical speciation of lead as a function of pH (redrawn after Ref. 183).

Three possible mechanisms may account for the binding of Pb onto Mn oxides: first, strong specific adsorption, second, a special affinity for Mn oxides as it has been found for Co [327, 328], and third, the formation of some specific Pb-Mn minerals such as coronadite. The presence of soil organic matter also plays an important role in Pb adsorption. Soil organic matter may immobilize Pb via specific adsorption reactions, while mobilization of Pb can also be facilitated by its complexion with dissolved organic matter or fulvic acids [329, 330]. Pb adsorption onto α-Al_2O_3 has been found to involve several mechanisms. In general, adsorption kinetics of Pb exhibit a biphasic behaviour. An initial fast reaction is followed by a slower reaction. The slow adsorption reaction is not caused by surface precipitation of Pb but may be due to diffusion to internal sites, adsorption onto sites that have slower reaction rates due to low affinity, and probably formation of additional adsorption sites due to the slow transformation of α-Al_2O_3 into the less reactive solid phase. The initial fast reaction is most likely caused by chemical reactions on readily accessible surface sites [331]. Pb has been shown to exhibit the strongest affinity to clays, peat, Fe oxides, and usual soils [332, 333]. Lead phosphates have been shown to be a very stable environmental form of Pb with low solubilities, a fact, which is widely used in remediation of Pb-contaminated soils (see Chapter 3.5 Solidification).

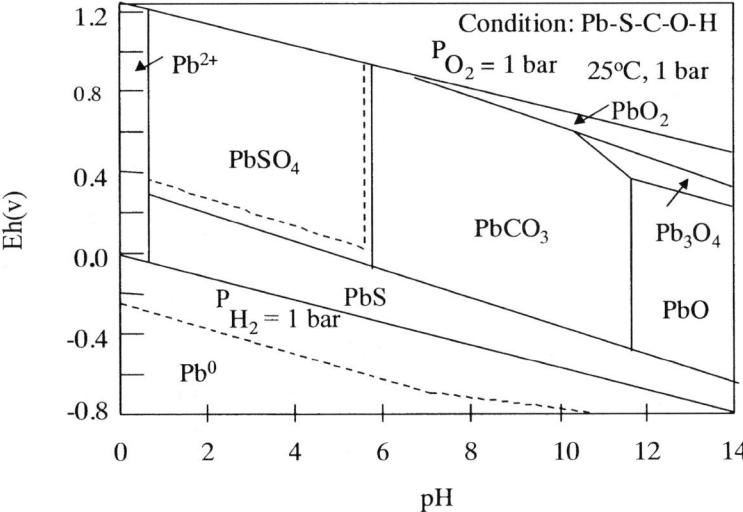

Fig. 37. Predicted Eh-pH-stability field for lead. The assumed activities of dissolved species are: Pb = 10^{-6}, S = 10^{-3}, C = 10^{-3} (redrawn after Ref. 183).

4.5.2. Sources and Applications of Lead

Pb has been used by man for more than 5000 years. It is vital for any industrial economy and is used mainly for large rechargeable batteries, pigments, rolled and extruded products, cable sheathing, alloys, shot and ammunition, and gasoline additives. Pb is also used as radiation sheeting and as a heat stabilizer in PVC. Pb-containing pesticides have been banned in the US, Austria, Belgium, and Germany, but are still in use in other countries. Pb has also been extensively used as gasoline additive, which has been banned in the US, Canada, and Europe since the 1070s, but is still in use in China, Russia, India, and other countries.

Pb is the most abundant element among the heavy metals with atomic numbers >60 with an average content of approximately 15 ppm [334]. The most common Pb minerals are galena (PbS), cerussite ($PbCO_3$), and anglesite ($PbSO_4$). Potassium feldspars and pegmatites are usually enriched in Pb. Table 14 gives the most commonly reported Pb values in various environmental media. Atmospheric deposition is one of the major inputs in the biogeochemical cycle of Pb, and it can be observed even in remote ecosystems.

The most common ways of contact with Pb is through the air and exposure to dust, and ingestion of food and water. The most significant source of Pb poisoning in children is considered to be dust contaminated by Pb-based paint. It is estimated that about 64 million U.S. housing units contain some Pb-based paint. The Pb is released when the paint deteriorates, is disturbed during renovation, or is abraded on surfaces such as window tracks or floors. This is a severe problem, which has been recognized in the United States and Europe, but is mostly ignored by the rest of the world.

Table 14
Common values for Pb concentrations (ppm) in various environmental media

Material	Average Concentration	Range
Igneous rocks	15	2 – 30
Sandstone	7	1-31
Limestone	9	-
Shale	20	16 - 50
Coal	16	up to 60
Fly ash	170	21 - 220
Sewage sludges	1832	136 - 7627
Soils (agricultural)		2-300
Freshwater (µg/L)	3	0.06 - 120
Seawater (µg/L)	0.03	0.03 – 13

Modified after Ref.183.

4.5.3. Ecotoxicological Effects of Lead

In general, Pb phytotoxicity is not encountered in ordinary environmental settings. Pb is not a serious problem in soils because there is a high affinity of Pb for soil organic matter. Exceptions are old mining and smelting sites, where plants are subjected to high Pb levels. Obviously microorganisms seem to be more sensitive to Pb than plants [335]. As for livestock, fish, and wildlife, Pb is a nonspecific toxin, which inhibits many enzymatic activities. Typical effects of Pb poisoning are effects on the hematological, central nervous system, and reproduction function.

As for humans, poisoning with Pb is a major environmental and public hazard, especially for infants and young children. The way by which Pb enters the body depends on its chemical and physical form. Inorganic Pb is mainly inhaled and ingested, and does not undergo biological transformation, while organic Pb such as tetraethyl Pb (which is used as an antiknock agent in gasoline) enters the body mainly by skin contact and inhalation, and is metabolized in the liver [183]. Once Pb is in the blood, it is distributed primarily among three compartments, viz. blood, soft tissue (such as kidney, bone marrow, liver, and brain), and mineralizing tissue (bones and teeth). Pb is accumulating in the body over a lifetime and releases it only slowly, thus Pb poisoning can be caused by small doses over time. Typical symptoms of Pb poisoning include general fatigue, tremor, headache, vomiting, seizures, blue-black lead line on gingival tissue, and colic. Pb also interferes with hemoglobin synthesis, and severely damages kidney functions. It affects the viability of the fetus and its development. Children under 6 years of age are especially affected by Pb poisoning, and there has no safe level been found so far.

4.6. Manganese
4.6.1. Chemical and Physical Character of Manganese

Manganese has atom number 25, atom weight 54.94, melting point 1244 ± 3 °C, and specific gravity of 7.2 gcm^{-3}. It is a member of group VII-A of the periodic table and resembles Fe, its next neighbour, in chemical behaviour. Mn has oxidation states of I, II, III, IV, VI, and VII, with oxidation state of II, IV, VI, and VII forming the most stable salts. Lower oxides (MnO and Mn_2O_3) are basic, the higher ones are acidic. Mn is a whitish-grey metal, very brittle, and oxidizes superficially in air.

The biogeochemistry of Mn in soils is very complex due to the following observations: Mn can exist in several oxidation states, Mn oxides can exist in several crystalline or pseudocrystalline states, the oxides can form coprecipitates with Fe oxides, Fe and Mn oxides exhibit amphoteric behaviour and interact both with cations and with anions, and oxidation-reduction reactions involving Mn are influenced by a variety of physical, chemical, and microbiological processes. Therefore, Mn adsorption is more complicated as it forms insoluble oxides in response to pH-Eh conditions. Fig. 38 displays the predicted Eh-pH-stability field for Mn. In most acid and alkaline soils, Mn^{2+} is the predominant solution species.

Adsorption of Mn has been shown to conform to the Langmuir or Freundlich isotherm [336]. Fig. 39 shows Mn adsorption by the B horizon of a highly weathered sand. The adsorption conforms to the Freundlich model. Adsorption in the Ao horizon near the surface (0 to 4 cm) has been found to be increased due to the higher CEC, higher soil organic matter, and higher content in amorphous Fe oxide. Adsorption enhances with increasing pH, which can be explained by the increased hydrolysis of Mn^{2+}, increased likelihood of Mn precipitation, and increased negative charge on the exchange complex. Manganese is strongly adsorbed by clay minerals. Adsorption has been found to increase with increasing pH [337]. In general, sorption of Mn onto soils can be facilitated by several mechanisms. First, the oxidation of Mn to higher-valence oxides and/or precipitation of insoluble compounds in soils are subjected to wetting and drying.

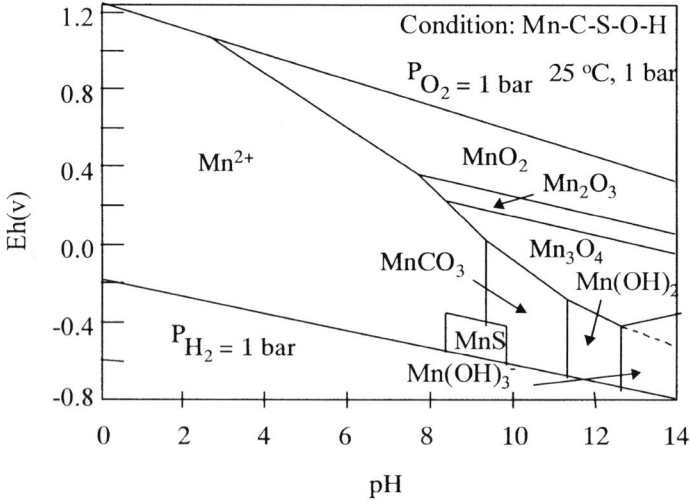

Fig. 38. Predicted Eh-pH-stability field for manganese; the assumed activities of dissolved species are: $Mn = 10^{-6}$, $C = 10^{-3}$, $S = 10^{-3}$ (redrawn after Ref. 183).

Second, absorption into the crystal lattice of clay minerals and adsorption on exchange sites may occur. In calcareous soils, chemisorption onto $CaCO_3$ and following precipitation of $MnCO_3$ may play an important role. Presence of chelating agents is not able to form stable Mn complexes in soils because Fe or Ca can substitute for Mn [338].

4.6.2. Sources and Applications of Manganese

Mn is frequently found in metamorphic, sedimentary, and igneous rocks. Its average content in the lithosphere is about 1000 ppm. As its ionic size is similar to Ca, the two elements can replace each other in silicate minerals. Mn also replaces Fe in magnetite. Although there are more than 100 Mn minerals such as sulfides, oxides, carbonates, silicates, phosphates, arsenates, tungstates, and borates, the most important Mn mineral is the native black manganese oxide, pyrolusite (MnO_2). Other main ores are rhodochrosite ($MnCO_3$), manganite ($Mn_2O_3 \cdot H_2O$), hausmannite (Mn_3O_4), braunite ($3Mn_2O_3 \cdot MnSiO_3$), and rhodonite ($MnSiO_3$).

Mn is mainly used in the metallurgical industry and is an essential ingredient of steel to improve strength, toughness, and hardness. It is also used for the production of alloys of steel, aluminum, and copper. Mn is necessary for the production of alkaline batteries, electrical coils, ceramics, matches, welding rods, glass, dyes, paints, and as a catalyst.

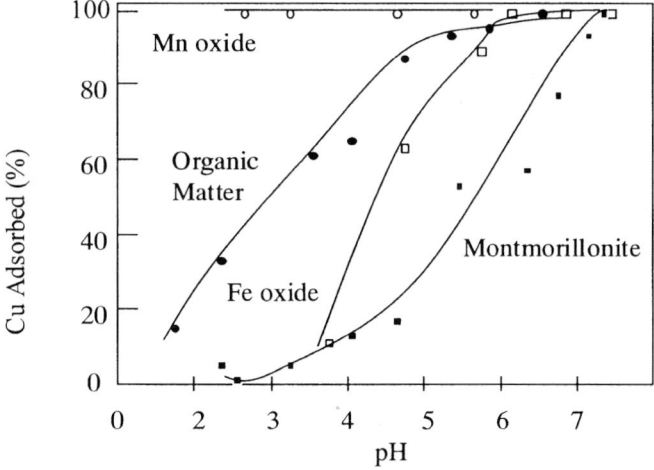

Fig. 39. Adsorption of Mn by soils from B horizons (modified after Ref. 336).

Table 15 shows commonly observed Mn concentrations in various environmental media. The main anthropogenic sources of Mn are industrial activities such as metal smelting and refining, agriculture (fertilizer use, sewage sludge, and animal waste disposal) and atmospheric deposition from fossil fuel combustion and waste incineration. In spite of high levels of Mn in certain sludges, Mn in sludges is not generally a problem in the food chain.

4.6.3. Ecotoxicological Effects of Manganese

Manganese is an essential micronutrient for plants, which activates many enzymes needed for metabolism of organic acids, P, and N. It is also required in photosynthetic O_2 evolution (Hill reaction) in the chloroplasts [183]. Mn is also a constituent of a variety of plant enzymes responsible for respiration and protein synthesis and functions along with Fe in chlorophyll formation. Mn is also important in the development of resistance of plants to root and foliar diseases of fungal origin. Mn deficiencies have been observed, especially in members of the bean family and oat, but Mn phytotoxicity is considered to be more important and occurs frequently with Al toxicity. Al and Mn toxicity are the most important growth-limiting factors in acid soils [339].

Mn is essential for animals and humans as well. If normal diets are consumed, there is hardly any Mn deficiency. Concentration differences between organs and between children and adults are small [340]. Tea has been reported to be exceptionally rich in Mn. One cup of tea contains 0.3 to 1.4 mg Mn. Mn serves as a cofactor for a number of enzymes such as hydrolases, kinases, decarboxylases, and several metalloenzymes.

Table 15
Commonly observed manganese concentrations (ppm) in various environmental media

Material	Average Concentration	Range
Igneous rocks	-	390 – 1620
Sandstone	460	-
Limestone	620	-
Shale	850	-
Coal	100	-
Fly ash	357	44 - 1332
Sewage sludges	-	60 - 1170
Soils (agricultural)	1000	20-10,000
Freshwater (µg/L)	8	0.02 - 130
Seawater (µg/L)	0.2	0.03 – 21

Modified after Ref.183.

Mn toxicity in humans is rare and is generally a result of chronic inhalation of airborne Mn. While ingested Mn is low in toxicity [341], inhaled Mn can be neurotoxic as well as toxic to the respiratory and reproductive tract, probably because inhaled Mn passes the brain first, while ingested Mn passes the liver, which is able to metabolize and eliminate it. Symptoms of Mn toxicity resemble Parkinson's disease in terms of impairment in movement control, facial expression, and certain neurochemical functions. Mn is also used as a gasoline additive and replaces organic Pb compounds such as tetraethyl and tetramethyl Pb.

4.7. Mercury
4.7.1. Chemical and Physical Character of Mercury

Mercury has atom number 80, atom weight of 200.6, melting point of -38.8 °C, specific gravity of 13.55 gcm^{-3}, and vapor pressure of $1.22 \cdot 10^{-3}$ mm at 20 °C. It is a silvery-white metal, which is liquid at room temperature and a good conductor of electricity, but a rather poor conductor of heat. It has three stable oxidation states of 0 (elemental Hg), I (mercurous), and II (mercuric). Its properties and chemical behaviour depend on the oxidation state. There are seven stable isotopes with the following relative abundances: ^{196}Hg (0.15%), ^{198}Hg (10.1%), ^{199}Hg (17.0%), ^{200}Hg (23.3%), ^{201}Hg (13.2%), ^{202}Hg (29.6%), and ^{204}Hg (6.7%). Most of the Hg found in the atmosphere is elemental Hg vapor, while most of Hg encountered in soil, water, sediments, or biota is in the form of inorganic salts and organic Hg complexes.

The fate of Hg in soils and sediments depends on the chemical form of Hg applied, reactivity of inorganic and organic soil colloids, soil pH, type of cations and anions in the exchange complex, soil organic matter, and redox potential. Among these factors controlling Hg adsorption, pH is the most important, as both surface charge of soil particles and metal speciation are mainly determined by pH. As the chloride ion has a strong affinity for Hg, Cl$^-$ is also important for Hg adsorption [342]. A variety of studies has been conducted on the inhibitory effects of Hg(II) complexation with Cl$^-$ on Hg mobility. The main effect is drastic reduction Hg(II) adsorption by inorganic colloids at high (>10^{-3} M) Cl$^-$ concentrations [343-347]. Hg adsorption is also affected by the presence of other ligands such as SO_4^{2-} [348]. Also the clay and organic matter content is of importance. Fig. 40 shows the predicted Eh-pH stability field for mercury. As pH or Cl$^-$ concentration increases, $Hg(OH)_2$ or $HgCl_2$ becomes dominant, while Hg(OH)Cl serves as a transitional species [349].

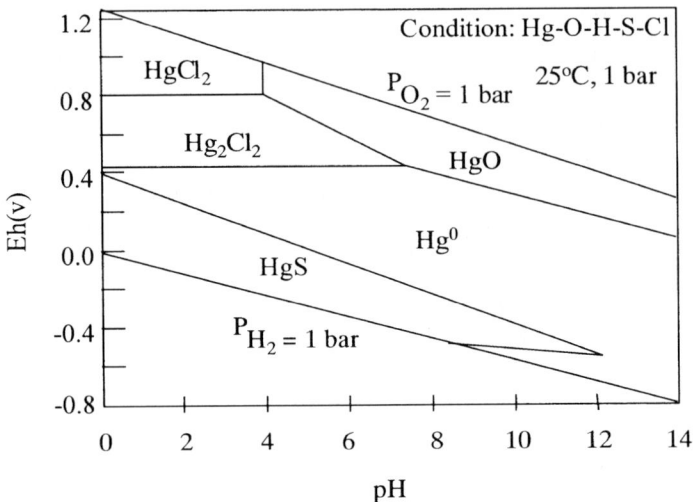

Fig. 40. Predicted Eh-pH stability field for mercury. The assumed activities for dissolved species are Hg = 10^{-8}, Cl = $10^{-3.5}$, S = 10^{-3} (redrawn after Ref. 183).

In general, Hg in soils and sediments is unstable and subjected to a variety of chemical, biological and photochemical reactions [350]. Fig. 41 shows a scheme for the environmental and biological cycling and interconversion of Hg in freshwater lakes [351]. Organic and inorganic Hg compounds decompose to elemental Hg, which can be converted to HgS, volatilizes into the atmosphere, or complexes with ligands. The transformation into toxic and volatile monomethyl- or dimethyl Hg depends largely on biological processes [352]. The Hg^{2+} ion can be methylated by aerobic and anaerobic bacteria [353]. Ref. 354 gives an overview over the different processes, by which Hg can be methylated chemically or biologically.

In terrestrial environments, Hg methylation can also be found [355] and has been shown to be a result of both chemical reactions [356] and microbial activity [357]. The methylation rates of Hg are influenced by several environmental factors such as complexation by dissolved organic matter, which has been found to decrease Hg methylation rates [358] and pH, which controls functional groups that would otherwise bind Hg [359]. The presence of HgS inhibits methylation, which in general decreases with increasing sediment sulfide content [360]. Methylation rates are affected by temperature, presence of organic compounds, and microbial inhibitors [361].

4.7.2. Sources and Applications of Mercury

The most common Hg minerals are cinnabar (α-HgS), which contains 86.2 % Hg, and metacinnabar (β-HgS), or both (polymorphs of HgS). Another Hg source is livingstonite (HgS · 2Sb$_2$S$_3$). The average content of the earth's crust is approximately 50 ppb Hg, mainly as sulfide. Table 16 shows commonly observed Hg concentrations in various environmental media. Hg is used in catalysts, paints, dental fillings, electrical equipment, and for several laboratory purposes. Hg is widely applied in industry and agriculture. The main industrial uses of Hg are chlor-alkali industry (electrolysis), electrical and control instruments industry, laboratory products, dentistry (dental amalgams), and pulp and paper industry (slimicides). In agriculture, Hg is used as a seed dressing in grain, potatoes, flower bulbs, sugar cane, etc., and as a foliar spray against plant diseases. This leads to the contamination of foodstuff and wildlife.

As Hg has an unusual high volatility, which increases with increasing temperature, it is ubiquitous in the atmosphere in appreciable contents. There are several sources of anthropogenic Hg emissions. First, combustion of fossil fuel, wood, wastes and sewage sludges are important sources of atmospheric Hg [362].

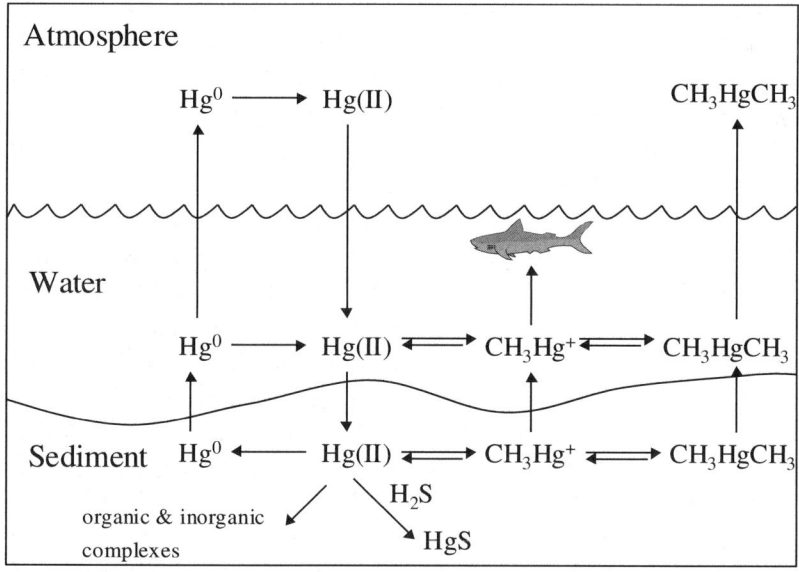

Fig. 41. Cycling and interconversion of various mercury species in freshwater lakes (modified after Ref. 351).

High temperature processes like smelting, coking, ore roasting, and cement and lime production, metal processing, waste sites, and mine spoils are important Hg sources as well [363]. Nowadays, 70 to 80% of total Hg in the atmosphere originates from man-made sources [364]. Total global atmospheric Hg burden has been estimated to be increased since the beginning of the industrialization by a factor between 2 and 5 [365].

Natural emissions of Hg are wind erosion and degassing of Hg from mineralized surface soils, volcanic eruptions and other geothermal activities, evasion from the subsurface crust, and reemissions from rivers, lakes, terrestrial and marine systems, and wetland areas.

4.7.3. Ecotoxicological Effects of Mercury

Hg and Hg compounds are absorbed by roots and translocated to other plant parts [366]. Bioavailability of soil Hg is rather low and Hg tends to accumulate in the roots. The use of Hg compounds such as mercuric oxide, mercuric chloride, or mercurous chloride for the control of plant diseases has lead to Hg transfer into the food chain. Toxic effects of Hg on plants include retardation of growth and premature senescence. Hg has been shown to inhibit synthesis of proteins in plant leaves [367] and to reduce photosynthetic activity [368], as it has a strong affinity for sulfhydryl or thiol groups, which are involved in enzymatic reactions.

In general, methylated organic Hg species are more toxic and bioaccumulate more readily due to their lipophilic nature relative to inorganic Hg, which results in the biomagnification of methyl Hg in the food chain [183]. The accumulation rate in aquatic biota is several orders of magnitude greater than Hg concentrations found in the water column.

Table 16
Commonly observed mercury concentrations (ppb) in various environmental media

Material	Average Concentration	Range
Igneous rocks	-	5 – 250
Sandstone	55	<10 - 300
Limestone	40	40 - 220
Shale	-	5 - 3250
Coal	-	10 - 8530
Fly ash	100	-
Soils (normal)	70	20- 250
Groundwater (μg/L)	0.05	0.01 – 0.10
Seawater (μg/L)	-	0.3 – 1.0
Rainwater, snow	0.2	0.01 – 0.48

Modified after Ref.183.

Therefore, predatory aquatic wildlife species such as the osprey and bald eagle are at greater risk of Hg toxicity [369]. Symptoms of acute Hg poisoning in birds and mammals include progressive weakness in wings and legs, an inability to coordinate muscle movements, neurological damages such as convulsions and fits and highly erratic movements. As for terrestrial ecosystems, bioaccumulation and toxicity of Hg are relatively low, as Hg is strongly sequestered by various soil components. Unlike the aquatic food chain, Hg does not biomagnify in the terrestrial food chain. Sublethal effects of Hg on birds include reproductive damages, liver and kidney damages, and neurobehavioural effects.

The best known example of the devastating effects of Hg on human health is that of Minamata Bay, a small bay on the southwestern shore of Kyushu Island in southern Japan. Between 1950 and 1968, Minamata Bay was severely contaminated with mercury from acetaldehyde and vinyl chloride plants, which lead to highly elevated Hg concentrations in seafood. Ref. 370 gives an excellent historical overview and describes the recovery steps such as sediment dredging and containment, which have been finished in 1990. As fish and fish-product consumption is the most important exposure for humans, the population of Minamata Bay was severely affected by Hg poisoning, which has since then be known worldwide as the Minamata disease. Toxicokinetics and effects of Hg in humans depends on the chemical form of Hg. Elemental Hg vapour is rapidly absorbed through the lungs, but only poorly absorbed through the GI tract. Hg is able to cross both placental and blood-brain barriers. Methyl Hg is rapidly absorbed through the GI tract. Typical symptoms of the Minamata disease include impairment of peripheral vision, disturbances in sensations such as numbness, incoordination of movements, impairment of speech, hearing, and mental disturbances. In general, the nervous system is damaged severely by Hg. Also kidney damages will occur upon long-term exposure to Hg. The US EPA has classified both inorganic and methyl Hg as possible human carcinogens.

Pregnant women, women of childbearing age, and children aged 14 and younger are potentially at increased risk from methyl Hg. Exposure to the developing embryo and foetus during pregnancy results in severe neurotoxicity. Those children can be expected to exhibit mental retardation, ataxia, blindness, and cerebral palsy. Methyl Hg persists in tissues; the half-life of methyl tissue in Hg is over two months. Methyl Hg poisoning epidemics have also been reported from Iraq, where children have been exposed *in utero*.

4.8. Molybdenum
4.8.1. Chemical and Physical Character of Molybdenum

Molybdenum has atom number 42 and atom weight 95.94, and is in the second row of the transition metal elements in group VI-B with Cr and W. Mo is a silvery white metal of great hardness with melting point of 2617 °C and a density of 10.22 gcm^{-3}. It occurs in the oxidation states II, III, IV, V, and VI, with oxidation states V and VI predominating in nature. In these oxidation states, it has an affinity for oxides and for sulfur- and oxygen-containing groups. The most important compound is the trioxide MoO_3. In dilute solutions as they can be found in natural environments, the predominant form is the molybdate anion MoO_4^{2-}. In soils, Mo can exist in different forms. It can be fixed within the crystal lattice of minerals, adsorbed by primary and secondary minerals and soil organic matter, in exchangeable, and in water-soluble form. In aqueous systems, Mo forms soluble anionic species. Below pH 4.2, $HMoO_4^-$ and H_2MoO_4 are the most common species; above this pH, MoO_4^{2-} becomes the major species [319]. Fig. 42 shows the predicted Eh-pH stability field for Mo. Mo is strongly absorbed in soils by Fe and Al oxides and to a lesser extent by Al oxide, halloysite, nontronite, and kaolinite [371, 372]. Adsorption is also strongly pH dependent and increases with decreasing pH [373]. Mo adsorption is also highly correlated with extractable Al [374] and Fe [375].

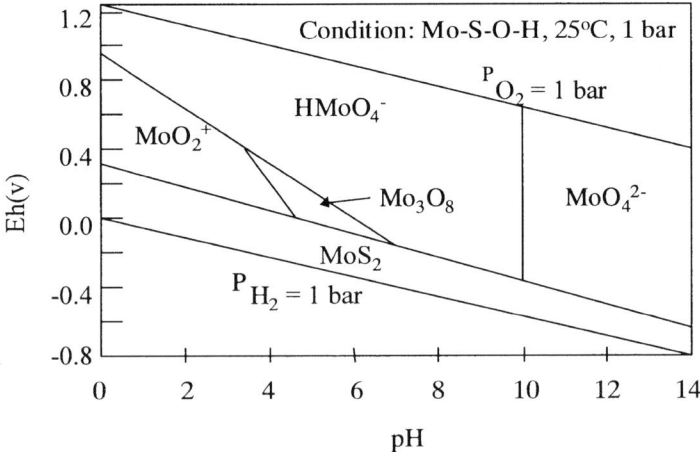

Fig. 42. Predicted Eh-pH stability field for molybdenum. The assumed activities for dissolved species are Mo = 10^{-8}, S = 10^{-3} (redrawn after Ref. 183).

4.8.2. Sources and Applications of Molybdenum

Mo is 53rd in crustal abundance and its average concentration varies from 1.0 to 2.3 ppm [260]. Table 17 shows commonly observed Mo concentrations in various environmental media. The most important primary minerals containing Mo are molybdenite (MoS_2), powellite [$(Ca(MoW)O_4)$], ferrimolybdenite [$Fe_2(MoO_4)_3$], wulfenite ($PbMoO_4$), ilsemanite (Mo oxysulfate), and jordisite (amorphous MoS_2). The most economically important Mo minerals are molybdenite, ferrimolybdenite, and jordisite.

Mo is nontoxic to humans and is therefore used widely as a substitution for Cr and other toxic metals in steel alloys, corrosion inhibitors, and pigments. The main uses of Mo are for the productions of various alloy steels and stainless steels. Mo is also used in catalysts, corrosion inhibitors, pigments, dyes, plastic and rubber parts, industrial gear oils, and high-pressure grease. In agriculture, Mo fertilizers are used.

The main sources of Mo in the environment are from the use of Mo fertilizers in agriculture, from sewage sludges, coal combustion, and mining and smelting. The fertilizers sodium molybdate ($Na_2MoO_4 \cdot 2H_2O$) and $(NH_4)_6Mo_7O_{24} \cdot 4H_2O$ are very soluble and are therefore used frequently in agriculture to correct Mo deficiency in plants. Mo is also frequently found in sewage sludges at levels above the natural soil concentration. The combustion of coal has been estimated to release as much as 1000 tons Mo per year in the United States [376]. Mining and processing of Mo also contributes to Mo release into the environment. Mo content of soils has been reported to be inversely correlated with distance from the Mo source [377] as well as high-Mo waters from the mine drainage system, which will increase Mo levels in rivers and streams [378].

Table 17

Commonly observed molybdenum concentrations (ppm) in various environmental media

Material	Average Concentration	Range
Igneous rocks	1.5	0.9 – 7
Sandstone	3	<3 - 30
Limestone	0.79	< 3 - 30
Shale	16	5 - 90
Coal	3	0.3 - 30
Fly ash	44	7 - 236
Soils (normal)	-	0.2- 5
Groundwater (µg/L)	0.5	0.3 – 10
Seawater (µg/L)	10	4 – 10

Modified after Ref.183.

4.8.3. Ecotoxicological Effects of Molybdenum

Mo is an essential nutrient for both plant and animal life as it is an important component of several enzymes that catalyze unrelated reactions. In plants, the most important functions of Mo are associated with N metabolism, e.g. with nitrogenase and nitrate reductase enzymes. Nitrogenase is extremely important for plants as it catalyzes the reduction of atmospheric N_2 to NH_3, the reaction, by which *Rhizobium* bacteria in root nodules supply N to the plant. Asymbiotic bacteria such as *Azotobacter*, *Rhodospirulum*, *Klebsiella*, and blue-green algae also carry out this reaction. Therefore, Mo-deficient plants often exhibit symptoms of N deficiency [379]. Another important molybdoenzyme is sulfite oxidase, which mediates the biochemical oxidation of sulfite to sulfate and has a wide range of indirect effects in regulating other enzymatic activities such as in carbohydrate metabolism, reproductive physiology, anion balance, root exudation, plant water relations, and disease control [380]. Mo toxicity in plants has not been observed under field conditions [183].

For animals and humans, Mo plays an important role as a catalyst of various enzymes. Toxic effects of Mo include increased blood xanthine oxidase, increased concentrations of uric acid in blood and urine, and a high incidence of gout. The primary source of Mo for humans is food such as milk and milk products, legume seeds, organ meats, cereal grains, and baked goods. As for grazing animals, the toxic range of Mo is 10 to 20 ppm [381]. Molybdenosis has been found to be a Mo-induced Cu deficiency or hypocuprosis, which can be avoided by a Cu/Mo ratio of 4.0 or greater in animal feed [382].

4.9. Nickel
4.9.1. Chemical and Physical Character of Nickel

Nickel has atom number 28, atom weight 58.71, specific gravity of 8.9 gcm^{-3}, and melting point of 1453 °C. It belongs to the socalled iron-cobalt group (group VIII) of the periodic table, and is a silvery-white, hard, malleable, ductile, ferromagnetic metal. It is insoluble in water, but soluble in dilute HNO_3, slightly soluble in HCl and H_2SO_4, and insoluble in NH_4OH. There are five stable isotopes with the following relative abundances: ^{58}Ni (68.27%), ^{60}Ni (26.10%), ^{61}Ni (1.13%), ^{62}Ni (3.59%), and ^{64}Ni (0.94%). It normally occurs in oxidation states 0 and II; the I and III oxidation states can exist under certain conditions, but are not stable in aqueous solutions. The most common Ni species in water-soluble I compounds is Ni^{2+}. Ni readily forms complexes with organic ligands, but complexes with inorganic ligands are only formed to a small degree in the order $OH^- > SO_4^{2-} > Cl^- > NH_3$.

In soils, Ni occurs in several chemical forms. In the soil solution, Ni occurs in the free ionic form (i.e. Ni^{2+}) or complexed with organic and inorganic ligands. Ni(II) is stable over a wide range of pH and redox conditions. Ni is readily sorbed by soils at low (<10 ppm) concentrations. The observed isotherms conform to the Freundlich equation, but multiphasic isotherms have also be observed [383]. Sorption of Ni in soils depends largely on pH.

4.9.2. Sources and Applications of Nickel

Ni ranks 23rd in crustal abundance with an average content of 80 ppm. Table 18 shows average Ni concentrations in various environmental media. The two most important Ni ores are pyrrhotite and pentlandite, a sulfide. Ni is mainly used for electroplating, alloy production, Ni-Cd batteries, electronic components, and catalysts for hydrogenation of fats and methanation. The largest application of Ni is in stainless steel production, and Ni is therefore found in a large variety of products, i.e. automobiles, batteries, coins, jewellery, surgical implants, kitchen appliances, sinks, and utensils. It is also used in Ni-Fe, Ni-Cu, Ni-Cr, and Ni-Ag alloys. Ni-steel alloys are used for armour plating and armaments as well as for turbine blades, jet engine components, and in nuclear reactors. The main sources of Ni in the environment are from mining and smelting, from sewage sludge, and from fuel oil and coal combustion. The latter process has been identified to be the primary source of Ni in the air, since petroleum contains Ni [384]. There is also a high correlation between Ni and V in the air [385].

Table 18
Commonly observed nickel concentrations (ppm) in various environmental media

Material	Average Concentration	Range
Igneous rocks	75	2 – 3600
Sandstone	2	-
Limestone	20	-
Shale and clay	68	20 - 250
Coal	15	3 - 50
Fly ash	141	23 - 353
Soils (world)	20	5 - 500
Freshwater (µg/L)	0.5	0.02 – 27
Seawater (µg/L)	0.56	0.13 – 43

Modified after Ref.183.

4.9.3. Ecotoxicological Effects of Nickel

Ni is an essential nutrient for both plants and animals and is a constituent in urease, methyl coenzyme M reductase, hydrogenase, and carbon monoxide dehydrogenase [386]. Symptoms of Ni deficiency in plants are various and include growth depression, premature senescence, decreased tissue Fe levels, chlorosis, and necrosis. In environments with ultrabasic substrate (e.g. peridotite or serpentinite) plants are known, which hyperaccumulate Ni in their tissues with Ni levels exceeding 1000 ppm. To date, over 240 Ni-hyperaccumulating taxa have been reported, of which 76 are in the family Brassicaceae, mostly in the genera *Alyssum* (48 taxa) and *Thlaspi* (23 taxa). Ni concentrations >50 ppm lead to manifest toxicity symptoms in plants.

As for fish, long-term exposure to low Ni levels may result in reduced skeletal calcification and asphyxiation [387]. There are several types of Ni poisoning in humans. Inhalation of Ni components such as [$Ni(CO)_4$], Ni_3S_2, NiO, and Ni_2O_3 leads to pneumonitis with adrenal cortical insufficiency, pulmonary oedema, and hepatic degeneration, cancer of the respiratory tract, pulmonary eosinophilia, and asthma. The most common effect of long-term Ni skin contact in humans is contact dermatitis [388]. Recently, the carcinogenity potential and Ni and Ni compounds has been of concern.

4.10. Platinum Group Elements (D. Stüben)
4.10.1. Introduction

Platinum group elements (PGE) belong to the group of precious metals, which comprises the rare metals such as Platinum (Pt), Palladium (Pd), Rhodium (Rh), Ruthenium (Ru), Iridium (Ir), and Osmium (Os). They are distributed in the earth's crust with an average content in the range of 0.001-0.005 mg/kg for Pt, 0.015 mg/kg for Pd, 0.0001 mg/kg for Rh, 0.0001 for Ru, 0.001 mg /kg for Ir and 0.005 mg/kg for Os [389, 390]. Higher concentrations are given in the earth core and in iron meteorites. PGE showing chalkophile element character occur in nickel, copper and iron sulphide deposits and are currently mined in Bushveld, South Africa, Norilsk, Siberia and Sudbury, Ontario. Worldwide production and use of PGE has been steadily increasing since 1970 [391]. The total worldwide supply in 2000 was 164 tonnes for Pt, 238 t for Pd and 16.2 t for Rh [392].

PGE are mainly introduced to the environment due to the emission of PGE bearing catalytic converters [393] and this amount is still increasing. Because PGE are not atmophile elements, once released to the air, they travel short distances due to their masses, and are then deposited along bordering traffic routes. Therefore they accumulate in neighbouring soils and enter vegetation and animals and hence, they enter the food chain. Regarding toxicology and sensitisation, Pt emitted in metallic or oxide form is considered to be biologically inert and non-allergenic [394]. Here in this chapter we mainly focus on platinum as data are only scarcely available for the other PGE.

4.10.2. Chemical and Physical Character of Platinum Group Elements

Platinum (Sp. Platina, silver) occurs native, accompanied by small amounts of Ir, Os, Pd, Ru and Rh, all belonging to the same group of metals called PGE. Natural Pt with an atomic weight of 195.09 consists of 6 isotopes with varying abundances: ^{190}Pt (0.01%), ^{192}Pt (0.79%), ^{194}Pt (32.9%), ^{195}Pt (33.8%), ^{196}Pt (25.3%), and ^{198}Pt (7.2%). Palladium (named after the asteroid Pallas – god of wisdom) with an atomic weight of 106.4 has 6 isotopes as well with higher abundances for 106 and 108 of around 25% and Rh has only one isotope at 103. This group of precious metals is known because of there specific physical and chemical characteristics, especially the very high melting points (Table 19).

Chemically the valence stage decreases from Os (VIII) via Ir (VI) and Pt (VI) to Pd (III), whereas Pt occurs as 0, +V and +VI [395]. Precious metals are known because of their low reactivity in regard to the low water and acid solubility. Platinum and Pd tend to form chloride complexes. Oxygen reacts with PGE under higher temperatures forming oxides. In natural environments, Pt occurs in elemental form as oxo- and chlorocomplexes (Table 19). Particularly the resistance to chemical corrosion over a wide temperature range, the high melting point, high mechanical strength, and the good ductility make various applications of these metals possible.

Platinum is a beautiful silvery-white metal, when pure, and is malleable and ductile. It has a coefficient of expansion almost equal to that of soda-lime-silica glass, and is therefore used to make sealed electrodes in glass systems. The metal does not oxidize in air at any temperature, but is corroded by halogens, cyanides, sulfur, and caustic alkalis. It is insoluble in hydrochloric and nitric acid, but dissolves when it is mixed as aqua regia, forming chloroplatinic acid (H_2PtCl_6), an important compound. Palladium is a steel-white metal, does not tarnish in air, and is the least dense and lowest melting of the PGE. When annealed, it is soft and ductile; cold working greatly increases its strength and hardness. Palladium is attacked by nitric and sulphuric acid.

Table 19
Physical and chemical characteristics of PGE

Characteristics	light PGE			heavy PGE		
Element	Ru	Rh	Pd	Os	Ir	Pt
atomic weight	101.07	102.90	106.42	190.2	192.22	195.09
density (g/cm^3)	12.45	12.41	12.02	22.61	22.65	21.45
melting point (°C)	2450	1960	1552	3050	2454	1769.3
boiling point (°C)	4150	3670	2930	5020	4530	3830
sublimation-enthalpy (kJ/mol)	+643.1	+557.3	+378	+791	+665.7	+565.6
atomic radius (Å)	1.246	1.252	1.283	1.260	1.265	1.295
metal atomic radius (Å)	1.325	1.345	1.376	1.338	1.357	1.388
ion-radius M^{2+} (Å)	-	-	0.80	-	-	0.80
" M^{3+}	-	0.68	-	-	-	-
" M^{4+}	0.67	-	0.65	0.69	0,68	0.65
hydratation enthalpy M^{2+} (kJ/mol)	-188	-203	-211	-186	-200	-2190
electrical-negativity	1.42	1.45	1.35	1.52	1.55	1.44
1. ionization-energy (eV)	7.364	7.46	8.33	8.7	9.1	9.0
normal potential M/M^{2+} (V)	+0.45	+0.6	+0.987	+0.85	+1.1	+1.2
" M^{2+}/M^{3+}	+0.23	+1.2	-	-	+1.15	-
" $M^{3+}/M(IV)$	+0.49	+1.4	-	-	+0.74	-
" MO_2/MO^{2-}_4	+2.01	+2.01	-	+1.61	+1.61	-
" MO^{2-}_4/MO^-_4	+0.59	-	-	-	-	-
" MO^-_4/MO_4	+0.95	-	-	-	-	-
Formation enthalpy MCl_2 (kJ/mol)	-	-151	-172	-	-179	-111
" MCl_3	-205	-2994	-	-190	-246	-174
" MCl_4	-	-	-	-255	-	-237
" MO	-	-90.9	-85.4	-	-	-
" MO_2	-305	-	-	-	-274	-172
" MO_4	-229	-	-	-394	-	-
valences	0,1,2,<u>3</u>,4,5,6,7,8	0,2,<u>3</u>,4,5,6	2,3,4	0,3,4,6,8	3,4	0,2,3,4
most frequent natural isotopes (Frequency, %)	101 (17.0) 102 (31.6) 104 (18.7)	103 (100) - -	105 (22.3) 106 (27.3) 108 (26.5)	189 (16.1) 190 (26.4) 192 (41.0)	191 (37.3) 193 (33.8) -	194 (32.9) 195 (33.8) 196 (25.3)
electronic ground-state configuration	[Kr]4d^7 5s	[Kr]4d^8 5s	[Kr]4d^{10}	[Xe]4f^{14} 5d^{6}6s^2	[Xe]4f^{14} 5d^{7}6s^2	[Xe]4f^{14} 5d^{9}6s

4.10.3. Sources and Applications of Platinum Group Elements

The occurrence of PGE in earth crust is very rare, the average concentration of Pt is 0.05g/t, and therefore it is the 76th most often element (Table 20). Higher enrichments of PGE are found in the earth core and meteorites and in rock-forming minerals such as chromite. PGE are enriched through mafic processes together with sulfide ores. They are mined in Norilsk, Siberia, Bushveld (Merensky Reef), South Africa, and Sudbury, Canada. Sperrylite ($PtAs_2$) occurring within nickel-bearing deposits are the major sources of considerable amount of these metals (1-3 ppm) (Table 21). Platinum is extensively used in jewellery, in wires and vessels for laboratory use, and in many valuable instruments, including thermocouple elements. It is also used for electrical contacts, corrosion-resistant apparatus, and in dentistry with increasing demand in the last 10 years (Fig. 43). Platinum resistance wires are used for constructing high-temperature electric furnaces. The metal is used for coating missile nose cones, jet engine fuel nozzles, etc., which must perform reliably for long periods of time at high temperatures.

Table 20
Abundance of platinum group elements in rock-forming minerals (ppb)

Sample description	Ru	Rh	Pd	Os	Ir	Pt	total PGE
Olivine	10		50			80	
Serpentine			80			50	
Chromite from Dunite							300
Chromite	500	500			5.000	1.000	
Peridotite						290	
Basalt			50		40	100	
Continental Basalt			35			27	
Oceanic Basalts			17			25	
Granodiorite			<3		<3		
Sand						7	
Kupferschiefer, Germany	3	0.2	20	0.3	0.4	50	
Soil, India						17	
Coal Ash						200	
Manganese Nodule						500	

Table 21
Compounds of the platinum metals found in nature

PGE Mineral	Formula	Structure
Sperrylite	$PtAs_2$	cubic pyrite type
Braggite	$(Pt, Pd, Ni)S$	tetragonal PdS type
Potarite	$PdHg$	tetragonal AuCu type
Froodite	$PdBi_2$	monoclinic, α $PdBi_2$ type
Niggliite	$Pt(Sn, Te)$	"hexagonal, NiAs type
Arsenopalladinite	Pd_3As	hexagonal, does not agree with artificial Pd_3As which has tetragonal Fe_3P type
Cooperite	PtS	tetragonal PtS type
Laurite	RuS_2	cubic pyrite type

Recently PGE gained importance as industrial catalysts due to their exceptional catalytic properties. PGE are widely used as catalytic converters and therefore the supply and demand for Pt has increased during the last years (Fig. 44). The large-scale introduction of the catalytic converters in car engines aimed at replacing leaded gasoline and for the reduction of automobile emissions, which was introduced into Europe in the late 1980s. From January 1993 on, all new cars sold within the EU were fitted with an automobile catalyst. Since 1975, vehicles have been equipped with converters in the US. Modern catalysts convert over 90% of CO, HCs and NO_x into CO_2, water, and nitrogen. Both NO_x and CO are eliminated together by a redox reaction on an Rh catalyst. NO_x oxidizes CO to CO_2 and is reduced to harmless nitrogen gas, whilst CO and HCs are oxidized by air on a Pt catalyst. Most converters consist basically of a monolithic honeycomb support, made of either aluminium or cordierite ($Mg_2[Al_4Si_5O_{18}]$) and are housed in a stainless steel box (Fig. 45). The honeycomb usually contains approximately 400 channels per square inch and is coated with an activated, high-surface aluminium layer, the so-called "washcoat". The washcoat consists of approximately 90% γ-Al_2O_3 and a mixture of base metal additives, mainly oxides of Ce, Zr, La, Ni, Fe, Ti, Y, W, and some alkaline earth metals acting as promoters of the desired catalytic reactions or serve as stabilizers against deterioration and aging. The noble metals are fixed on the surface of the washcoat usually by impregnation of hexachloroplatinic (IV) acid ($H_2PtCl_6 \cdot 6H_2O$), palladium chloride ($PdCl_2$) and rhodium chloride ($RhCl_3$) precursor salts, respectively. Reduction of the precursors in H_2 stream at 500 °C results in catalytically active highly dispersed metallic PGE particles with the diameter in the 1-10 nm range.

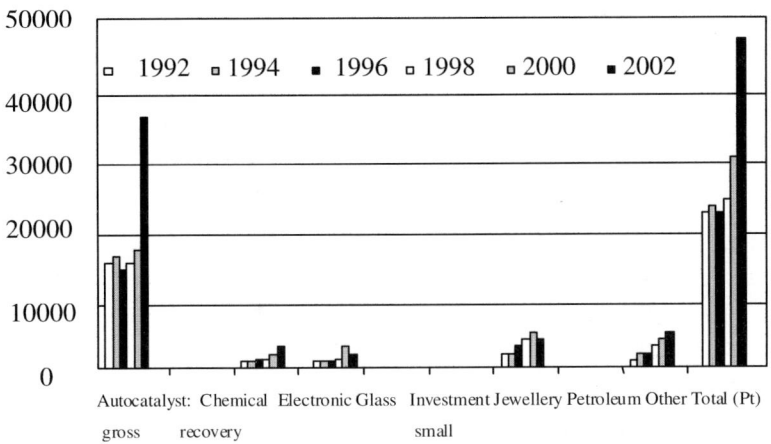

Fig. 43. Platinum demand by application in Europe (data from Ref. 393).

Nowadays, there is a wide range of possible combinations and concentrations of the noble metals mentioned above with base metals in the catalysts, which can be utilized to achieve different performance characteristics that are required in the various vehicle models. In European gasoline cars, the usual Pt/Rh or Pd/Rh ratio has been approximately 5; in the US, this ratio is approximately 10 [396-398]. Conventional converters typically contain 0.08% Pt, 0.04% Pd and 0.005-0.007% Rh [399]. Since Rh was introduced into catalytic converters, approximately 73% of the Rh production of the world has been consumed by the catalyst manufacturing industry until the end of the 1980s [400], which increased to 99% until 2000. Demands by application in 1999 and 2000 for these metals are shown in Fig. 46. Apart from catalytic converters, additional major uses of PGE are in the glass, chemical, electrical, electronic and petroleum industries, the manufacture of jewellery, in medicine as cancer treatment drugs, and in dentistry as alloys. Pd absorbs large volumes of hydrogen, retaining it at ordinary temperatures but giving it up at red heat. In the finely divided state, Pt is an excellent catalyst, having long been used in the contact process for producing sulphuric acid. It is also used as a catalyst in cracking petroleum products. Palladium is a good catalyst as well and is used for hydrogenation and dehydrogenation reactions. It is alloyed and used in jewellery trades. White gold is an alloy of gold decolorized by the addition of Pd.

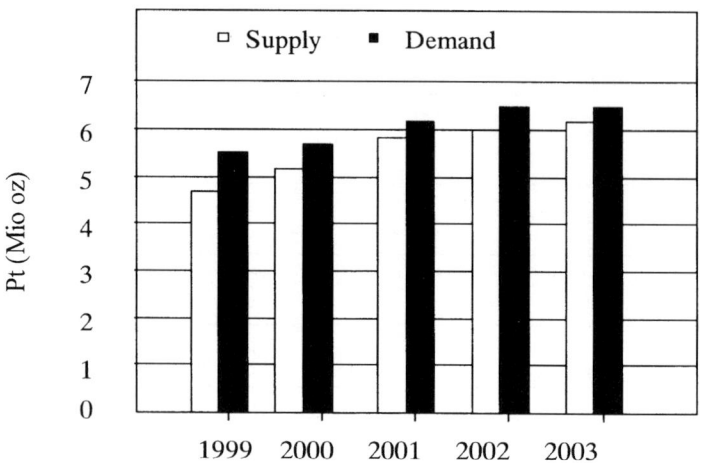

Fig. 44. Supply and demand for Platinum 1999-2003 (redrawn after Ref. 392).

The metal is used in dentistry, watch making, and in marking surgical instruments and electrical contacts. Its use in catalytic converters is increasing.

4.10.4. Platinum Group Elements Emission by Car Catalytic Converters

Due to chemical and physical stress by fast changing oxidative/reductive conditions, high temperature and mechanical abrasion, PGE are emitted into the environment [401-406]. Despite the multitude of studies on the contamination of different environmental compartments by PGE, e.g. in soil and dust [407-409], in sediments [410], in snow [411], in plants [412, 413], and in wine [414], except of some experimental data on the solubility of PGE in different solvents [398, 414-418], still little is known about the mobility and speciation of this group of elements in the environment [416, 419]. Also only few data exist on the emission mechanisms and on the characteristics of the emitted particles [420-422]. The amount and rate of PGE emission are affected by the speed of the automobile, the type of the engine, the type and age of the catalyst and the type of fuel additives [406, 423]. Emission can be intensified by the unfavourable operating conditions (misfiring, excessive heating), which can even destroy the converter.

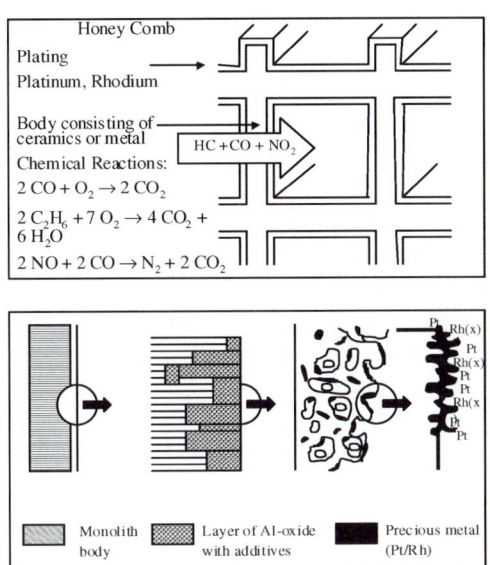

Fig. 45. Design of the automobile catalytic converter, cross section of the honeycomb and the catalytic active surface.

The quantity of PGE released from converters has been evaluated by direct (sampling in the exhaust fume: [402, 406] and indirect methods (sampling of street sediments, street dust: e.g. [407]). The emission rate of Pt ranged from 7 to 123 ng m^{-3}, where the values correspond to the emission factor of 9-124 ng km^{-1} [406, 412, 424, 425]. Increasing Pt emission was pointed out with increasing car speed and operating conditions [426].

4.10.5. Platinum Group Elements in Environmental Matrices

A large number of studies have reported the dispersion and accumulation of precious metals in various environmental compartments. Studies in Germany have shown that the total Pt concentration in airborne particulate matter ranged from 0.02 to 5.1 pg m^{-3} in urban areas, whilst along a highway it reached 30 pg m^{-3}, with Pt mainly present in the small (0.5–8µm) particle size fraction (Fig. 47). The proportion of soluble Pt in airborne PM varied from 30 to 43% [427]. Ref. 418 observed a 46-fold increase in the Pt levels of air over a 10-year period up to 1998. A lot of studies show enrichments of PGE in traffic-related soil, dust, and vegetation (Fig. 47). Ref. 419 calculated the mean Pt concentrations in highway run-off ranging from 0.1 to 0.7 ng l-1.

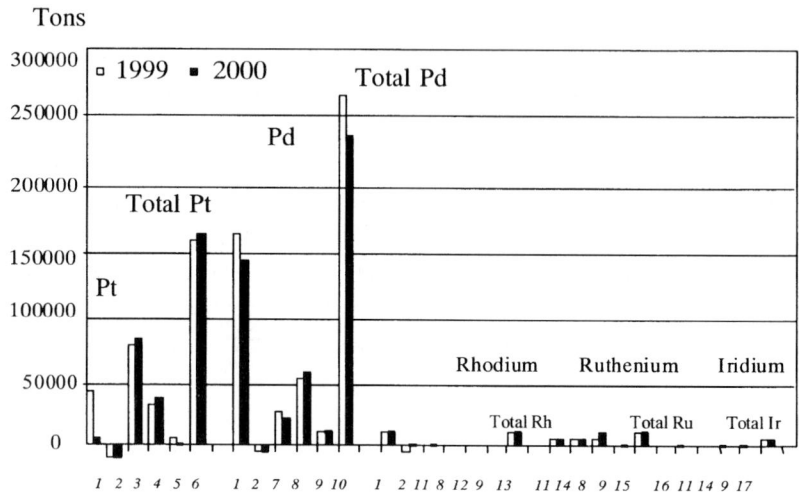

Fig. 46. PGE demand by application (worldwide): *1*: Autocatalysts, gross; *2*: Autocatalysts, recovery; *3*: Jewellery; *4*: Industrial; *5*: Investment; *6*: Total demand (Pt); *7*: Dental; *8*: Electronics; *9*: Other; *10*: Total demand (Pd); *11*: Chemical; *12*: Glass; *13*: Total demand (Rh); *14*: Electrochemical; *15*: Total demand (Ru); *16*: Automotive; *17*: Total demand (Ir), (data after Ref. 402)

In Ref. 423, a strong correlation of Pt, Pd, and Rh with traffic-related elements such as Ni, Cu, Zn, and Pb has been observed in soils from US roadsides. In heavily polluted Mexico City, the PGE contamination of the soil exposed to high traffic densities exceeded the natural background values by up to two orders of magnitude and was also strongly dependent on the traffic conditions [428].

In Germany, at monitoring stations with around 100.000 vehicles per day, the Pt concentration in surface soil adjacent to the road reached several hundred nanograms per gram [429] decreasing with distance from the road (Fig. 47). Dust samples collected in a street tunnel showed an increase in Pd concentrations from 21.8 ng g^{-1} in 1994 to 100.5 ng g^{-1} in 1998 [430]. Studies on vegetation showed that elevated Pt and Rh levels are given in grass with increasing trends for the last years [413] (Fig. 48). Tree barks are effective substrates for the accumulation of atmospheric aerosols and airborne PM as well [431]. Accumulation of PGE has been shown in pine needles collected in Italy, reaching higher concentrations than two orders of magnitude than the crustal abundance [432].

Plotting 374 Pt and Rh analyses on soil, dust, street sediments and fallen leave samples (Fig. 49) the Pt/Rh ratio of 5:1 followed well those applied in catalytic converters, which allowed the tracking of traffic-related PGE in the environment [426]. The concentrations of Pt and Pd in sewage sludge and effluent particles from different cities reflect the anthropogenic emission situation. Ref. 433 implied that the PGE enrichment is ubiquitous through the industrialized areas. However, the higher Pd content of the sludge samples compared to Pt suggested an additional anthropogenic source of Pd and/or its more particle reactive behaviour than Pt. Daily Pt loads indicated that in large industrial areas, such as Munich, the traffic is likely not the dominant source of Pt in municipal sewage [434]. Ref. 435 explained the increased Pd content in the sludge with an enhanced emission from the German dental-alloy industry. Other possibilities are the electroplating waste associated with jewellery and electrical industry, and most importantly, the chemical industry, which releases dissolved and/readily dissolvable compounds of PGE [410].

4.10.6. Transformation of Platinum Group Elements and Bioaccumulation in the Environment

According to the size distribution of the emitted particles, the larger particles (>10 μm) with 62-67% abundance dominated over the medium (3.1-10 μm), and the small (<3.1μm) sized fraction of ~21% and ~13%, respectively [399]. The soluble and the volatile Pt species were tracked by the application of double-layered glass-fibre filters and condensate traps, respectively. It was found that the soluble Pt was on average approximately 1% of the total Pt emitted, also in the Dimroth-type condenser, whereas no volatile Pt was found in the liquid nitrogen trap [436]. As indicated by the results, most of the PGE released from converters are found to be in particulate form (Pt >95%, Pd >85%, and Rh>90%) and are dispersed into the environment at a rate of up to 200 ng PGE per km per car [436]. Examination of the released particulates showed that approximately 99% of Pt is in metallic state with approximately 1 % present as oxidized Pt, presumably in the form of Pt^{4+} [420, 437]. The particulate Pt exists in the form of surface oxidized metal nanoparticles attached to larger aluminium particles (substrate), and hence, the ablated washcoat particles are considered to be the carrier of the precious metals [438]. A similar observation on the existence of carrier particles and also the association of PGE with Ce fingerprint originating from the washcoat was made by Ref. 422 and 439.

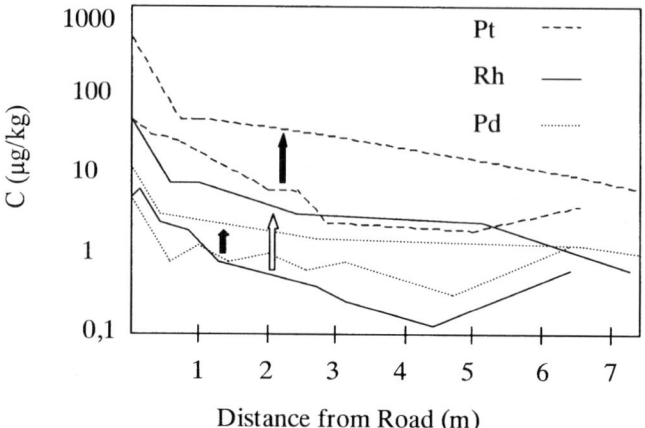

Fig. 47. Element content and variation of PGEs of road-side soils between 1994 and 1996.

Very little is known on the uptake of traffic-related PGE by the biosphere. Calculating the transfer coefficient (TC) defined by Ref. 440, the TC of Pt, Rh, and Pd was found to be within the range from immobile to moderately mobile, and decreased in the order of Pd>Pt>Rh; i.e., Pd is the most biologically available in this group [426]. Further studies indicate that the uptake depends on the type of vegetation [441-444]. Ref. 445 examined the biological availability of Pd, Pt, and Rh by freshwater isopods and demonstrated a time-dependent bioaccumulation and a higher accumulation from matrices of higher PGE content. Further studies showed the biological availability of Pt, Rh, and Pd for earthworms [426], rats [406], zebra mussels, and European eels [446, 447]. Laboratory experiments revealed that the aquatic organisms bioaccumulate anthropogenic PGE to a considerable extent in different organs (Fig. 50) [446, 447]. The parasite of eels took up and accumulated 1600 times higher Rh and 50 times higher Pt levels than the concentration in the water exposed to catalytic converter materials [448]. A number of studies have reported on the PGE levels in human body fluids and tissues. Ref. 393 gives an overview on the data available. Studies in Australia have suggested that a major pathway of Pt into the human body is via the diet with an average dietary intake of 1.44 µg Pt per day for adults.

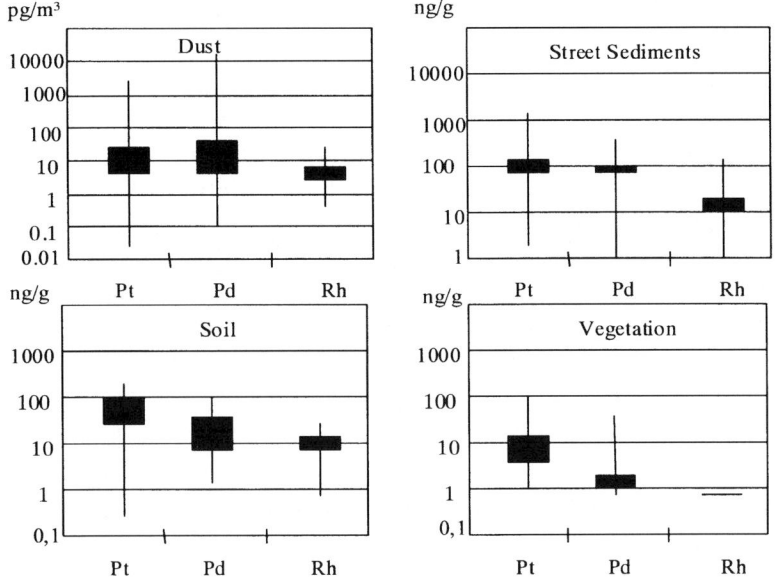

Fig. 48. Average Pt content in various environmental compartments.

Toxic effects of metallic Pt, as emitted by catalytic converter are only expected if the Pt is bioavailable. The scarcity of information on the speciation and solubility of PGE under real environmental conditions and the possible toxicological effects of inhaled micron-sized particles and the uptake within the food chain impose the necessity of further investigations in order to estimate the ecotoxicological and health effects of the emitted PGE.

4.11. Zinc (H.B. Bradl)
4.11.1. Chemical and Physical Character of Zinc

Zinc has atom number 30, atom weight of 65.38, a melting point of 420 °C and density of 7.13 gcm^{-1}. It belongs to group II-B of the periodic table and occurs divalent in all its compounds. Zinc is a bluish white soft metal, which comprises of five stable isotopes with the following relative abundances: ^{64}Zn (48.89%), ^{66}Zn (27.81%), ^{67}Zn (4.11%), ^{68}Zn (18.65%), and ^{70}Zn (0.62%). The six radioactive isotopes are ^{62}Zn, ^{63}Zn, ^{65}Zn, ^{69}Zn, ^{72}Zn, and ^{73}Zn. The oxidation state in nature is II, the Zn^{2+} ion is colourless. In alkaline solutions, the hydroxide is precipitated. The zincate ion, Zn(OH)$_4^{2-}$, is formed with excess base. As it is amphoteric in nature, Zn forms water-soluble chlorates, chlorides, sulfates, and nitrates, while the oxides, carbonates, phosphates, and silicates are relatively insoluble in water. Sorption is an important factor governing Zn concentration in soils and is influenced by several factors, such as pH, clay mineral content, CEC, soil organic matter, CEC, and soil type. Clay minerals show variations in their adsorbing capacity due to their different CEC, specific surface area, and basic structural makeup. 2:1 clays such as montmorillonite and illite exhibit greater fixing capacities for Zn than 1:1 clays such as kaolinite. This fact can be explained by entrapment of Zn^{2+} in the interlattice wedge zones of the clay when the zones expanded due to wetting and contracted upon drying [449]. Clay-bound Zn was characterized as dominantly reversible in association with clay surface groups, while the rest exists in an irreversible nonexchangeable form associated with lattice entrapment [450]. In calcareous and alkaline soils, Zn unavailability is due to sorption of Zn by carbonates, precipitation of Zn hydroxide or carbonates, or formation of insoluble calcium zincate [183]. The surface charge on hydrous oxides depends highly on pH and increases with increasing pH. Zn retention is partly due to the presence of oxide surfaces in soils whose clay fractions are dominated by layer silicates [451]. Chelating agents, either natural or synthetic, play an important role in Zn mobility in soils. Zn also forms complexes with Cl^-, PO_4^-, NO_3^-, and SO_4^{2-} [452].

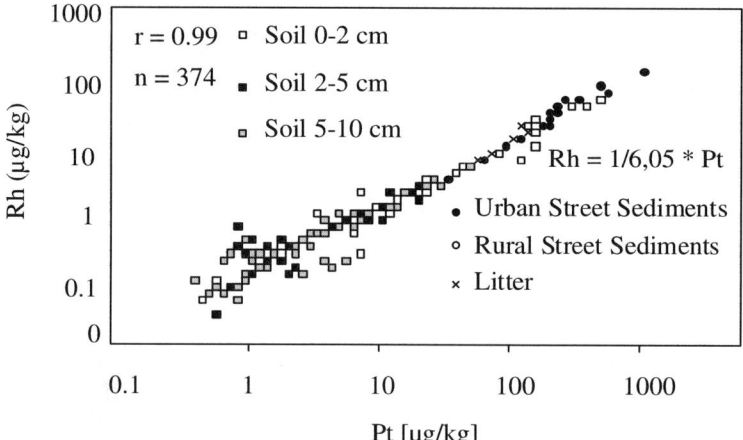

Fig. 49. Platinum and rhodium content of contaminated environmental samples (urban street sediments, rural street sediments, vegetation, soils).

As the presence of EDTA in soil suspension can decrease Zn sorption by soils, Zn is believed to form strong complexes with EDTA thus decreasing its affinity for sorption sites [453]. In contrast, complex formation of Zn with Cl^-, NO_3^-, and SO_4^{2-} did not have significant effects on Zn sorption. Thus, the presence of synthetic chelates maintains most of the Zn in mobile form.

4.11.2. Sources and Applications of Zinc

Zinc is fourth among metals of the world in annual consumption (behind Fe, Al, and Cu) [183]. It is extensively used in the automobile industry, for the production of protective coatings for iron and steel, in cosmetics, powders, ointments, antiseptics, paints, varnishes, rubber, and linoleum. Zinc is also needed for manufacture of parchment papers, glass, automobile tires, television screens, dry cell batteries, and electrical equipment.

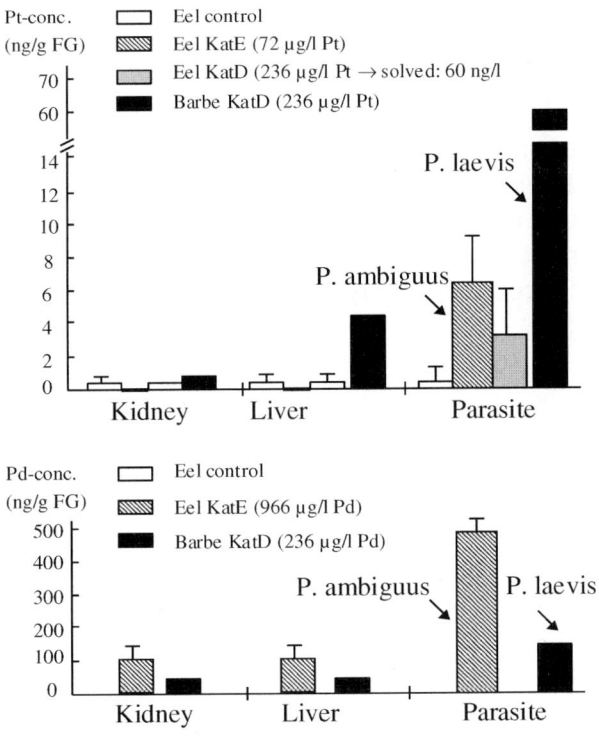

Fig. 50. Platinum content of fish organs (redrawn after Ref. 447).

In agriculture, Zinc is an important micronutrient fertilizer, a wood preservative, and an insecticide. Znc is the 24th most abundant element in the earth's crust (70 ppm average value). The most important commercial Zn ores are the sulfides sphalerite and wurzite and their weathering products, smithsonite ($ZnCO_3$) and hemimorphite [$Zn_4Si2O_7(OH)_2 \cdot 2H_2O$]. Table 22 shows commonly observed Zn concentrations in various environmental media.

The main sources of Zn in the environment are Zinc fertilizers, sewage sludges, and mining and smelting. In agriculture, four main classes of Zn-bearing fertilizers are used: inorganic, synthetic chelates, natural organic complexes, and inorganic complexes [454]. The most commonly Zn fertilizers are ZnO, $ZnCO_3$, $Zn(NO_3)_2$, $ZnCl_2$, and Zn-ammonia-complex. Sewage sludges may contain high concentrations of Zn. It has been shown in long-term experiments that Zn is the most bioavailable metal in soils treated with sewage sludges [183]. Also mining and smelting is an important input source of Zn [455]. Other sources include vehicle exhausts and tire wear, corrosion of galvanized steel, and storm water runoff in agricultural areas [456].

4.11.3. Ecotoxicological Effects of Zinc

Zinc is essential for both plants and animals. In plants, Zn is involved in various metalloenzymes, in the stability of cytoplasmic ribosomes, and of root cell plasma membrane. It also catalyzes oxidation and protein synthesis as well as the transformation of carbohydrates. If Zn concentrations > 100 ppm are encountered, phytotoxicity symptoms similar to chlorosis may occur [457].

Table 22
Commonly observed zinc concentrations (ppm) in various environmental media

Material	Average Concentration	Range
Igneous rocks	65	5 – 1070
Sandstone	30	5 - 170
Limestone	20	< 1 - 180
Shale	97	15 - 1500
Fly ash	449	27 - 2880
Sewage sludge	2250	1000 - 10000
Soils (normal)	90	1- 900
Freshwater	15	< 1 – 100
Seawater	5	< 1 – 48

Modified after Ref.183.

Zn is essential for humans, as more than 200 Zn enzymes and proteins have been identified so far [458]. Zn deficiency is a result of inadequate dietary intake especially during periods of growth, pregnancy, and lactation. Clinical symptoms of Zn deficiency include dermatitis, anaemia, poor wound healing, hypogonadism, and neuropsychological dysfunction [459]. Zn toxicity in humans is very rare. High Zn intake may affect cholesterol metabolism.

4.12. Other Heavy Metals
4.12.1. Cobalt

Cobalt is a silvery white metal with atom number 27, atom weight 58.93, and specific gravity of 8.9 gcm^{-3}. It is resistant to corrosion and to alkali, can be solved in acids, and occurs in oxidation state of II or III. The most important Co minerals are cobaldite, smaltite ($CoAs_3$), and erythrite [$Co_3(AsO_4)_2 \cdot 2H_2O$]. Co is commonly obtained from Ag, Ni, Pb, Cu, and Fe ores as a by-product. The main industrial use of Co is in the production of high-grade steels, alloys, superalloys, and magnetic alloys. Co is also used as a pigment, glass decolourizer, and as a drying agent in paints, varnishes, enamels, and inks.

The predicted Eh-pH stability field for Co is presented in Fig. 51. Acid conditions increase the availability of Co as Co^{2+}. Co was found to accumulate in hydrous oxides of Fe and Mn in soils [460, 461]. It was also found that Co adsorption by certain soils was increased by removal of Fe, which is believed to expose clay mineral surfaces that were more reactive than previously exposed Fe oxide surfaces [462]. Co sorption capacity of soils was found to highly correlate with Co content and surface area and to a lesser extent with Mn and clay contents and pH [463]. Almost all of the Co in soils could be accounted for by that present in Mn minerals, indicating that these minerals can be an important sink for Co in soil [464]. Sorption of Co by Fe and Mn oxides as a function of pH is shown in Fig. 52. Cryptomelane ($K_2Mn_8O_{16}$) has a point of zero charge below 3 and a high surface area of 200 m²/g. It sorbed significant amounts of Co even at relatively low pH. On the other side, goethite, which has a relatively small surface area of 90 m²/g and a point of zero charge of 8.7, shows significant Co sorption only at pH values above 6.0 [465]. Two forms of bound Co in montmorillonite have been identified [466]. The first form, which is characterised as being slowly dissociable, seems to be bound in a monolayer by chemisorption and would exchange with Zn^{2+}, Cu^{2+}, or other Co^{2+} ions but not with a Ca^{2+}, Mg^{2+}, or NH_4^+ ions.

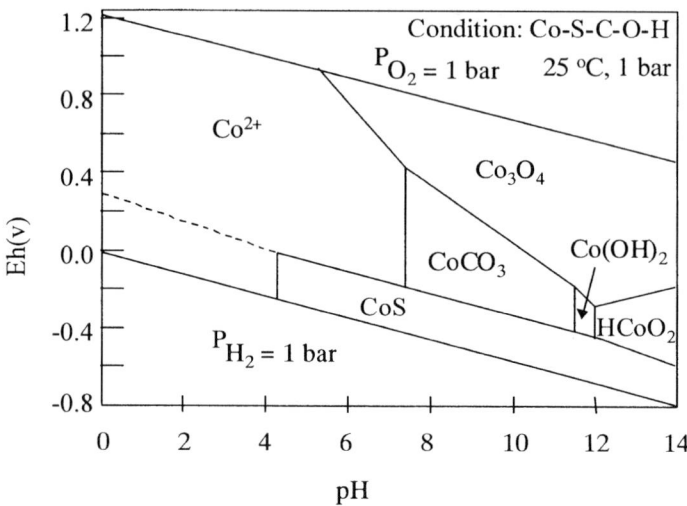

Fig. 51. Predicted Eh-pH stability field for cobalt. The assumed activities for dissolved species are Co = 10^{-6}, C = 10^{-3}, S = 10^{-3} (redrawn after Ref. 183).

The second form of Co is not dissociable and is believed to either enter the crystal lattice or become occluded in the precipitates of another phase.

Co is essential to certain blue-green algae, free-living bacteria, and symbiotic systems [467], e.g., it is required by *Rhizobium*, the symbiotic bacterium that fixes N_2 in the root nodules of legumes. It is not essential to plants, but it is a constituent of vitamin B_{12}, which is required by all animals. A level of about 0.07 ppm of Co in foodstuff is necessary to maintain health of animals. Sheep and cattle feeding lower levels of Co show various symptoms including reduced growth, loss in body weight, then extreme emaciation, anaemia, and death.

4.12.2. Silver

Silver is a lustrous, brilliant white metal, which is very ductile and malleable. It has atom number 47, atom weight 107.87, and specific gravity of 10.5 gcm^{-3}. It occurs in four oxidation states in nature: 0, I, II, and III, with O and I being the most common, while II and III species are rare. The most important Ag sources are metallic Ag, argentite (Ag_2S), and hornsilver (AgCl). It is also encountered in Pb, Pb-Zn, Cu, Ag, and Cu-Ni ores.

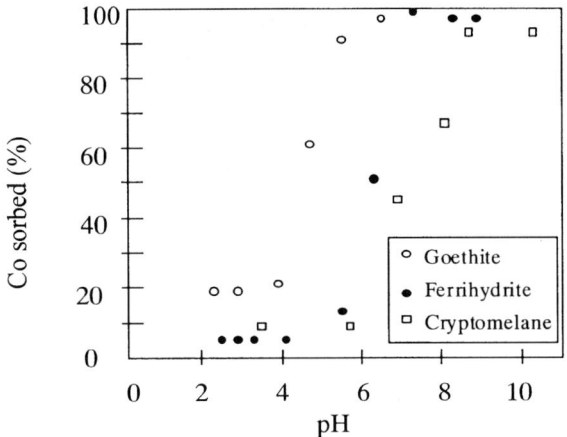

Fig. 52. Adsorption of Cu by different soil constituents as a function of pH (redrawn after Ref. 183).

Another source of Ag is the recovery during electrolytic Cu refining and smelting of Ni ores. The main industrial uses for Ag are for photographic manufacturing, which amounts to more than 40% of the world's total Ag demand. It is also used for electrical contacts and conductors, brazing alloys and solders, catalysts, batteries, sterlingware, jewellery, silverplate, mirrors, and others.

The average content of Ag in the earth's crust is about 0.1 ppm [468]. In soils, Ag is usually found as sulfides in association with Fe, Pb, Au, or tellurides. Fig. 53 shows the predicted Eh-pH stability field for silver. Ag^+ is very reactive and forms stable complexes with negative binding sites in suspended soils and sediments. Various investigations showed the relative unavailability of Ag to plants [469, 470]. Ag is of environmental concern with respect to aquatic ecosystems. Silver ions are one of the most toxic heavy metal ions to microorganisms, particularly to the heterotrophic bacteria [471]. As Ag^+ easily forms insoluble compounds, it is relatively harmless in terrestrial environments. In aquatic environments, it is very toxic to aquatic species such as fish, mollusks, and crustaceans, with a median lethal concentration value (LC_{50}) of 6.5 to 70 μgL^{-1} for freshwater fish. The bioavailability and toxicity of Ag is mainly influenced by is chemical speciation, as well as water hardness, pH, alkalinity, Cl^-, and DOC.

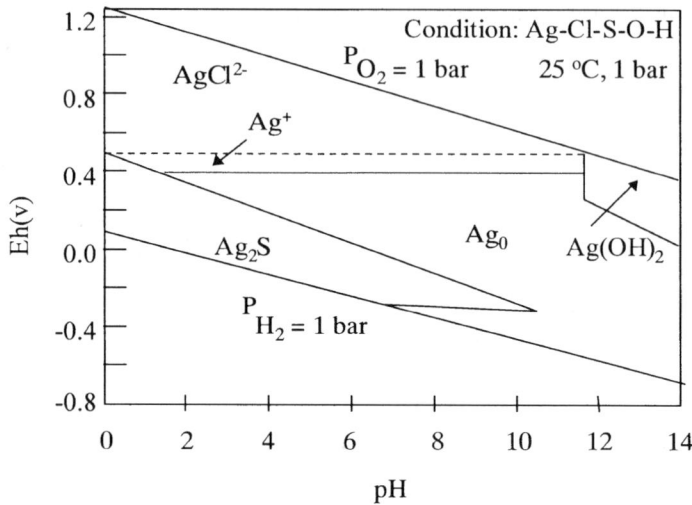

Fig. 53. Predicted Eh-pH stability field for silver. The assumed activities for dissolved species are Ag = 10^{-8}, S = $10^{-3.5}$ (redrawn after Ref. 183).

The toxicity of Ag in aquatic systems can also be influenced by the presence of ligands that form Ag complexes such as Cl$^-$ and DOC [472]. To humans, Ag is relatively nontoxic.

4.12.3. Thallium

Thallium has atom number 81, atom weight 204.4, specific gravity of 11.85 gcm^{-3}, and melting point of 303 °C. It is a grey-white, soft, ductile metal, which occurs in nature in the primary oxidation state I (thallous species), although the III oxidation state (thallic species) can also be found. The main Tl minerals are crooksite [(Cu, Tl, Ag)$_2$Se], lorandite (TlAsS$_2$), and hutchinsonite [(PbTl)2(Cu, Ag)As$_5$S$_{10}$]. Its geochemical behaviour is similar to Rb and other alkali metal cations. The main use of Tl is in Pb-, Ag-, or Au-based alloys, which are characterized by low friction coefficients, high endurance limits, and high resistance to acids. Its agricultural use as a rodenticide and insecticide was banned in the 1970s. Further industrial uses of Tl include manufacturing of Hg vapour lamps, deep-temperature thermometers, semiconductors, highly-refractive optical glasses, cardiac imaging, as a catalyst in olefin and hydrocarbon synthesis, and in the electrical and electronic industry.

Tl is widely distributed in the environment and has become of increasing concern due its toxicity to organisms in small amounts, the potential biomethylation in the environment, and its release from coal combustion and from cement production [473].

In soils, data indicate that although Tl is mobile it does not migrate substantially in the soil profile. In waters, Tl has been found in trace amounts only, while in agricultural products, Tl concentrations can be in the ppb to ppm level [183]. The predominant Tl species in soils and natural waters is Tl (Fig. 54). Tl also forms stable complexes with fulvic acid and chloride as $TlCl^0$. Phytotoxicity of Tl is known and produced impairment of chlorophyll synthesis and seed germination, reduced transpiration, growth reduction, and leaf chlorosis. In general, not much is known about the environmental biogeochemistry of Tl, especially of Tl in the food chain pathway.

4.12.4. Tin

Tin is a soft, silvery-white, pliable metal, which has atom number 50, atom weight 118.7, specific gravity of 5.5.7 gcm^{-3}, and melting point of 232 °C. It occurs in two oxidation states, II and IV. Sn is ranked 49th in crustal abundance with 2.5 ppm average crustal concentration.

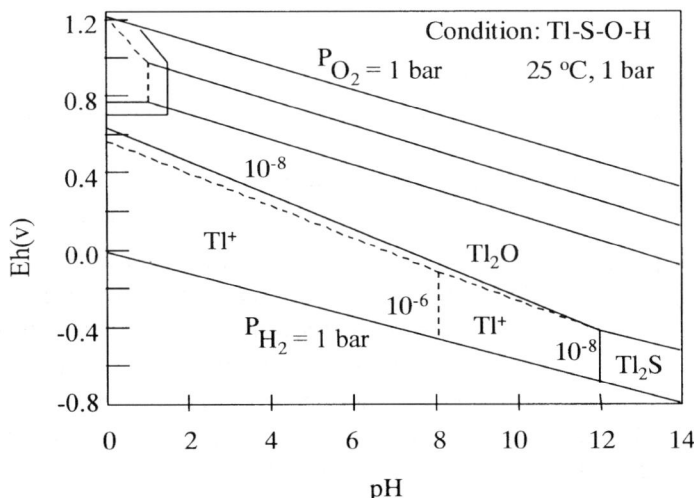

Fig. 54. Predicted Eh-pH stability field for thallium. The assumed activities for dissolved species are Tl = 10^{-8}, S = 10^{-3} (redrawn after Ref. 183).

The most important Sn mineral is cassiterite (SnO_2). Sn is mainly used as a protective coating agent and as an alloying metal, mostly in tinplate, bronze, and solder. Tinplate is widely used for cans, Zn-Sn and Cd-Sn alloys are needed in the automotive and aerospace industry. Organotin compounds such as mono-, di-, and tributyltin (MBT, DBT, TBT) have been used as biocides or as stabilizers, e.g., in the heat stabilization of polyvinyl chloride polymers or as catalysts in the manufacture of silicone and polyurethane, or as an antifouling agent in paints for ships. This extended use has led to serious concern on its environmental biogeochemistry, as various data imply that butyltin contamination has already reached deep waters [474].

Organotin compounds are of concern because of their known toxicological effects on both animals and humans [475]. Sn can also be methylated and therefore be more bioavailable [387]. In aquatic systems, TBT is the most toxic of all butyltin compounds [476]. More recently, butyltin compounds have been found in human blood, which indicates the exposure of humans to these substances [477]. The toxicological consequences of these findings are not known to date.

REFERENCES

[1] P.J. Potts, A Handbook of Silicate Rock Analysis, Blackie Academic & Professional, London, 1987.
[2] H. Rump, Laborhandbuch für die Untersuchung von Wasser, Abwasser, Boden, Wiley-VCH, Weinheim, 1998.
[3] M. Koksoy, P.M.D. Bradshaw and J.S. Thomas, Inst. Min. Metall., Trans., Sect B, 76 (1967)121.
[4] W.K. Fletcher, Analytical methods in geochemical exploration, Elsevier, Amsterdam, 1981.
[5] A.M. Ure, P. Quevauviller, H. Munteau and B. Griepink, Int. J. Environ. Anal. Chem., 51 (1993)135.
[6] A. Tessier, P.G.C. Campbell and M. Bisen, Anal. Chem., 51 (1979) 844.
[7] W. Baeyens, F. Monteny, M. Leermakers and S. Bouillon, Anal. Bioanal. Chem., 376 (2003) 890.
[8] J.R. Dean and D.J. Ando, Atomic Absorption and Plasma Spectroscopy, Wiley & Sons, Chichester, 2000.
[9] J. Dedina and D.L. Tsalev, Hydride Generation Atomic Absorption Spectrometry, Wiley & Sons, Chichester, 1995.
[10] L. Ebdon, E.H. Evans, A. Fisher and S.J. Hill, An Introduction to Analytical Atomic Spectrometry, Wiley & Sons, Chichester, 1998.
[11] J.M. Harnly, Fres. J. of Analyt. Chem., 355 (1996) 501.
[12] S.J. Haswell, Atomic Absorption Spectrometry, Theory, Design and Applications, Elsevier, Amsterdam, 1991.

[13] S.J. Hill, J.B. Dawson, W.J. Price, I.L. Shuttler, C.M.M. Smith and J.F. Tyson, Analyt. Atomic Spectr., 13 (1998) 131.
[14] J.C. Lindon, G.E. Trainter and J.L. Holmes, Encyclopedia of Spectroscopy and Spectrometry, Academic Press, London, 2000.
[15] A.L. Gray, in" Inductively coupled Plasma Mass Spectrometry" (A.R. Date and A.L. Gray, eds.), p. 1-39, Blackie & Son, Glasgow, 1989.
[16] S. Hill, Inductively Coupled Plasma Spectrometry and its Applications, Blackwell Publishing, Oxford, 1999.
[17] H. Taylor, Inductively Coupled Plasma-Mass Spectrometry, Elsevier, Amsterdam, 2000.
[18] C. Vandecasteele and C.B. Block, Modern Methods for Trace Element Determination, Wiley & Sons, Chichester, 1993.
[19] B.K. Argawal, X-ray Spectroscopy, Springer, Berlin, Heidelberg, New York, 1979.
[20] E.P. Bertin, Introduction to X-ray Spectrometric Analysis, Plenum Press, New York, London, 1978.
[21] P. Hahn-Weinheimer, A. Hirner and K. Weber-Diefenbach, Röntgenfluoreszenzanalytische Methoden. - Grundlagen und praktische Anwendung in den Geo-, Material- und Umweltwissenschaften, Vieweg, Braunschweig, Wiesbaden, 1995.
[22] J. Heckel , J. Trace Microprobe Tech., 13 (1995) 97.
[23] R. Jenkins, R.W. Gould and D. Gedecke, Quantitative X-Ray spectrometry, Marcel Dekker, New York, 1981.
[23] R. Jenkins, X-ray fluorescence Spectrometry, Wiley & Sons, New York, 1999.
[24] R. Klockenkämper, Total reflection X-ray Fluorescence Analysis, Wiley & Sons, New York, Chichester, Brisbane, Toronto, Singapore, Weinheim, 1997.
[25] G.R. Lachance, and F. Claisse, Quantitative X-ray Fluorescence Analysis, Wiley & Sons, New York, Chichester, Brisbane, Toronto, Singapore, Weinheim, 1995.
[26] C. Whiston, X-Ray Methods. - Analytical Chemistry by Open Learning, Wiley & Sons, Chichester, New York, Brisbane, Toronto, Singapore, 1987.
[27] K.L. Williams, Introduction to X-Ray Spectrometry, Allen & Unwin, London, 1987.
[28] G. Gillen, R. Lareau, J. Bennett and F. Stevie, (eds.), Secondary Ion Mass Spectometry SIMS XI, Wiley & Sons, New York, 1998.
[29] S. Johansson, J. Campbell and K. Malmqvist, (eds.), Particle induced X-ray emission Spectometry, Wiley & Sons, New York, 1995.
[30] D.C. Koningsbergerer and R. Prins, X-ray Absorption: Principles, Applications, Techniques of EXAFS, SEXAFS and XANES, Wiley & Sons, New York, 1988.
[31] V.M. Goldschmidt, Fortschr. Mineral., 17 (1932) 112.
[32] V.M.Goldschmidt, J. Chem. Soc., (1937) 655.
[33] K.B. Krauskopf, Geochim. Cosmochim. Acta, 10 (1956) 1.
[34] G.E.J. Brown and G.A. Parks, Int. Geol. Rev., 43 (2001) 963.
[35] G.E.J. Brown, in "Mineral-Water Interface Geochemistry" (M.F. Hochella and A.F. White, eds.), p. 309-363, Mineralogical Society of America, Washington, D.C., 1990.
[36] G. Sposito, Surface Chemistry of Soils, Oxford University Press, Oxford, UK, 1984.

[37] W. Stumm, Chemistry of the Solid-Water Interface: Processes at the Mineral-Water and Particle-Water Interface in Natural Systems, John Wiley & Sons, New York, 1992.
[38] G.A. Parks, in "Mineral-Water Interface Geochemistry: (M.F. Hochella and A.F. White, eds.), p. 133-175, Mineralogical Society of America, Washington, D.C., 1990.
[39] P.W. Schindler and W. Stumm, in " Aquatic Surface Chemistry - Chemical Processes at the Particle-Water Interface" (W. Stumm, ed.), p. 83-125, Wiley-Interscience, New York, NY, 1987.
[40] K.F. Hayes, Equilibrium, Spectroscopic, and Kinetic Studies of Ion Adsorption at the Oxide/Aqueous Interface, Department of Civil Engineering, Stanford University, Stanford, CA., 1987.
[41] W. Stumm and J.J. Morgan, Aquatic Chemistry, John Wiley, New York, NY, 1996.
[42] J.D. Ostergren, G.E. Brown, Jr., G.A. Parks and P. Persson, J. Colloid Interface Sci., 225 (2000) 483.
[43] J.D. Ostergren, T.P. Trainor, J.R. Bargar, G.E. Brown., Jr. and G.A. Parks, J. Colloid Interface Sci., 225 (2000) 466.
[44] L. Gunneriusson and S. Sjoberg, J. Colloid Interface Sci., 156 (1993) 121.
[45] C.S. Kim, J.J. Rytuba and G.E. Brown, Jr., J. Colloid Interface Sci., 270 (2004) 9.
[46] P. W. Schindler, in "Mineral-Water Interface Geochemistry" (M.F. Hochella and A.F. White, eds.), p. 281-307, Mineralogical Society of America, Washington, D.C., 1990.
[47] J.M. Zachara, S.C. Smith and L.S. Kuzel, Geochim. Cosmochim. Acta, 59 (1995) 4825.
[48] J.M. Zachara, C.T. Resch and S.C. Smith, Geochim. Cosmochim. Acta, 58 (1994) 553.
[49] T. Zuyi, C. Taiwei, D. Jinzhou, D. XiongXin and G. Yingjie, Appl. Geochem., 15 (2000) 145.
[50] C.-H. Wu, C.-F. Lin, H.-W. Ma and T.-Q. Hsi, Water Res., 37 (2003) 743.
[51] S.F. Cheah, G.E. Brown, Jr. and G.A. Parks, Am. Mineral., 85 (2000) 118.
[52] C.J. Chisholm-Brause, J.E. Brown, Jr. and G.A. Parks, G.A., Physica B: Condensed Matter, 158 (1989) 646.
[53] C.J. Chisholm-Brause, P.A. O'Day, G.E. Brown, Jr. and G.A. Parks, Nature, 348 (1990b) 528.
[54] C.J. Chisholm-Brause, K.F. Hayes, A.L. Roe, J. Brown, E. Gordon, G.A. Parks and J.O. Leckie, Geochim. Cosmochim. Acta, 54 (1990) 1897.
[55] L. Charlet and A. Manceau, Int. J. Environ. Anal. Chem., 46 (1992) 97.
[56] S.E. Fendorf, G.M. Lamble, M.G. Stapleton, M.J. Kelley and D.L. Sparks, Environ. Sci. Technol., 28 (1994) 284.
[57] P.A. O'Day, J. Brown, E. Gordon and G.A. Parks, J. Colloid Interface Sci., 165 (1994) 269.
[58] P.A. O'Day, G.A. Parks and G. E. Brown, Jr., Clays Clay Min., 42 (1994) 337.
[59] A.M. Scheidegger, G.M. Lamble and D.L. Sparks, J. Colloid Interface Sci., 186 (1997) 118.
[60] J.-B. d'Espinose de la Caillerie, M. Kermarec and O. Clause, J. Am. Chem. Soc., 117 (1995) 11471.

[61] S.N. Towle, J.R. Bargar, G.E. Brown, G.A. Parks, Jr. and T.W. Barbee, Jr., Physica B: Condensed Matter, 208-209 (1995) 439.
[62] M.J. Eick and S.E. Fendorf, Soil Sci. Soc. Am. J., 62 (1998) 1257.
[63] A.C. Scheinost and D.L. Sparks, J. Colloid Interface Sci., 223 (2000) 167.
[64] H.A. Thompson, G.A. Parks and G.E. Brown, Jr., J. Colloid Interface Sci., 222 (2000) 241.
[65] T.P. Trainor, G.E. Brown., Jr. and G.A. Parks, G.A., J. Colloid Interface Sci., 231 (2000) 359.
[66] R.G. Ford, P.M. Bertsch and K.J. Farley, Environ. Sci. Technol., 31 (1997) 2028.
[67] G.E. Brown, Jr. and G.A. Parks, Rev. Geophys., 27 (1989) 519.
[68] C.S. Kim, J.J. Rytuba and G.E. Brown, Jr., J. Colloid Interface Sci., 271 (2004) 1.
[69] G.E. Brown, Jr. and N.C. Sturchio, in "Applications of Synchrotron Radiation in Low-Temperature Geochemistry and Environmental Science" (P.A. Fenter, M. Rivers, N. Sturchio and S. Sutton, eds.), p. 1-115, Mineralogical Society of America, Washington, D.C., 2002. pp. 1-115.
[70] M.L. Peterson, G.E. Brown, Jr. and G.A. Parks, Colloids Surf., 107 (1996) 77.
[71] S. Fendorf, M.J. Eick, P. Grossl and D.L. Sparks, Environ. Sci. Technol., 31 (1997) 315.
[72] R.R. Patterson, S. Fendorf and M. Fendorf, Environ. Sci. Technol., 31 (1997) 2039.
[73] J.P. Fitts, G.E. Brown, Jr. and G.A. Parks, Environ. Sci. Technol., 34 (2000) 5122.
[74] L. Charlet and A. Manceau, J. Colloid Interface Sci., 148 (1992b) 443.
[75] A. Manceau and L. Charlet, J. Colloid Interface Sci., 148 (1992) 425.
[76] G.A. Waychunas, J.A. Davis, and R. Reitmeyer, J. Synch. Rad., 6 (1999) 615.
[77] P.A. O'Day, C.J. Chisholm-Brause, S.N. Towle, G.A. Parks and G.E. Brown, Jr., Geochim. Cosmochim. Acta, 60 (1996) 2515.
[78] S.N. Towle, J.R. Bargar, G.E. Brown, Jr. and G.A. Parks, J. Colloid Interface Sci., 187 (1997) 62.
[79] G.M. Lamble, R.J. Reeder and P.A. Northrup, J. Phys., 7 (1997) 793.
[80] M.L. Schlegel, L.Charlet and A. Manceau, J. Colloid Interface Sci., 220 (1999) 392.
[81] M.L. Schlegel, A. Manceau, D. Chateigner and L. Charlet, L., J. Colloid Interface Sci., 215 (1999)140.
[82] C.-C. Chen and K.F. Hayes, Geochim. Cosmochim. Acta, 63 (1999) 3205.
[83] R.G. Ford, K.M. Kemner and P.M. Bertsch, Geochim. Cosmochim. Acta, 63 (1999) 39.
[84] P. Trivedi, L. Axe and T.A. Tyson, J. Colloid Interface Sci., 244 (2001) 230.
[85] S.-F. Cheah, J. Brown, E. Gordon and G.A. Parks, G.A., J. Colloid Interface Sci., 208 (1998) 110.
[86] L. Bochatay, P. Persson, L. Lovgren and G.E. Brown, Jr., J. Phys., 7 (1997) 819.
[87] A.C. Scheinost, S. Abend, K.I. Pandya and D.L. Sparks, D.L., Environ. Sci. Technol., 35 (2001) 1090.
[88] A. Manceau, B. Lanson and V.A. Drits, V.A., Geochim. Cosmochim. Acta, 66 (2002) 2639.
[89] J.D. Morton, J.D. Semrau and K.F. Hayes, Geochim. Cosmochim. Acta, 65 (2001) 2709.

[90] G.A. Waychunas, C.C. Fuller and J.A. Davis, Geochim. Cosmochim. Acta, 66 (2002) 1119.
[91] G.A. Waychunas, C.C. Fuller, J.A. Davis and J.J. Rehr, Geochim. Cosmochim. Acta, 67 (2003) 1031.
[92] F. Juillot, G. Morin, P. Ildefonse, T.P. Trainor, M. Benedetti, L. Galoisy, G. Calas and G.E. Brown, Jr., Am. Min., 88 (2003) 509.
[93] R.G. Ford and D.L. Sparks, Environ. Sci. Technol., 34 (2000) 2479.
[94] D.R. Roberts, R.G. Ford and D.L. Sparks, J. Colloid Interface Sci., 263 (2003) 364.
[95] L. Spadini, A. Manceau, P.W. Schindler and L. Charlet, J. Colloid Interface Sci., 168 (1994) 73.
[96] A. Manceau, K.L. Nagy, L. Spadini and K.V. Ragnarsdottir, J. Colloid Interface Sci., 228 (2000) 306.
[97] S.R. Randall, D.M. Sherman, K.V. Ragnarsdottir and C.R. Collins, Geochim. Cosmochim. Acta, 63 (1999) 2971.
[98] L. Bochatay and P. Persson, J. Colloid Interface Sci., 229 (2000) 593.
[99] I. Berrodier, F. Farges, M. Benedetti and G.E. Brown, Jr., J. Synch. Rad., 6 (1999) 651.
[100] C.R. Collins, D.M. Sherman and K.V. Ragnarsdottir, J. Colloid Interface Sci., 219 (1999) 345.
[101] C.J. Chisholm-Brause, A.L. Roe, K.F. Hayes, G.E. Brown, Jr., G.A. Parks, and J.O. Leckie, Physica B: Condensed Matter, 158 (1989) 646.
[102] F.J. Weesner and W.F. Bleam, J. Colloid Interface Sci., 205 (1998) 380.
[103] J.R. Bargar, G.E. Brown, Jr. and G.A. Parks, Geochim. Cosmochim. Acta 61, (1997) 2639.
[104] R.J. Reeder, G.M. Lamble and P.A. Northrup, Am. Min., 84 (1999) 1049.
[105] C.P. Huang and W. Stumm, J. Colloid Interface Sci., 43 (1973) 409.
[106] H. Hohl and W. Stumm, J. Colloid Interface Sci., 55 (1976) 281.
[107] M.A. Anderson, J.F. Ferguson and J. Gavis, J. Colloid Interface Sci., 54 (1976) 391.
[108] E. Osthols, A.A. Manceau, F. Farges, L. Charlet, J. Colloid Interface Sci., 194 (1997) 10.
[109] C. Papelis, G.E. Brown, Jr., G.A. Parks and J.O. Leckie, Langmuir, 11 (1995) 2041.
[110] F.J. Wessner, W. F. Bleam, J. Colloid Interface Sci., 196 (1997) 79.
[111] S. Fendorf, P.M. Jardine, R.R. Patterson, D.L. Taylor and S.C. Brooks, Geochim. Cosmochim. Acta, 63 (1999) 3049.
[112] L. Bonneviot, O. Clause, M. Che, A. Manceau and H. Dexpert, Catalysis Today, 6 (1989) 39.
[113] T.E. Alcacio, D. Hesterberg, J.W. Chou, J.D. Martin, S. Beauchemin and D.E. Sayers, Geochim. Cosmochim. Acta, 65 (2001) 1355.
[114] S.F. Cheah, G.E. Brown, Jr. and G.A. Parks, Geochim. Cosmochim. Acta, 63 (1999) 3229.
[115] J.P. Fitts, P. Persson, G.E. Brown, Jr. and G.A. Parks, J. Colloid Interface Sci., 220 (1999) 133.
[116] D. Hesterberg, D.E. Sayers, W. Zhou, G.M. Plummer and W.P. Robarg, J. Phys. IV Colloque, C2 (1997) 833.

[117] R.A.D. Pattrick, K.E.R. England, J.M. Charnock and J. F.W. Mosselmans, Int. J. Min. Proc., 55 (1999) 247.
[118] S. Lin, H.-C. Kao, C.-H. Cheng and R.-S. Juang, Colloids Surfaces A: Physicochemical and Engineering Aspects, 234 (2004) 71.
[119] R.A. Alvarez-Puebla, C. Aisa, J. Blasco, J.C. Echeverria, B. Mosquera and J.J. Garrido, Appl. Clay Sci., 25 (2004) 103.
[120] C.R. Collins, K.V. Ragnarsdottir and D.M. Sherman, D.M., Geochim. Cosmochim. Acta, 63 (1999a) 2989.
[121] J.R. Bargar, P. Persson and G.E. Brown, Jr., J. Phys. IV Colloque C2, (1997) 825.
[122] J.R. Bargar, G.E. Brown, Jr. and G.A. Parks, Geochim. Cosmochim. Acta, 62 (1998) 193.
[123] J.R. Bargar, P. Persson and G.E. Brown, Jr., Geochim. Cosmochim. Acta, 63 (1999) 2957.
[124] E.J. Elzinga, D. Peak and D.L. Sparks, Geochim. Cosmochim. Acta, 65 (2001) 2219.
[125] J.J. Lenhart, J.R. Bargar and J.A.Davis, J. Colloid Interface Sci., 234 (1001) 448.
[126] M.P. Papini and M. Majone, in "Encyclopedia of Surface and Colloid Science" (A. Hubbard, ed.), p. 3483-,3498, Marcel Dekker, New York, 2000.
[127] M. Majone, M.P. Papini and E.J. Rolle, J. Colloid Interface Sci., 179 (1996) 412.
[128] S. M. Toth, in "Encyclopedia of Surface and Colloid Science" (A. Hubbard, ed.), p.212-224, Marcel Dekker, New York, 2002.
[129] J. Langmuir, J. Am. Chem. Soc., 38 (1916) 2267.
[130] M.Z. Volmer, Z. Phys. Chem., 115 (1925) 23.
[131] J.W. Gibbs, Scientific Papers, Longmans Co., London, 1906.
[132] S. Brunauer, P.H. Emmett and E. Teller, J. Am. Chem. Soc., 60 (1938) 309.
[133] J. Langmuir, J. Am. Chem. Soc., 40 (1918) 1361.
[134] R.D. Harter and D.E. Baker, Soil Sci. Soc. Am. J., 41 (1977) 1077.
[135] J.A. Veith and G. Sposito, Soil Sci. Soc. Am. J., 41 (1977) 697.
[136] A.M. Elprince and G. Sposito, Soil Sci. Soc. Am. J., 45 (1981) 277.
[137] G. Sposito, Soil Sci. Soc. Am. J., 43 (1979) 197.
[138] R.A. Griffin and A.K. Au, Soil Sci. Soc. Am. J., 41 (1977) 880.
[139] G. Sposito, Soil Sci. Soc. Am. J., 46 (1982) 1147.
[140] B.A. Manning and S. Goldberg, Environ. Sci. Technol., 31 (1997) 2005.
[141] D.G. Kinniburgh, J.A. Barker and M.A. Whitfield, J. Colloid Interface Sci., 95 (1983) 370.
[142] J. Lützenkirchen, in "Encyclopedia of Surface and Colloid Science" (A. Hubbard, ed.), p. 5028-5046, Marcel Dekker, New York, 2002.
[143] F. Boily, in "Encyclopedia of Surface and Colloid Science" (A. Hubbard, ed.), p. 3223-3238, Marcel Dekker, New York, 2002.
[144] S. Goldberg, in "Structure and Surface Reactions of Soil Particles" (P.M. Huang, N. Senesi and J. Buffle, eds.), p. 377-412, John Wiley, New York, 1998.
[145] P.W. Schindler and H.R. Kamber, Helv. Chim. Acta, 51 (1968) 1781.
[146] P.W. Schindler and H. Gamsjager, Kolloid Z. Z. Polym., 250 (1972) 759.
[147] H. Hohl and W. Stumm, J. Colloid Interface Sci., 55 (1976) 281.
[148] P.W. Schindler, P. Liechti and J.C. Westall, Neth. J. Agric. Sci., 35 (1987) 219.
[149] O. Stern, Z. Elektrochem., 30 (1924) 508.

[150] W. Stumm, C.P. Huang and S.R. Jenkins, Croat. Chem. Acta, 42 (1970) 223.
[151] D.A. Dzombak and F.M.M. Morel, Surface Complexation Modelling, John Wiley, New York, 1990.
[152] D.E. Yates, S. Levine and T.W. Healy, J. Chem. Soc. Faraday Trans., 70 (1974) 1807.
[153] J.A. Davies, R.O. James and J.O. Leckie, J. Colloid Interface Sci., 63 (1978) 480.
[154] K.F. Hayes and J.O. Leckie, J. Colloid Interface Sci., 115 (1987) 564.
[155] W.H. Van Riemsdijk, G.H. Bolt, L.K. Koopal and J. Blaakmer, J. Colloid Interface Sci., 109 (1986) 219.
[156] G.H. Bolt, Soil Chemistry, Part B: Physicochemical Methods, Elsevier, Amsterdam, 1982.
[157] W.H. van Riemsdijk, J.C.M. de Wit, L.K. Koopal and G.H. Bolt, J. Colloid Interface Sci., 116 (1987) 511.
[158] T. Hiemstra, W.H. van Riemsdijk and G.H. Bolt, J. Colloid Interface Sci., 133 (1989) 91.
[159] T. Hiemstra, J.C.M. de Wit and W.H. van Riemsdijk, J. Colloid Interface Sci., 133 (1989) 105.
[160] J. Westall and H. Hohl, H., Adv. Colloid Interface Sci., 12 (1980) 265.
[161] N.J. Barrow, Adv. Agron., 38 (1985) 183.
[162] M.H. Bradbury and B. Baeyens, J. Contam. Hydrol., 27 (1997) 223.
[163] M. Majone, M.P. Papini and E. Rolle, Ing. Sanit., 2 (1993) 59.
[164] Q. Zang, Z. Xu and J.A. Finch, J. Colloid Interface Sci.,169 (1995) 414.
[165] J. Garcia-Miragaya and M. Davalos, Water, Air, Soil Pollut., 27 (1986) 217.
[166] M. Stadler and P.W. Schindler, Clays Clay Miner., 42 (1994) 148.
[167] N.J. Barrow, J.W. Bowden, A.M. Posner and J.P. Quirk, Aust. J. Soil Res., 18 (1980) 37.
[168] S. Bank, J.F. Bank and P.D. Ellis, J. Phys. Chem., 93 (1989) 4847.
[169] A.M. Scheidegger, G.M. Lamble and D.L. Sparks, Environ. Sci. Technol., 30 (1996) 548.
[170] H. Kerndorf and M. Schnitzer, Geochim. Cosmochim. Acta, 44 (1980) 1701.
[171] A. Heidemann, A. Geochim. Cosmochim. Acta, 15 (1959) 305.
[172] R. Apak, in"Encyclopedia of Surface and Colloid Science" (A. Hubbard, ed.), p. 385-417, Marcel Dekker, New York, 2002.
[173] A. Tessier, R. Carignan, B. Dubreuil and F. Rapin, Geochim. Cosmochim. Acta, 53 (1989) 1511.
[174] Z.S. Kooner, Environ. Geol., 21 (1998) 242.
[175] C.E. Cowan, J.M. Zachara and C.T. Resch, Environ. Sci. Technol., 25 (1991) 437.
[176] J.J. Higgo and L.C.V. Rees, Environ. Sci. Technol., 20 (1986) 483.
[177] M.M. Benjamin and J.O. Leckie, J. Coll. Interface Sci., 79 (1981) 209.
[178] L.L. Hendrickson and R.B. Corey, Soil Sci., 131 (1981) 163.
[179] R.J. Crawford, I.H. Harding and D.E. Mainwaring, Langmuir, 9 (1993) 3050.
[180] K. Kalbitz and R. Wennrich, Sci. Total Environ., 209 (1998) 27.
[181] N.H. Neal and G. Sposito, Soil Sci., 146 (1986) 164.
[182] L.J. Lund, E.E. Betty, A.L. Page and R.A. Elliott, J. Environ. Qual., 10 (1981) 551.

[183] D.C. Adriano, Trace Elements in Terrestrial Environments, Springer, Berlin, Heidelberg, 2001.
[184] A.C.M. Bourg, in" Heavy Metals – Problems and Solutions" (W. Salomons, U. Förstner and P. Mader, eds.), p. 19-31, Springer, Berlin, Heidelberg, 1995.
[185] M.M. Benjamin and J.O. Leckie, Environ. Sci. Technol., 15 (1981) 1050.
[186] P.W. Schindler, in" Mineral-Water Interface Geochemistry Reviews in Mineralogy" (M.F. Hochella, Jr. and A.F. White, eds.), p. 281- 293, Mineralogical Society of America, Washington, D.C., 1990.
[187] U. Hoins, L. Charlet and H. Sticher, Water, Air, Soil Pollut., 68 (1993) 241.
[188] K.R. Nordquist, M.M. Benjamin and J.F. Ferguson, Water Res., 22 (1988) 837.
[189] J.A. Davis and J.O. Leckie, Environ. Sci. Technol., 12 (1978) 1309.
[190] C.R. Collins, K.V. Ragnarsdottir and D.M. Sherman, Geochim. Cosmochim. Acta, 63 (1999) 2989.
[191] A.M. Schlautman and J.J. Morgan, Geochim. Cosmochim. Acta, 58 (1994) 4293.
[192] A. P. Davis and V. Bhatnagar, Chemosphere, 30 (1995) 243.
[193] E.M. Murphy and J.M. Zachara, Geoderma, 67 (1995) 103.
[194] J.B. Fein and D. Delea, Chem. Geol., 161 (1999) 375.
[195] J.B. Fein, J.-F. Boily, K. Güclü and E. Kaulbach, Chem. Geol., 162 (1999) 33.
[196] A.R. Bowers and C.P. Huang, J. Colloid Interface Sci., 110 (1986) 575.
[197] B. Nowack and L. Sigg, J. Colloid Interface Sci., 177 (1996) 106.
[198] J.-K. Yang and A. P. Davis, J. Colloid Interface Sci., 213 (1999) 77.
[199] A. P. Davis and M. Upadhyaya, Water Res., 30 (1996) 1894.
[200] J.M. Zachara, S.C. Smith and L.S. Kuzel, Geochim. Cosmochim. Acta, 59 (1995) 4825.
[201] J.E. Szecsody, J.M. Zachara and P.L. Bruckhardt, Environ. Sci. Technol., 28 (1994) 1706.
[202] B. Nowack, J. Lützenkirchen, P. Behra and L. Sigg, Environ. Sci. Technol., 30 (1996) 2397.
[203] D.C. Girvin, P.L. Gassman and H. Bolton, Jr., Clays Clay Miner., 44 (1996) 757.
[204] M.S. Vohra and A. P. Davis, J. Colloid Interface Sci., 198 (1998) 18.
[205] B.E. Reed and S.R. Cline, Sep. Sci. Technol., 29 (1994) 1529.
[206] J. Cairns and D.I. Mount, Environ. Sci. Technol., 24 (1990) 154.
[207] M.C. Newman, Quantitative Methods in Aquatic Ecotoxicology, CRC Press, Boca Raton, FL, 1995.
[208] J.M. Pacyna and G.J. Keeler, Water, Air, Soil Poll., 80 (1995) 621.
[209] F.R. Siegel, Environmental Geochemistry of Potentially Toxic Metals, Springer, Berlin, Heidelberg, 2002.
[210] W.H. Rulkens, J.T.C. Grotenhuis and R. Tichy, in" Heavy Metals – Problems and Solutions" (W. Salomons, U. Förstner and P.Mader, eds.), p. 165-191, Springer, Berlin, Heidelberg, 1995.
[211] A. Kabata-Pendias, Appl. Geochem., 2 (1993) 3.
[212) A. Kabata-Pendias, Trace Substances in Environmental Health, 25 (1992) 53.
[213] G.M. Rand, Fundamentals of Aquatic Toxicology, Taylor & Francis, Washington D.C., 1995.
[214] P.F. Landrum, Environ. Sci. Technol., 23 (1989) 588.

[215] M. Gibaldi, Biopharmaceutics and Clinical Pharmacokinetics, Lea & Febiger, Philadelphia, PA, 1991.
[216] M.V. Ruby, R. Schoof, W. Brattin, G. Goldack, P. Post, M. Harnois, D.E. Mosby, S.W. Casteel, W. Berti, M. Carpenter, D. Edwards, D. Cragin and W. Chappell, Environ. Sci. Technol., 33 (1999) 3697.
[217] A.J.M. Baker, New Phytol., 106 (1987) 93.
[218] A. Tessier and D.R. Turner, Metal Speciation and Bioavailability in Aquatic Systems, Wiley and Sons, New York, 1995.
[219] D.J.H. Phillips and P.S. Rainbow, Mar. Environ. Res., 28 (1989) 207.
[220] H.E. Allen, R.H. Hall and T.P. Brisbin, Environ. Sci. Technol., 14 (1980) 441.
[221] L.M. Walsh, M.E. Sumner and D.R. Keeney, Environ. Health Perspect., 19 (1977) 67.
[222] Wauchope and L.L. McDowell, J. Environ. Qual., 13 (1984) 499.
[223] J.M. Harrington, S.E. Fendorf and R.F. Rosenzweig, Environ. Sci. Technol., 32 (1998) 2425.
[224] F.F. Orumwense, J. Chem. Technol. Biotechnol., 65 (1996) 363.
[225] J.F. Ferguson and J. Gavis, J., Water Res., 6 (1972) 1259.
[226] Maeda, A. Ohki, T. Tokuda and M. Ohmine, Appl. Organomet. Chem., 4 (1990) 251.
[227] S. Maeda, A. Ohki, K. Kusadome, T. Kuroiwa, I. Yoshifuku and K. Naka, Appl. Organomet. Chem., 6 (1992) 213.
[228] S. Tamaki and W. Frankenberger, Rev. Environ. Contam. Toxicol., 124 (1992) 79.
[229] C.C. Tanner and J.S. Clayton, NZ J. Mar. Freshwater Res., 24 (1990) 173.
[230] J.O. Nriagu, Arsenic in the Environment Part 1: Cycling and Characterization, John Wiley & Sons, New York, 1994.
[231] J.T. Hindmarsh, and R.F. McCurdy, CRC Crit. Rev. Clin. Lab. Sci., 23 (1986) 315.
[232] D. Das, G. Sawanta, and D. Chakraborti, Environ. Geochem. Health, 18 (1996) 5.
[233] R.T. Nickson, J.M. McArthur, W. Burgess, K.M. Ahmed, P. Ravenscroft and M. Rahman, Nature, 395 (1998) 338.
[234] R.T. Nickson, J.M. McArthur, P. Ravenscroft, W.G. Burgess and K.M. Ahmed, Appl. Geochem.,15 (2000) 403.
[235] C.J. Chen and C.J. Wang, Cancer Res., 50 (1990) 5470.
[236] H.Y.Chiou, Y.M. Hsueh, and Y.M. Liaw, Cancer Res., 55 (1995) 1296.
[237] M.G. Hickey and J.A. Kittrick, J. Environ. Qual., 13 (1984) 372.
[238] T. Asami, M. Kubota and K. Orikasa, Water Air Soil Pollut. 83/84 (1995) 187.
[239] A. Chlopeka, J.R. Bacon, M.J. Wilson and J. Kay, J. Environ. Qual., 25 (1996) 69.
[240] L. Ramos, L.M. Hernandez and M.J. Gonzalez, J. Environ. Qual., 23 (1994) 50.
[241] X. Xian, Plant Soil, 113 (1989) 257.
[242] D. Hirsch and A. Banin, J. Environ. Qual., 19 (1990) 366.
[243] J. Garcia-Miragaya and A.L. Page, Soil Sci. Soc. Am. J., 40 (1976) 658.
[244] M.J. McLaughlin and K.G. Tiller, Aust. J. Soil Res., 34 (1996) 1.
[245] M.J. McLaughlin and K.G. Tiller, Trans. 15th World Congress Soil Science, 36 (1994) 195.

[246] P.E. Holm, T.H. Christensen, J.C. Tjell and S.P. McGrath, J. Environ. Qual., 24 (1995) 183.
[247] W.E. Emmerich, L.J. Lund, A.L. Page and A.C. Chang, J. Environ. Qual., 11 (1982) 182.
[248] P.E. Holm, B.B.H. Andersen and T.H. Christensen, Soil Sci. Soc. Am. J., 60 (1996) 775.
[249] J.D. Hem, USGS Water Supply Report 2254, US Geological Survey (USGS), Alexandria, VA, 1985.
[250] K.M. Krupka, D.I. Kaplan, G. Whelan, S.V. Martigod and R.J. Serne, Understanding Variation in Partition Coefficient K_d Values, US EPA 402-R-99-004A, B, US EPA, Washington, D.C., 1999.
[251] J.D. Hem, Water Resources Res., 8 (1972) 661.
[252] T.H. Christensen, Water Air Soil Pollut., 21 (1984) 105.
[253] J. Santillan-Medrano and J.J. Jurinak, Soil Sci. Soc. Am. J., 29 (1975) 851.
[254] J.J. Street, B.R. Sabey and W.L. Lindsay, J. Environ. Qual., 7 (1978) 286.
[255] M.B. McBride, Soil Sci. Soc. Am. J., 44 (1980) 26.
[256] H.A. Elliott and C.M. Denneny, J. Environ. Qual. 11 (1982) 658.
[257] R.P. Milberg, D.L. Brower, and J.V. Lagerwerff, Soil Sci. Soc. Am. J., 42 (1978) 892.
[258] C.E. Cowan, J.M. Zachara and C.T. Resch, Environ. Sci. Technol., 25 (1991) 437.
[259] J.A. Davis and J.O. Leckie, J. Coll. Interface Sci., 74 (1980) 32.
[260] K.B. Krauskopf, Introduction to Geochemistry, McGraw-Hill, New York, 1979.
[261] J.J. Mortvedt, J. Environ. Qual., 16 (1987) 137.
[262] A.S. Jeng and B.R. Singh, Plant Soil, 175 (1995) 67.
[263] B.G. Wixson, N.L. Gale and K. Downey, Trace Subs. Environ. Health, 11 (1977) 455.
[264] J. Kobayashi, Trace Subs. Environ. Health, 5 (1971) 117.
[265] P. Little and M.H. Martin, Environ. Pollut., 3 (1972) 241.
[266] J.R. Preer and W.G. Rosen, Trace Subs. Environ. Health, 11 (1977) 399.
[267] Y. Takijima and F. Katsumi, Soil Sci. Plant Nutr.,19 (1973) 29.
[268] M. Chino and A. Baba, J. Plant Nutr., 3 (1981) 203.
[269] D.H. Khan and B. Frankland, Plant Soil, 70 (1983) 335.
[270] Y. Waisel, U. Kafkani and A. Eshel, Plant Roots: The Hidden Half, Marcel Dekker, New York, 1991.
[271] L. Marchiol, L. Leita, M. Martin, A. Peressotti and G. Zerbi, J. Environ. Qual., 25 (1996) 562.
[272] H.Van Assche and F.H. Clijsters, Plant Cell Environ., 13 (1990) 195.
[273] C.H.R. DeVos, H. Schat, M. deWaal, R. Youijis and W.H.O. Ernst, Physiol. Plant, 82 (1991) 523.
[274] A. Lagriffoul, B. Macquot, M. Mech and J. Vangronsfeld, Plant Soil, 200 (1998) 241.
[275] S.M. Ross, Toxic Metals in Soil-Plant System, Wiley & Sons, New York, 1994.
[276] L. Friberg, M. Piscator and G. Nordberg, Cadmium in the Environment, CRC Press, Cleveland, OH, 1971.
[277] J. Artiole and W.H. Fuller, J. Environ. Qual., 8 81979) 503.
[278] R.A. Griffin, A.K. Au and P.P. Prost, J. Environ. Sci. Health A12, (1977) 431.

[279] D. Rai, L.E. Eary and J.M. Zachara, Sci. Total Environ., 86 (1989) 15.
[280] J.M. Zachara, D.C. Girvin, R.L. Schmidt and C.T. Resch, Environ. Sci. Technol., 21 (1987) 589
[281] J.M. Zachara, C.E. Cowan, R.L. Schmidt and C.C. Ainsworth, Clays Clay Miner., 36 (1988) 317.
[282] J.M. Zachara, C.C. Ainsworth, C.E. Cowan, C.T. Resch, Soil Sci. Soc. Am. J., 53 (1989) 418.
[283] A. Davis and R.L. Olsen, Groundwater, 33 (1995) 759.
[284] C.D. Palmer and P.R. Wittbrodt, Environ. Health Perspect., 92 (1991) 25.
[285] K.G. Stollenwerk and D.B. Grove, J. Environ. Qual., 14 (1985) 150.
[286] J.H. Grove, B.G. Ellis, Soil Sci. Soc. Am. J., 44 (1980) 238.
[287] M. Stoeppler, Hazardous Elements in the Environment, Elsevier, Amsterdam, 1992.
[288] C.H. Williams, J. Austr. Inst. Agric. Sci., (Sept.-Dec.) (1975) 99.
[289] L.E. Sommers, J. Environ. Qual., 6 (1977) 225.
[290] W. Mertz, Physiol. Rev., 49 (1969) 163.
[291] W. Mertz, J. Nutr., 123 (1994) 626.
[292] H.D. Chapman, Diagnostic Criteria for Plants and Soils, Quality Printing, Abilene, TX, 1966.
[293] E.W.D. Huffman, Jr. and W.H. Alloway, Plant Physiol., 53 (1973) 72.
[294] K. Schwartz and W. Mertz, Arch. Biochem. Biophys., 85 81959) 292.
[295] M.D. Cohen, B. Kargacin, C.B. Klein and M. Costa, Crit. Rev. Toxicol., 23 (1993) 255.
[296] J.O. Nriagu and E. Nieboer, Chromium in the Natural and Human Environments, Wiley and Sons, New York, 1988.
[297] H.J.M. Bowen, Environmental Chemistry of the Elements, Academic Press, New York, 1979.
[298] J.O. Nriagu, Copper in the Environment, Wiley and Sons, New York, 1979.
[299] S.P. McGrath, J.R. Sanders and M.H. Shalaby, Geoderma, 42 (1998) 177.
[300] E.A. Jenne, Adv. Chem., 73 (1968) 337.
[301] J. Wu, A. Liard and M.L .Thompson, J. Environ. Qual., 28 (1999) 334.
[302] R.G. McLaren and D.V. Crawford, J. Soil Sci., 24 (1973) 443.
[303] D.C. Adriano, G.M. Paulsen and L.S. Murphy, Agron. J., 63 (1971) 36.
[304] R.D. Harter, Soil Sci. Soc. Am. J., 47 (1983) 47.
[305] J.F. Hodgson, W.L. Lindsay and J.F. Trierweiler, Soil Sci. Soc. Am. Proc., 30 (1966) 723.
[306] C. Bloomfield and J.R. Sanders, J. Soil Sci., 28 (1977) 435.
[307] J.F. Loneragan, A.D. Robson, and R.D. Graham, Copper in Soils and Plants, Wiley and Sons, New York, 1981.
[308] L.M. Walsh, W.H. Erhardt and H.D. Seibel, J. Environ. Qual., 1 (1972) 197.
[309] J.S. Lima, Agric. Ecosys. Environ., 48 (1994) 19.
[310] A. Cordero and G.F. Ramirez, Agron. Costa., 3 (1979) 63.
[311] G.L. Cromwell, V.W. Hays and T.L. Clark, J. Anim. Sci., 46 (1978) 692.
[312] I.H.M. Lucas, R.M. Livingston, A.W. Boyne and I. McDonald, J. Agric. Sci., 58 (1962) 201.
[313] T.S. Stahly, G.L. Cromwell and H.J. Monegue, J. Anim. Sci., 51 (1980) 1347.

[314] A.C. Dalgarno and C.F. Mills, J. Agric. Sci., 85 (1975) 11.
[315] K.L. Blaxter, Vet. Rec., 92 (1973) 383.
[316] S. Dudka and D.C. Adriano, J. Environ. Qual., 26 (1997) 590.
[317] A.L. Sommer, Plant Physiol., 6 (1931) 339.
[318] C.B. Lipman and G. MacKinney, Plant Physiol., 6 (1931) 593.
[319] J.J. Mortvedt, P.M. Giordano and W.L. Lindsay, Micronutrients in Agriculture, Soil Sci. Soc. Am., Madison, WI, 1972.
[320] H.E. Allen and D.J. Hansen, Water Environ. Res., 68 (1996) 42.
[321] H. Ma, S.D. Kim, D.K. Cha and H.E. Allen, Environ. Toxicol. Chem., 18 (1999) 828.
[322] X.J.H.X. Stouthart, J.L.M. Haans, R.A.C. Lock and S.E.W. Bonga, Environ. Toxicol. Chem. 15 (1996) 376.
[323] L.W. Hall, M.C. Scott and W.D. Killen, Environ. Toxicol. Chem., 17 (1998) 1172.
[324] G.F. Soldatini, R. Riffaldi and R. Levi-Minzi, Water Air Soil Pollut., 6 (1977) 111.
[325] E.A. Elkhatib, G.M. Elshebiny and A.M. Balba, Environ. Pollut., 69 (1991) 269.
[326] R.M. McKenzie, Aust. J. Soil Res., 18 (1980) 61.
[327] R.M. McKenzie, Aust. J. Soil Res., 8 (1970) 97.
[328] R.M. McKenzie, Aust. J. Soil Res., 13 (1975) 177.
[329] J.P. Pinheiro, A.M. Mota and M.F.Benedetti, Environ. Sci. Technol., 33 (1999) 3398.
[330] A. Liu and R.D. Gonzalez, J. Colloid Interface Sci., 218 (1999) 225.
[331] D.G. Strawn, A.M. Scheidegger and D.L. Sparks, Environ. Sci. Technol., 32 (1998) 2596.
[332] N.T. Basta and M.A. Tabatabai, Soil Sci., 153 (1992) 331.
[333] S. Sauve, C.E. Martinez, M.B. McBride and W. Hendershot, Soil Sci. Soc. Am. J., 64 (2000) 595.
[334] H. Heinrichs and R. Mayer, J. Environ. Qual., 9 (1980) 111.
[335] C.L. Strojan, Oikos, 31 (1978) 41.
[336] I.R. Willett and W.J. Bond, J. Environ. Qual., 24 (1995) 834.
[337] M.R. Reddy and H.F. Perkins, Soil Sci., 121 (1976) 21.
[338] W.A. Norvell and W.L. Lindsay, Soil Sci. Soc. Am. J., 36 (1972) 778.
[339] C.D. Foy and T.A. Campbell, J. Plant Nutr., 7 (1984) 1365.
[340] C.E. Casey and M.F. Robinson, Br. J. Nutr., 39 (1978) 639.
[341] L.W. Chang, Toxicology of Metals, CRC Press, Boca Raton, FL, 1996.
[342] N.J. Barrow and V.C. Cox, J. Soil Sci., 43 (1992) 305.
[343] A. Anderson, Oikos Suppl., 9 (1967) 13.
[344] S. Aomine and K. Inoue, Soil Sci. Plant Nutr., 13 (1967) 195.
[345] D.W. Newton, R. Ellis, Jr. and G.M. Paulsen, J. Environ. Qual., 5 (1976) 251.
[346] D.G. Kinniburgh and M.L. Jackson, Soil Sci. Soc. Am. J., 42 (1978) 45.
[347] R.A. Lockwood and K.Y. Chen, Environ. Lett., 6 (1974) 151.
[348] D. Sarkar, M.E. Essington and K.C. Misra, Soil Sci. Soc. Am. J., 63 (1999) 1626.
[349] J.T. Gilmour, Environ. Lett., 2 (1971) 143.
[350] O. Hutzinger, Handbook of Environmental Chemistry, Springer, New York, 1980.
[351] M.R. Winfrey and J.W.M. Rudd, Environ. Toxicol. Chem., 9 (1990) 853.

[352] J.A. Jay, F.M.M. Morel and H.F. Hemmond, Environ. Sci. Technol. 34 (2000) 2196.
[353] J.M. Wood, F.S. Kennedy and C.G. Rosen, Nature, 220 (1968) 173.
[354] J.O. Nriagu, The Biogeochemistry of Mercury in the Environment, Elsevier, Amsterdam, 1979.
[355] W.F. Beckert, A.-A. Moghissi, F.H.F. Au, E.W. Bretthauer and J.C. McFarlane, Nature, 249 (1974) 674.
[356] F.Y. Toribara, C.P. Shields and L. Koval, Talanta, 17 (1970) 1025.
[357] R.D. Rogers and J.C. McFarlane, J. Environ. Qual., 8 (1979) 255.
[358] B.M. Mimmiskin, J.W.M. Rudd and C. A. Kelly, Can. J. Fish Aquat. Sci., 49 (1992) 17.
[359] C.G. Gilmour and E.A. Henry, Environ. Poll., 71 (1991) 131.
[360] C.G. Gilmour and D.G. Capone, EOS Trans. Geophys. Union, 68 (1987) 1718.
[361] J.K. King, F.M. Saunders, R.F. Lee and R.A. Jahnke, Environ. Toxicol. Chem., 18 (1999) 1362.
[362] D.B. Porcella, P. Chu and M.A. Allen, in"Global and Regional Mercury Cycles: Sources, Fluxes, and Mass Balances", (W. Baeyens, R. Ebinghaus and O. Vasiliev, eds.), p. 179–190, NATO-ASI Series 2, Environment Vol. 21, Kluwer, Dordrecht, 1996.
[363] R. Ebinghaus, R.M. Tripathi, D. Wallschläger and S.E. Lindberg, in" Mercury Contaminated Sites (R. Ebinghaus, R.R. Turner, L.D. de Lacerda, O. Vasiliev and W. Salomons, eds.), p. 3-50, Springer, Berlin, Heidelberg, New York, 1999.
[364] R.P. Mason, W.F. Fitzgerald and F.M.M. Morel, Geochim. Cosmochim. Acta, 58 (1994) 3191.
[365] F. Slemr and E. Langer, Nature, 355 (1992) 424.
[366] N.A. Smart, Residue Rev., 23 (1968) 1.
[367] W. Beauford, J. Barber and A.R. Barringer, Plant Physiol., 39 (1977) 261.
[368] M. Bolli, Ann. Fac. Agrar. Univ. Stud. Perugia, 4 (1947) 180.
[369] A.M. Scheuhammer, Environ. Pollut., 71 (1991) 329.
[370] A. Kudo and R.R. Turner, in" Mercury Contaminated Sites (R. Ebinghaus, R.R. Turner, L.D. de Lacerda, O. Vasiliev and W. Salomons, eds.), p. 143-158, Springer, Berlin, Heidelberg, New York, 1999.
[371] L.H.P. Jones, Science, 123 (1956) 1116.
[372] W.M. Jarrel and M.D. Dawson, Soil Sci. Soc. Am. J., 42 (1978) 412.
[373] H.M. Reisenauer, A. Tabikh and P.R. Stout, Soil Sci. Soc. Am. Proc., 26 (1962) 23.
[374] N.J. Barrow, Soil Sci., 109 (1970) 282.
[375] N. Karimian and F.R. Cox, Soil Sci. Soc. Am. J., 42 (1978) 757.
[376] I. Straughan, A.A. Elseewi and A.L. Page, Trace Subs. Environ. Health, 12 (1978) 389.
[377] R. Parada and L. Covarrubias, Arch. Med. Vet., 12 (1980) 229.
[378] P.A. Krenkel, Heavy Metals in the Aquatic Environment, Pergamon Press, New York, 1975.
[379] G.R. Hagstrom, Fert. Solutions, July-August (1977) 18.
[380] U.C. Gupta, Molybdenum in Agriculture, Cambridge University Press, Cambridge, 1997.

[381] J. Kubota, V.A. Lazar, G.H. Simonson and W.H. Hill, Soil Sci. Soc. Am. Proc., 31 (1967) 667.
[382] B.J. Alloway, J. Agric. Sci., 80 (1973) 521.
[383] R.D. Harter, Soil Sci. Soc. Am. J., 47 (1983) 47.
[384] J.O. Nriagu, Nickel in the Environment, Wiley and Sons, New York, 1980.
[385] H.A. Schroeder, Environment, 13 (1971) 18.
[386] E. Nieboer and J.O. Nriagu, Nickel and Human Health, Wiley and Sons, New York, 1992.
[387] J.W. Moore, Inorganic Contaminants in Surface Waters, Springer, New York, 1991.
[388] F.W. Sunderman, S.M. Hopfer, K.R. Sweeney, A.H. Marcus, B.M. Most and J. Cuanson, Proc. Soc. Biol. Med., 191 (1989) 5.
[389] N.N. Greenwood and A. Earnshaw, Chemistry of the Elements, Pergamon Press, Oxford, 1989.
[390] R. Hartley, Chemistry of the Platinum Group Metals, Amsterdam, Elsevier, 1991.
[391] WHO, Environmental Health Criteria 125 – Platinum. Geneva: World Health Organization, International Programme on Chemical Safety, Geneva, 1991.
[392] M. Johnson, Precious Metal Division, Johnson Matthey Publishing Company, London, 2001. (http://www.Matthey.com/divisions/precious)
[393] K. Ravindra, L. Bencs and R. Van Grieken, Sci. Total Environ., 318 (2004) 1.
[394] M.E. Farago, P. Kavanagh, R. Blanks, J. Kelly, G. Kazantzis, I. Thornton, P.R. Simpson, J.M. Cook, H.T. Deleves and G.E.M. Hall, Analyst, 123 (1998) 451.
[395] A.F. Holleman and E. Wiberg, Lehrbuch der anorganischen Chemie, De Gruyter, Berlin, New York, 1995.
[396] R.J. Farrauto, R.M. Heck and B.K. Speronello, Chem. Engin. News, 70 (1992) 5.
[397] R.M. Heck and R.J. Farrauto, Appl. Catal. A., 221 (2001) 443.
[398] M. Moldovan, M.M. Gomez and M.A. Palacios, J. Anal. At. Spectrom., 14 (1999) 1163.
[399] J. E. Hofman, in" Precious and Rare Metal Technologies, Proceedings of a Symposium on precious and rare metals", (A.E. Torma, ed.), p. 345-364, Albuquerque, NM, 1989.
[400] L. Manziek, in "Precious Metal Recovery and Refining", Texas: Historic Publications, 7, 1990.
[401] H.F. Hertel, H.P. König, O. Inacker and R. Malessa, in "Nachweis der Freisetzung und Identifizierung von Edelmetallen im Abgasstrom von Katalysatorfahrzeugen", p. 16-21, Edelmetallemissionen, Zwischenbericht München: Gesellschaft f. Strahlen- und Umweltforschung, 1990.
[402] H.P. König, R.F. Hertel, W. Koch and G. Rosner, Atmospheric Environ., 26A (1992) 741.
[403] S. Knobloch, H. König and G. Wünsch, in " Application of plasma source mass spectrometry II " (G. Holland and A.N. Eaton, eds.), p. 108-114, Cambridge, Royal Society of Chemistry, 1993.
[404] F. Alt, H.R. Eschnaur, B. Mergler, J. Messerschmidt and G. Tölg, Fresenius J. Anal. Chem., 357 (1997) 1013.
[405] R.R. Barefoot, Environ. Sci. Technol., 31 (1997) 309.

[406] S. Artelt, H. Kock, H.P. König, K. Levsen and G. Rosner, Atmos. Environ., 33 (1999) 3559.
[407] J. Schäfer, J.D. Eckhardt, Z.A. Berner and D. Stüben, Environ. Sci. Technol., 33 (1999) 3166.
[408] G. Köllensperger, S. Hann and G. Stingeder, J. Anal. At. Spectrom., 15 (2000) 1553.
[409] S. Lustig, R. Schierl, F. Alt, E. Helmers and K. Krümmerer, Z. Umweltchem. Ökotox., 91(1997) 149.
[410] C.B. Tuit, G.E. Ravizza and M.H. Bothner, Environ. Sci. Technol., 34 (2000) 927.
[411] C. Barbante, A. Veysseyre, C. Ferrari, K. Van de Velde, C. Morel, G. Capodoglio, P. Cescon , G. Scarponi and C. Boutron, Environ. Sci. Technol., 35 (2001) 835.
[412] E. Helmers, Environ. Sci. Pollut. Res., 4 (1997) 100.
[413] E. Helmers and N. Mergel, Fresenius J. Anal. Chem., 362 (1998) 522.
[414] K.E. Jarvis, S.J. Parry and J.M. Piper, Environ. Sci. Technol., 35 (2001) 1031.
[415] J.F.W. Bowles, A.P. Gize, D.J. Vaughan and S.J. Norris, Chronique de la Recherche miniere, 520 (1995) 65.
[416] S. Lustig, B. Michalke, W. Beck and P. Schramel, Fresenius J. Anal. Chem., 360 (1998) 18.
[417] S. Lustig, S. Zang, W. Beck and P. Schramel, Mikrochim. Acta, 129 (1998) 189.
[418] F. Zereini, B. Skerstupp, K. Rankenberg, F. Dirksen, J.M. Beyer, T. Claus and H. Urban, J. Soil Sediments, 1 (2001a) 44.
[419] C. Wei and G.M. Morrison, Sci. Total Environm., 146, (1994a) 169.
[420] R. Schlögl, G. Indlekofer and P. Oelhafen, Angew. Chemie Int. Ed., 26 (1987) 309.
[421] F. Zereini, C. Wiseman, F. Alt, J. Messerschmidt, J. Müller and H. Urban, Environ. Sci. Technol., 35 (2001b) 1996.
[422] S. Rauch, G.M. Morrison, M. Motelica-Heino, O.F.X. Donard and M. Murris, Environ. Sci. Technol., 34 (2000) 3119.
[423] J.C. Ely, C.R. Neal, C.F. Kulpa, M.A. Schneegurt, J.A. Seidler and J.C. Jain, Environ. Sci. Technol., 35 (2001) 3816.
[424] M. Cubelic, R. Pecoroni, J. Schäfer, J.D. Eckhardt, Z. Berner and D. Stüben, Z. Umweltchem. Ökotox., 9 (1997) 249.
[425] F. Zereini, B. Skerstupp. F. Alt, E. Helmers and H. Urban, Sci. Total Environm., 206 (1997) 137.
[426] J. Schäfer, D. Hannker, J.D. Eckhardt and D. Stüben, Sci. Total. Environ., 215 (1998) 59.
[427] F. Alt, A. Bambauer, K. Hoppstock, B. Mergler and G. Tölg, Fresenius J. Anal. Chem., 346 (1993) 693.
[428] O. Morton, H. Puchelt, E. Hernandez and E. Lounejeva, J. Geochem. Explor., 72 (2001) 223.
[429] J. Schäfer and H. Puchelt, J. Geochem. Explor., 64 (1998) 307.
[430] M. Schuster, M. Schwarzer and G. Risse, in " Anthropogenic platinum-group element emissions. Their impact on man and environment " (F. Zereini, F. Alt, eds.), p. 173-182, Berlin, Springer-Verlag, 2000.

[431] Walkenhorst, J. Hagemeyer and S.W. Breckle, in "Plants as Biomonitors-Indicators for Heavy Metals in the Terrestrial Environment" (B. Markert, ed.), p. 523-540, VCH, Weinheim, 1993.
[432] G. Dongarra, D. Varrica and G. Sabatino, Appl. Geochem., 18 (2003) 109.
[433] B.G. Lottermoser, Int. J. Environ. Stud., 46 (1994) 167.
[434] D. Laschka and M. Nachtwey, Chemosphere, 34 (1997) 1803.
[435] E. Helmers, M. Schwarzer and M. Schuster, Environ. Sci. Pollution Res., 5 (1998) 44.
[436] K. Ravindra, L. Bencs and R. Van Grieken, Sci. Total Environ., 318 (2004) 1.
[437] S. Artelt, K. Levsen, H.P. König and G. Rosner, in "Anthropogenic platinum-group element emission: Their impact on man and environment" (F. Zereini and F. Alt, eds.), p. 33-44, Springer Verlag, Berlin, Heidelberg, 2000.
[438] T. Rühle, H. Schneider, J. Find, D. Herein, N. Pfänder, U. Wild, R. Schlögl, D. Nachtigall, S. Artelt and U. Heinrich, Appl. Catal. B., 14 (1997) 69.
[439] S. Rauch, G.M. Morrison and M. Moldovan, Sci. Total Environ., 286 (2002) 243.
[440] D. Sauerbeck, in " Beurteilung von Schwermetallkontaminationen im Boden " (D. Behrens, J. Wiesner, eds.), p. 281-316, Dechema-Fachgespräche, Frankfurt/Main, 1989.
[441] R. Djingova, P. Kovacheva, G. Wagner and B. Markert, Sci. Total Environ., 308 (2003) 235.
[442] T. Hess, B. Wenclawiak, S. Lustig, P. Schramel, M. Schwarzer, M. Schuster, D. Verstraete, R. Dams and E. Helmers, Environ. Sci. and Pollution Res., 5 (1998) 105.
[443] S. Lustig, S. Zang, B. Michalke, P. Schramel and W. Beck, Fresenius J. Anal. Chem., 357 (1997) 1157.
[444] S. Lustig, R. Schierl, F. Alt, E. Helmers and K. Krümmerer, Z. Umweltchem. Ökotox., 9 (1997) 149.
[445] M. Moldavan, S. Rauch, M. Gomez, M.A. Palacios and G.M. Morrison, Water Res., 35 (2001) 4175.
[446] B. Sures, F. Thielen and S. Zimmermann, Z. Umweltchem. Ökotox., 14 (2002) 30.
[447] B. Sures, S. Zimmermann, J. Messerschmidt and A. von Bohlen, Ecotoxicol., 11 (2002) 385.
[448] B. Sures, S. Zimmermann, C. Sonntag, D. Stüben and H. Taraschewski., Environ. Pollut., 122 (2003) 401.
[449] M.R. Reddy and H.F. Perkins, Soil Sci. Soc. Am. J., 38 (1974) 229.
[450] K.G. Tiller and J.F. Hodgson, Clays Clay Miner., 9 (1962) 393.
[451] M.B. McBride and J.J. Blasiak, Soil Sci. Soc. Am. J., 43 (1979) 866.
[452] W.L. Lindsay, Chemical Equilibria in Soils, Wiley and Sons, New York, 1979.
[453] M.A. Elrashidi and G.A. O`Connor, Soil Sci. Soc. Am. J., 46 (1982) 1153.
[454] A.D. Robson, Zinc in Soils and Plants, Kluwer, Dordrecht, 1993.
[455] G.D. Hogan and D.L. Wotton, J. Environ. Qual., 14 (1984) 377.
[456] W.M. Stigliani, P.R. Joffe and S. Anderberg, Environ. Sci. Technol., 27 (1993) 786.
[457] N.R. Benson, Soil Sci., 101 (1966) 171.
[458] K.M. Hambidge, C.E. Casey and L.B. McLeod, Trace Elements in Human and Animal Nutrition, Academic Press, Orlando, 1986.
[459] A.S. Prasad, J.A. Halsted and M. Nadimi, J. Med., 31 (1961) 532.

[460] J. Kubota, Soil Sci., 99 (1965) 166.
[461] R.M. McKenzie, Aust. J. Soil Res., 16 (1978) 209.
[462] J.F. Hodgson, K.G. Tiller and M. Fellows, Soil Sci., 108 (1969) 391.
[463] K.G. Tiller, J.L. Honeysett and E.G. Hallsworth, Aust. J. Soil Res., 7 (1969) 43.
[464] R.M. Taylor and R.M. McKenzie, Aust. J. Soil Res., 4 (1966) 29.
[465] C.A. Backes, R.G. McLaren, A.W. Rate and R.S. Swift, Soil Sci. Soc. Am. J., 59 (1995) 778.
[466] J.F. Hodgson, Soil Sci. Soc. Am Proc., 24 (1960) 165.
[467] S. Palit, A. Sharma and G. Talukder, Bot. Rev., 60 (1994)149.
[468] T.W. Purcell and J.J. Peters, Environ. Toxicol. Chem., 17 (1998) 539.
[469] P.J. Peterson, M.A.S. Burton, M. Gregson, S.M. Nye and E.K. Porter, Trace Subs. Environ. Health, 10 (1976) 123,
[470] C.T. Horovitz, H.H. Schock and L.A. Horovitz-Kisimova, Plant Soil, 40 (1974) 397.
[471] H.T. Ratte, Environ. Toxicol. Chem., 18 (1999) 89.
[472] N.R. Bury, J.C. McGreer and C.M. Wood, Environ. Toxicol. Chem., 18 (1999) 49.
[473] J.O. Nriagu, Thallium in the Environment, Wiley and Sons, New York, 1998.
[474] S. Takahasi, S. Tanabe and T. Kubodera, Environ. Sci. Technol., 31 (1997) 3103.
[475] R.D. Cardwell, M.S. Brancato and L. Tear, Environ. Toxicol. Chem., 18 (1999) 567.
[476] M.C. Newman and A. McIntosh, Metal Toxicology: Concepts and Applications, Lewis Publishers, Boca Raton, 1991.
[477] K. Kannan, K. Senthilkumar and J.P. Giesy, Environ. Sci. Technol., 33 (1999) 1779.

Heavy Metals in the Environment
H.B. Bradl (editor)
© 2005 Elsevier Ltd. All rights reserved.

Chapter 3

Remediation Techniques

H. Bradl[a] and A. Xenidis[b]

[a]Department of Environmental Engineering, University of Applied Sciences Trier, Umwelt-Campus Birkenfeld, P.O. Box 301380, D-55761 Birkenfeld, Germany
[b]Laboratory of Metallurgy, National Technical University of Athens, 9, Iroon Polytechneiou Str., GR-157 80 Zografos, Greece

1. INTRODUCTION

To date, a variety of methods for the remediation of both solid and liquid media such as water, groundwater, industrial wastewaters, soils, sediments, and sludges is known. To give a general overview over different technologies it is most convenient to divide them into three major categories: first, physical methods such as soil washing, encapsulation, or electrokinesis, second chemical methods such as solidification, precipitation, or ion exchange, and third, biological methods, which use plants to remove heavy metals [1]. In the past years, innovative approaches such as passive treatment technologies for soil and groundwater contaminations have been developed. The following chapters introduce the various remediation technologies.

2. PHYSICAL REMEDIATION TECHNIQUES

Physical remediation techniques include soil washing, vitrification, encapsulation of contaminated areas by impermeable vertical and horizontal layers, electrokinesis, and permeable barrier systems.

2.1. Soil Washing

Soil washing is a technique widely used for removing heavy metals and organic pollutants from soils. Most of the process steps in soil washing plants have not been developed for the remediation of contaminated soils but have been used for a long time in the mineral processing industry.

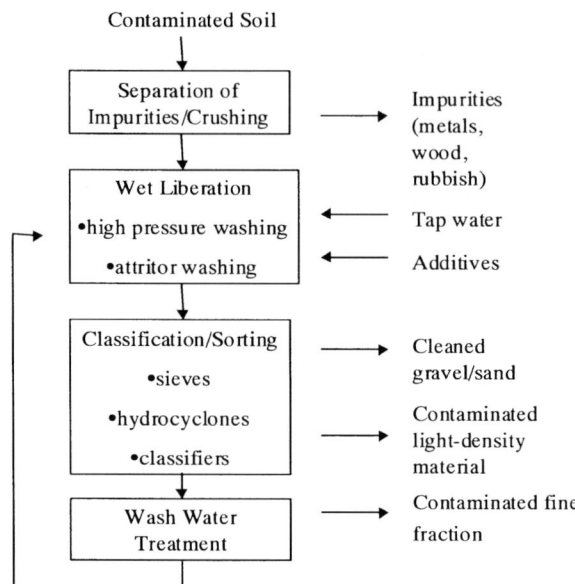

Fig. 1. General flow sheet of the soil washing process.

These processes have been then adapted to the special requirements of soil washing [2]. The main principle of soil washing is a selective classification of highly contaminated fines followed by the solid/liquid phase separation of the remaining suspension. For the cleaning of fines alternative processes like flotation, leaching, or high-gradient magnetic separation can be used. Soil washing does not attack the pollutants directly but separates different soil fractions with high contaminant content from soil fractions with low contaminant content. In general, contaminants concentrate in the fine particle fraction. The lowly contaminated coarse fraction can be re-used, while the highly contaminated fraction must undergo additional treatment. Soil washing plants consist of two principal steps, the wet liberation and the classification unit. Fig. 1 shows the general flow sheet of the soil washing process.

The first process step is the removal of coarse impurities such as wood, rubbish, metals, and other materials. These impurities are removed by sieving or manual sorting. Magnetic metals are removed by using magnets. In the next step, water and additives are added to the soil. Both cleaned process water and fresh water can be used. The soil is then liberated by mechanical energy input. Most soil washing plants use drums or attritors for the wet liberation step. It is also possible to use high-pressure steel jet tubes for this step. The high mechanical input is needed to detach the pollutants bound to the surface from the coarse soil particles. Contaminant detachment may be supported by the addition of additives such as surfactants, oxidizing agents, acids, and others depending on the nature of the contaminants. Then the cleaned soil fractions (particularly sand and gravel) are separated from the wash water, which contains the highly-contaminated fine-grained soil fraction. The separation of the cleaned material is carried out with the help of different classification and sorting devices such as sieves, hydrocyclones, up-current classifiers, spiral classifiers, and jigging machines.

The polluted wash water must be treated afterwards to separate both the fine particles and the dissolved pollutants from the water phase. Different techniques such as precipitation and flocculation of solid matter, filtration, flotation, or adsorption of organic pollutants on activated carbon are used for this purpose. The cleaned wash water can then be returned to the washing process. The highly-contaminated fine fraction has to be disposed either by deposition on a waste deposit or by thermal treatment in an incineration plant. As soil washing plants work with a closed wash water circuit there is no need for wastewater disposal.

2.1.1. Particle-Size Dependent Distribution of Pollutants

The importance of increasing surface can be simply demonstrated [3]. Imagine a spherical particle of some unspecified material with radius R. Then reapport this fixed quantity of material by subdividing it into an array of sphere, each with a radius half that of the original sphere. In a second subdivision, the radius of each of theses spheres will be cut in half again and so on. As for a sphere the volume of a particle varies with the cube of its radius R, therefore halving R decreases the volume by a factor of 8. Since the total amount of material is unchanged, the number of spheres must be increased by a factor of 8 by this cut.

The area varies with R^2; therefore halving R decreases the area per sphere by a factor of 4. Since the number of particles is increased eightfold by the same cut, however, the total area of the array is increased by a factor of 2.

The contaminants are mostly attached to the soil particle surfaces. The fine fraction contents soil particles with large surface reactivities and large surface areas such as clay minerals, iron and manganese oxyhydroxides, and others and displays enhanced adsorption properties, which result in increased contaminant content. The relationship between surface area and volume of a spherical particle is connected to its particle diameter d by:

$$S = \frac{\pi \ d^2 \ 6}{\pi d^3} = \frac{6}{d} \tag{1}$$

Fig. 2 shows this relationship for the different soil fractions clay, silt, sand and gravel. Most of the pollutants (> 90%) can be found in the silt and clay fraction. Particles with a diameter above 100 µm do not contribute to the total soil surface area even if coarse grained soils are considered. If the soil contents humic matter, the accumulation of pollutants is even enhanced as it coats the surface of the fine particles and strongly adsorbs both organic and inorganic pollutants. The knowledge of the particle-size dependent pollutant distribution is of vital importance for the process of soil washing. For the separation of the highly contaminated fine-grained fraction, classification processes are necessary.

2.1.2. Wet Liberation

Wet liberation is the first principal step in soil washing, which aims at removal of contaminants and highly contaminated fines from large particles. This process is mainly brought about by three physical mechanisms. The first process is abrasion of contaminants and fine particles from coarser particles. The second process is the destruction of agglomerates, and the third process is the disintegration of particles [4].

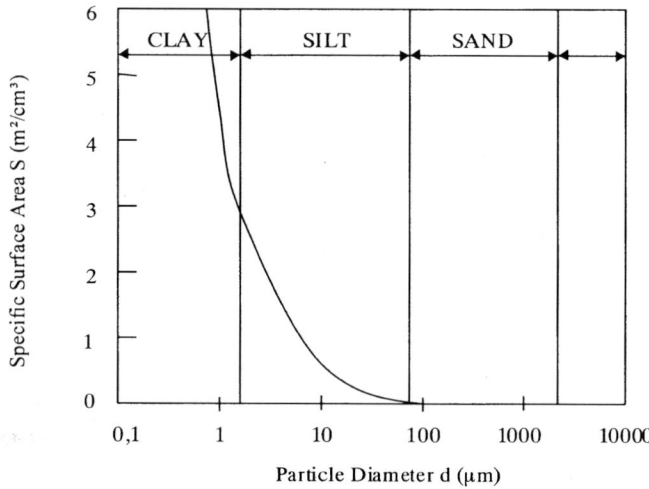

Fig. 2. Specific surface area of spherical particles dependent on the particle diameter.

These processes can be induced by using different methods such as hydraulic transport for dredged sludges, impact stress in high pressure water jets, attrition in rotation drums, and attrition with agitators for soil. Attrition cells usually consist of two stirrers counter-rotating in a steel vessel to induce shear stress. In a high pressure liberation unit, high pressure jets are focused into the suspension to accelerate it against a deflector plate. Comparisons of different methods showed that the higher the energy amount introduced into the suspension, the higher is the extent of shifting of contaminants into the fines fraction [4]. For equal energy input, attrition has been found to work more efficiently than high pressure jet liberation. This result could be caused by the increased particle residence time in the attrition cell [5].

2.1.3. Classification of Fine Particles

The second principal step in soil washing consists of the classification of the coarse soil fraction, which is characterized by no or very low contamination, from the fine grained soil fraction, which contains the majority of contaminants. For particles of the sand fraction, sieves can be used without any problems. If the particles become smaller, sieving is no longer possible. For further classification, devices such as a hydrocyclone or a hydro deflector wheel classifier can be used. Particles can also be separated according to their different surface properties, a process called flotation. Flotation has been developed in the mineral processing industry for the separation of the valuable minerals from tailings using different flotation agents [6]. The use of flotation for soil washing is rarely discussed in the scientific literature [7].

The basic principle of flotation is the separation of particles according to their different surface properties. Hydrophobic particles are caught by rising air bubbles, while hydrophilic particles remain in the pulp. The hydrophobic particles are then removed from the surface in a froth, which can be stabilized by adding different frothing agents. The surface properties of the particles can be altered in order to enhance the flotation effectiveness. Collectors are used to provide the particles, which have to be removed from the pulp, with a hydrophobic surface, while depressants enhance the wettability of those particles, which have to remain in the pulp. Modifying agents, which alter the chemical nature of the particle surfaces, can be used to promote or inhibit collector action [8].

If applied to soil washing, the use of flotation is limited by the particle size of the material. If the particles become too big, they cannot be transported upwards by the air bubbles. If they become too small, particles will unselectively adsorb flotation reagents due to their large specific surface area, thus severely disturbing the flotation process. Therefore, flotation is used for the material, which has been subjected to the removal of the fine and the coarse fraction, respectively. Investigations using flotation for the soil washing of soils contaminated with copper and zinc showed that the process can be tailored down to the appropriate flotation conditions to enhance the cleaning efficiency of the process significantly [4].

2.2. Encapsulation

Encapsulation of contaminated areas is commonly used for remediation by containment or pollution prevention [9]. Most of these techniques have been adapted to the use in the field of environmental engineering from the water-tight encapsulation of construction pits [10]. The basic principle is the underground construction of an impermeable vertical barrier to allow the containment of gases and liquids. A variety of construction methods such as cut-off slurry walls using mainly cement-bentonite-water slurries, thin walls, sheet pile walls, bored-pile cut-off walls, jet grouting curtains, injection walls, and frozen barriers has been developed [11].

2.2.1. Slurry Walls

Slurry walls use the supporting ability of bentonite slurries, which support a trench during excavation. Bentonite contains more than 80% montmorillonite, which is a three-layer clay of the smectite group (see subchapter 2.4.2. Clay Minerals). Bentonite is widely used as a slurry wall suspension. The stability of the suspension is of utmost importance for its workability. Stability is defined as the resistance of the suspension against flocculation of the solid particles, i.e. the separation between a solid and a liquid phase (see subchapter 3.3.1 Colloidal Systems). The determining parameters for the stability of a suspension are the electrolyte content of the fluid phase on the one hand and the interactions of the suspended particles on the other hand [12]. These parameters as well as the rheological properties of the slurry have to be controlled continuously during execution.

Slurry walls can be executed in two different ways as single-phase or single-phase walls. For both execution methods, a trench is excavated in the ground to the desired depth using grab buckets, clamshells, or vertical trench cutters (hydrofraise). The open trench is stabilized with the help of a slurry, mostly a suspension of bentonite and water, although polymer slurries are also in use. The wall consists of single panels, which can be from

0.4 up to 1 m thick. If the single-phase method is used, the supporting slurry remains in the trench after excavation and solidifies. It thus fulfils both supporting and containing functions. Single-phase slurries usually consist of cement, bentonite and water, but several additives can be added to improve certain properties of the slurry such as adsorption capacity, chemical stability, and others [13].

When using the twin-phase method, the supporting bentonite slurry is replaced after trench excavation by a second cut-off sealing slurry, which remains in the trench. For displacement, the contractor method with tremie pies is used. To achieve proper displacement, care has to be taken that the density difference between bentonite slurry and cut-off sealing slurry is high enough (at least 500 kg/m^3 are recommended by Ref. 11). If the density difference is not high enough, the two slurries may mix, thus impairing the integrity and hydraulic conductivity of the wall.

To ensure the long-term integrity of the wall, two parameters are most important: first, the quality of the slurry, and second, the contact between individual panels, which can be ensured by correct execution. Slurry walls are usually executed in the socalled pilgrim's pace method, i.e. the primary panels 1, 3, 5 etc. are excavated and filled with slurry first. After a period of time (mostly up to 2 days) the slurry in the primary panels has solidified. Then the secondary panels can be executed with an overlapping into the already hardened primary panels. This method ensures close contact between the panels, which is a prerequisite for the structural integrity of the wall and its containment function. Also, the verticality of the panels is of utmost importance. To enhance the strength and impermeability of a slurry wall, additional barrier elements can be inserted into the still liquid slurry. Barrier elements, which are used frequently, are HDPE membranes [14], sheet piles [15], geomembranes [16], and glass tiles. Execution of composite slurry walls is more complicated as special construction apparatuses are needed for placement of the additional barrier elements; also special lock constructions are necessary to ensure watertight joints between the single barrier elements [17, 18].

2.2.2. Thin Walls

Thin walls are a quick and cost-effective way of encapsulation. A heavy steel beam is vibrated into the ground, which is equipped with a high pressure jet. If the desired depth is reached, the beam is retracted slowly and a clay-cement-water suspension is injected simultaneously thus creating a watertight thin wall. To provide sufficient overlap and impermeability, the panels have to cut into the adjacent ones [10]. The impermeability is improved by the compaction of the soil and the penetration of the slurry into the soil pores. Thin walls can reach a thickness up to twice the nominal thickness, which can be up to 0.3 m.

2.2.3. Sheet Pile Walls

Sheet pile walls offer an attractive way of encapsulation as they can be constructed using a variety of materials such as steel, precast concrete, aluminium, or wood piles. These piles are driven into the ground and connected to each other by locks [19, 15]. The material used mostly is steel as they are easy to construct and is able to carry heavy loads. Sheet pile walls can be executed in short time with low costs and there is no need to dispose contaminated soil.

2.2.4. Bored-pile Walls and Jet Grouting

Bored-pile walls can be executed using contiguous or secant piles with or without casing. The execution method is similar to the pilgrim's pace method used with slurry walls. Jet grouting or high-pressure injection is a common execution method. Soilcrete columns are constructed underground. In the first step, a borehole is executed with the help of a rotary casing equipped with a high-pressure jet. If the desired depth is reached, the rotating casing is retracted and a cement-water suspension is injected into the ground under high pressure through the jet [14, 20]. The resulting void is filled with the suspension thus forming underground soilcrete columns.

2.2.5. Injection Walls

Injection is the placement of a solidifying material into the underground pores and fissures. Therefore, ground permeability is reduced. A variety of injection materials and injection techniques is available [14]. The injection material is injected into the ground through boreholes. The distance between the boreholes is a function of the ground permeability, the viscosity of the injected material, and the highest permissible injection pressure [21, 22]. A large variety of injection materials such as cement suspensions, water-glass based materials, and artificial resins are available. Fig. 3 shows the application of different injection materials as a function of the particle diameter of the ground to be treated. To prevent sedimentation of the cement particles, bentonite or other clay minerals are added to the suspension. Impermeability and plasticity of the solidified injection mass are determined by the bentonite and cement content. If the ground shows large permeabilities, additives such as sand, fly ash, or others can be used in order to increase the density of the suspension.

2.2.6. Artificial Ground Freezing

Artificial ground freezing is a frequently used technique for groundwater control, tunnelling, mine shafts, and other civil engineering construction for many years [23, 24]. Artificial ground freezing is also applied in soil remediation and hazardous waste management. Sometimes, the term cryogenic barrier is used. Freeze pipes are inserted into the ground, in which a heat exchange fluid (e.g. $CaCl_2$) or liquid nitrogen circulates. During the freezing process, the pore water in the soil pores is transformed into ice and fills the pores, which results in decreased soil permeability. The frozen soil turns into a solid watertight body, which prevents advective and diffusive contaminant transport. Although previously considered to be a temporary measure of isolating the contaminants until other treatment methods can be applied, recent investigations indicate that permeabilities of cryogenic barriers can be as low as 10^{-12} m/s and diffusivities are about 10^{-9} cm²/s [25]. The construction of frozen barriers of arbitrary thickness, shape, and depth is possible by choosing different pipe configurations.

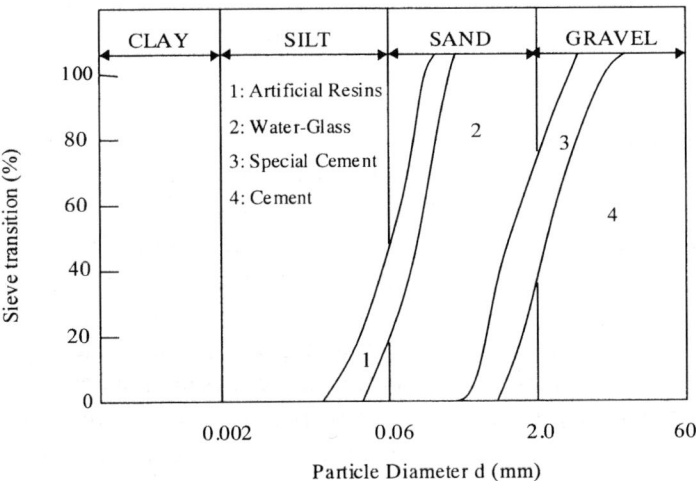

Fig. 3. Application of different injection materials as a function of particle diameter (modified after Ref.14).

It is also possible to contain brines and other antifreeze liquids where the temperature of the barrier is higher than the freezing point of the mixture. Barrier performance can be checked by using ground penetration radar [26, 27].

Artificial ground freezing offers several advantages. First, little or no excavation of contaminated soil is required; second, frozen walls can simply be removed by stopping the cooling, leaving no wastes, which have to be disposed and third, they can be executed to depths of a few hundred meters both in fine-grained and coarse-grained soils. Therefore, cryogenic barriers are a proven technology, which is both economically and environmentally sound.

2.3. Vitrification

Vitrification is a process, by which materials are converted into glass or glass-like substances [23]. Glass is characterized by its noncrystallinity and rigidity as well as its very limited porosity. For soil and waste remediation, vitrification can be used both as an in situ and as an ex situ technique. The processing and heating of excavated soil or waste is easier to control than the in situ process but it is disadvantageous due to greater exposure if radioactive or dispersive contaminants are treated. Vitrification uses heat produced by different sources, which destroys organic contaminants through pyrolysis or combustion, and fuses inorganic metals (including radioactive elements) into the glass structure. Glass formation requires the availability of component elements, which might not be always the case in contaminated media. In these cases, additives for glass formation improvement may be added to the deficient media.

The primary constituents of glass are oxides of silicon, boron, aluminium, and alkali and alkaline earth elements. Primary glass types used in vitrification are sodium silicates, borosilicates, and aluminosilicates. In industrial glass production, mostly sand (SiO_2), limestone ($CaCO_3$), dolomite ($CaMg(CO_3)_2$), and soda ash (Na_2CO_3, $KAlSi_3O_8$) are used as raw materials. The contaminated media are heated in a temperature range of 1000-2000°C and will melt. Upon cooling, the hot liquid solidifies into an amorphous mass. The glass consist of a three dimensional network of silicon-oxygen tetrahedra. The lack of crystallinity is caused by the irregularity and randomness of the silicon-oxygen bonds of the network. Some metals are able to replace silicon atoms in the interior of the tetrahedral or interrupt linkages among neighbouring tetrahedra.

The heat required for the melting process can be generated by different energy sources. The primary types used are joule heating, plasma heating, and microwave heating. Joule and plasma heating are technologies from the metalworks and glassmaking industries, while microwave heating stems from materials and incineration industries.

The most common heating process for in situ vitrification is joule heating where an electric current, which is applied through electrodes embedded in the ground flows through the contaminated medium. As the medium has an internal resistance to current flow, power will be reduced and heat is transferred to the medium. The power dissipated in the media can be estimated using Joule's law:

$$P = I^2 R \tag{2}$$

$$R = \frac{V}{I} \tag{3}$$

with
P: dissipated power (W)
I: current flow through the media (A)
R: resistance of the material (Ω)
V: voltage across the material (V)

In general, soils show a very high resistance due to the low electric conductivity of SiO_2, which is their main constituent. The resistance is a function of temperature and decreases with increasing temperature. Therefore, high voltages between 2000-4000 V are applied first to compensate for the high resistance of the soil at low temperatures. To initiate soil heating, enhancers such as flaked graphite and glass frit are placed between the electrodes on order to reduce initial soil conductivity. With increasing temperatures, these enhancers are destroyed and the soil can be heated more efficiently by reducing the voltage by as much as 10-fold.

The second heat generation method is plasma heating, which is mostly used ex situ. Heat is produced through conversion of a gas into plasma. Energy is supplied by an electric arc which is generated by direct or alternating current. Plasma is ionized gas, which is mostly produced using materials such as oxygen, nitrogen, air, noble gases, or mixtures of these. As ionized materials are good electricity conductors, heat can be transferred to the soil through convection, radiation, and electrical resistance.

The third heat generation method is microwave heating, which is currently used ex situ. The contaminated media are heated by absorption of electromagnetic radiation in the effective microwave range, which is 3000-30 000 MHz. The principle mechanism of soil heating by microwaves is described by Ref. 28. Soil is a dielectric material which means that its heat conduction properties are not very efficient. If an electric field is applied, dielectric materials exhibit some polarization. If the electric field is alternated, distortion of the molecules of the dielectric material will result in the alternation. The main vitrification treatment mechanisms for heavy metal contaminants are the covalent bonding of metal ions to oxygen atoms of tetrahedra through replacement of silicon, ionic bonding to oxygen atoms outside tetrahedra such that the glass network is interrupted, and the encapsulation by surrounding material without their being a part of the glass structure.

Fractions of initially bound contaminants from vitrified masses can be released by several deterioration processes such as matrix dissolution of glass and interdiffusion of elements, which are usually termed leaching. Leaching is exacerbated by the presence of aqueous media of suitable chemistry and temperature. Table 1 gives approximate ranges of solubility of elements in silicate glasses, which have been developed by Ref. 29. For a detailed discussion of leaching in connection with soil stabilization and solidification processes, see subchapter 3.5.3.

2.4. Electrokinetic Techniques

Electrokinetic decontamination or electroremediation of polluted sites is a promising in situ treatment technology especially for fine-grained soils [30]. Electrokinetic phenomena have been applied to environmental purposes since the 1990s [31-33].

Table 1
Approximate Ranges of Solubility of Elements in Silicate Glasses

Less than 0.1 wt%:	Ag, Ar, Au, Br, H, He, I, Kr, N, Ne, Pd, Pt, Rh, Rn, Ru, Xe
Between 1 and 3 wt%:	As, C, Cl, Cr, S, Sb, Se, Sn, Tc, Te
Between 3 and 5 wt%:	Bi, Co, Cu, Mn, Mo, Ni, Ti
Between 5 and 15 wt%:	Ce, F, Ge, Gd, La, Nd, Pr, Th
Between 15 and 25 wt%:	Al, B, Ba, Ca, Cs, Fe, Fr, K, Li, Mg, Na, Ra, Rb, Sr, U, Zn
Greater than 25 wt%:	P, Pb, Si

Reprinted with permission from: L.N. Reddi, and H.I. Inyang, Geoenvironmental Engineering, Marcel Dekker, New York, 2000, p. 306.

Electrodes are placed into the soil and a direct electrical current is applied, which induces movements of ions to their respective electrodes (Fig. 4). The three principle electrokinetic phenomena occurring are electroosmosis, electromigration, and electrophoresis [34], while the major physical mechanisms, which transport contaminants in the electric field, are electromigration and electroosmosis in fine grained soils, and electromigration and electrophoresis in coarse grained soils.

2.4.1. Principle Electrokinetic Transport Processes

The first principle electrokinetic transport phenomenon is electroosmosis, which is defined as the movement of a liquid relative to a stationary charged surface. In fine grained soils, these charged surfaces are provided mainly by clay minerals with diffuse double layers around their negatively charged surfaces. If an electrical field is applied tangentially along the charged surface, it will exert a force on the charges of the diffuse double layer. If the force of the applied electrical field exceeds the electrostatic attraction of ions to the surface, the ions will move along the field and will drag water by viscous interaction through the pores towards the electrodes. As there are more cations than anions in the diffuse double layer, there will be a net water flow towards the cathode.

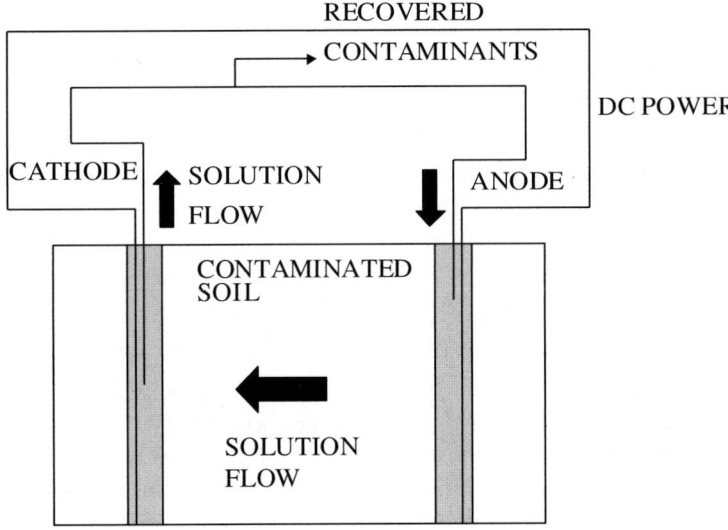

Fig. 4. Schematic principle of electrokinetic soil remediation (redrawn after Ref. 30).

This water transport is proportional to the electrical gradient and to the zeta potential of the charged surface and is independent of pore size distribution. In fine grained soils, electroosmotic velocities of several centimetres per day have been achieved by applying electric gradients of 100 V/m [30].

The second principle electrokinetic transport phenomenon is electromigration. Electromigration is defined as the mass transfer of charged ions and molecules, which are dissolved in the pore fluid of a soil when an electrical field is applied to the soil. Anions are moved towards the anode, and cations are moved towards the cathode. The migration velocity is proportional to the ionic charge, the local electrical field, and the ion mobility. In fine grained soils, transport velocities by electromigration are 5-40 times higher than those transport velocities by electroosmosis [35]. In coarse grained soils, however, there is only a low content of charged surfaces, and electromigration dominates clearly over electroosmosis.

The third principle electrokinetic transport phenomenon is electrophoresis, which is defined as the movement of charged particles relative to a stationary fluid under the influence of an electrical field [34]. As charged particles are of distinct size, electrophoresis can only take part if the pore sizes are large enough for the particle which is not the case in fine grained soils. Therefore, electrophoresis can be neglected in fine grained soils.

2.4.2. Electrode Reactions

The system depicted in Fig. 3 behaves like an electrolytic cell with current flow occurring from the positive anode to the negative cathode. Electron flow occurs the opposite way. Oxidation-reduction reactions will occur at the electrodes to maintain electrical neutrality. The principle chemical reaction observed is the electrolysis of water, which is reduced at the cathode and oxidized at the anode.

Table 2 gives an overview on the electrochemical processes occurring at the cathode and the anode. A front of high pH is produced at the cathodes, whereas a front of low pH is produced at the anode, and the two fronts move towards the opposite electrode. Thus the soil is divided in between the electrodes into a low and a high pH zone with a sharp jump of pH. This pH gradient will have a significant effect on electroosmosis, solubility, ionic state and charge, and level of adsorption of contaminants [33]. Dissolved heavy metal ions will precipitate at the cathode as oxides, hydroxides, carbonates, and others.

Table 2
Electrochemical processes during electrokinetic remediation

	Cathode (-)	Anode (+)
process	reduction	oxidation
redox processes	$4\,H_2O + 4\,e^- \rightarrow 2\,H_2\,(g) + 4\,OH^-$	$2\,H_2O \rightarrow 2\,O_2\,(g) + 4\,e^-$
pH	alkaline	acidic
heavy metals	precipitation as (hydr)oxides, carbonates etc.	dissolution

Modified after Ref. 30.

2.4.3. Applications

Electroremediation is used to treat both organic [36, 37] and inorganic contaminants such as Pb, Cr, Cd, and U [1]. Electroremediation is advantageous in low permeability soils because the transport rate which is induced by the electrical field is not affected by the low permeability. It can be applied on site, off site, and in situ depending on the conditions encountered at the site. Another advantage of electroremediation is its application to contaminations located at great depths or out of reach (e.g., below buildings). Also, inhomogeneous soil conditions such as alternating layers of fine and coarse grained soil can be controlled by this method. The combination of electroremediation with hydraulic and microbiological techniques and in situ passive permeable reactive barriers (see 2.6.) is discussed [30, 37]. The basic idea is to apply an electrical field upstream of the barrier and thus to reduce the amount of groundwater constituents flowing into the barrier, which might impair barrier function by coating the reactive material or clogging the pores of the reactor by precipitates.

It is also possible to combine electrokinetics with treatment zones within fine grained soils, a process which is called "lasagne" due to its layered configuration of electrodes and treatment zones [38-40]. The contaminants are transported into the treatment zones by electrokinetic processes. The lasagne process has been reported to be successful for the removal of organic substances such as *p*-nitrophenol (PNP) and trichloroethylene (TCE) by using activated carbon and zero-valent iron [40].

2.5. Permeable Reactive Barrier Systems

Usually contaminated groundwater is remediated by taking the groundwater off the aquifer, treating the polluted water in a groundwater treatment plant, and then feeding the treated water into the aquifer again or discharging it (pump-and-treat technology). Yet practical experience shows that this technology is very inefficient especially when treating groundwater contaminated by organic contaminants such as chlorinated hydrocarbons, mineral oils, polycyclic aromatic hydrocarbons, or BTEX [41]. This inefficiency is caused by natural underground inhomogeneity, the inhomogeneous flow through and the irregular contaminant distribution as well as by low contaminant solubility and the slow contaminant rediffusion of the contaminants which have been intruded the rock matrix for long times [42]. As a consequence, hydraulic remediation by active systems, i.e. systems which require a permanent energy input, is only effective in high permeability regions. The low permeability regions of an aquifer are not influenced by this procedure.

2.5.1. Permeable Walls

Therefore alternative methods have been developed for groundwater remediation by using permeable reactive barriers (PBR) or passive technologies, which remove contaminants in situ from the aquifer [43]. Especially two procedures have been proven to be very efficient: permeable reactive walls and funnel and gate systems. The principle of these systems and their advantages in comparison to usual methods are explained shortly. Fig. 5 shows the basic principle of a permeable reactive wall. Contaminants can be Light Non-Aqueous Phase Liquids (LNAPL, e.g. mineral oils and light tar oils), which have a lower density than water and swim on the groundwater surface. Also Dense Non-Aqueous Phase Liquids (DNAPL, e.g. heavy tar oils and chlorinated hydrocarbons) can occur, which have a higher density than water and therefore tend to sink at the aquifer bottom. Fig. 5 also gives an idea, which kinds of mechanisms are active in soil and groundwater. Contaminants are solved in the groundwater and are transported away from the contaminated area by advection. At the same time, dispersion and sorption result in a mixing and retardation of the contaminants. These substances are also able to intrude into low permeability regions due to diffusion processes. As a rule, times for diffusion processes are very long in comparison to transport processes, which are caused by advection and dispersion. This fact is of utmost importance in all cases where contaminants remain in the underground for a long time (often more than 10-50 years), e.g. gas works or industrial production plants. If these cases were to be treated by hydraulic remediation, which in principle only causes an increased flowthrough through the contaminated area, very long remediation times can be expected.

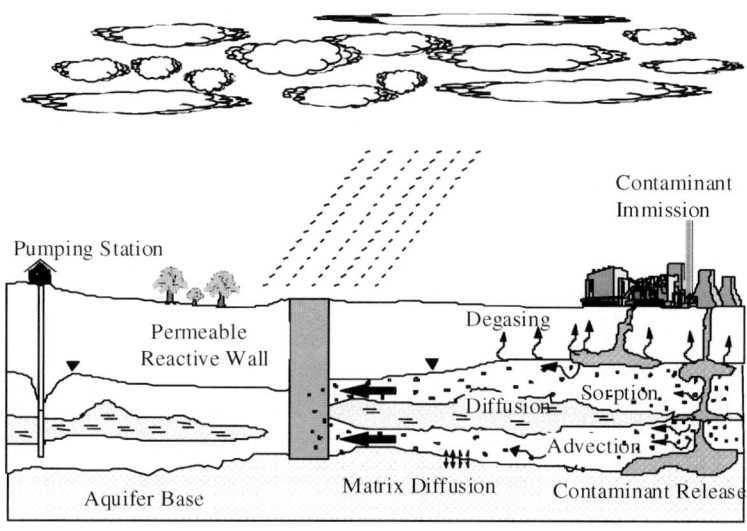

Fig. 5. Principle of a permeable reactive wall (redrawn after Ref. 42).

Calculations on the base of laboratory experiments resulted in remediation times of several hundreds of years for organic contaminants of very low solubility [44]. A scenario as shown in Fig. 5 can be treated by a permeable reactive wall. In this case, a permeable wall perpendicular to the groundwater flow direction is executed, which is filled with reactive material over the total width and depth of the contamination. The whole wall acts as the process reactor. The wall dimensions are calculated in such a way that the reactive material does not have to be replaced during the remediation. The wall permeability has to be in the same order of magnitude as the aquifer permeability, the wall depth depends on parameters such as the depth of the contamination, groundwater flow direction, aquifer properties and anisotropy of aquifer permeability. For execution, methods from the field of foundation engineering can be used [45]. Passive systems also offer some economical advantages when compared to active pump-and-treat [46-49]. Fig. 6 shows a comparison between costs for treating groundwater contaminated by chlorinated hydrocarbons using permeable barrier technology on the one side and pump-and-treat method on the other side. The cost saving has been estimated around 50-70 % for the long term. Barrier maintenance costs of US $ 268 000 every 10 years have been estimated, which was equivalent to 25% of the iron medium costs.

2.5.2 Funnel and Gate Systems

Funnel and gate systems are an alternative to permeable reactive walls. Fig. 7 shows the construction principle in comparison to a permeable reactive wall. This concept has been developed in Canada [50]. Funnel and gate systems consist of the combination of two system components with contrary hydraulic properties. The major part of the system consists of funnels of low permeability, which direct the contaminated groundwater flow through high-permeability areas, the gates, which are executed as reactors [43]. The advantage of the system is the possibility of replacing the reactive material in the reactors, as its volume is only small in comparison to the total volume of the system. In addition, several reactors can be placed successively, and thus the treatment of several kinds of contaminants is possible. Sometimes only a small reactor volume is needed, e.g., when rapid reactions are used. Yet there is the disadvantage of changing the natural groundwater flow conditions, which may result in significant water rise upstream the funnel. Several criteria have to be taken into account when planning a funnel and gate system. An important point is volume, position and permeability of the reactor. To guarantee an effective streaming into the gate, it is of importance that the gate permeability exceeds the natural aquifer permeability by a factor of 10.

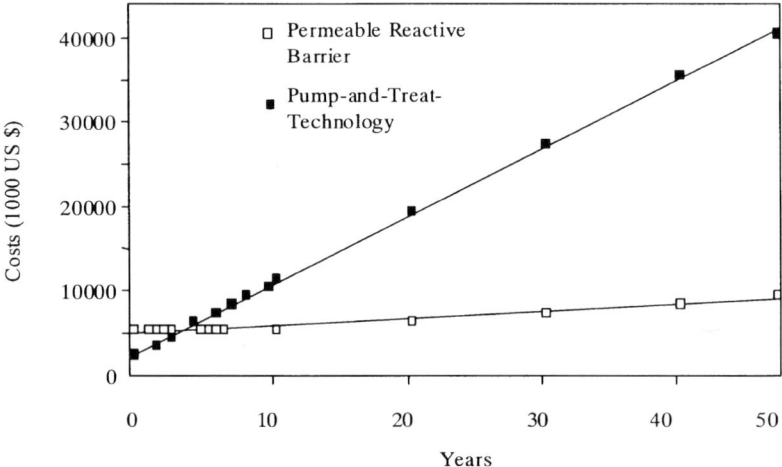

Fig. 6. Cost comparison of groundwater remediation between pump-and-treat technology and passive remediation systems (redrawn after Ref. 46).

On the other hand, the resident time of the contaminated groundwater in the reactor has to be taken into account. Furthermore, the design and the length of the low permeable funnel are of importance. Hydraulic considerations of different configurations showed that a rectilinear run of the funnel walls perpendicular to the groundwater flow direction results in the largest on-stream conditions [43]. This is only valid if homogeneous and isotropic conditions in a two-dimensional horizontal flow are assumed. Under natural conditions these assumptions are rather unreal. A more realistic assumption is to consider three-dimensional heterogeneous flow systems [43]. Therefore numeric flow models which take into account stochastic model approximations for heterogeneous aquifer systems have to be used for the design and optimisation of funnel and gate systems.

2.5.3 Reactor Technologies for Removal of Heavy Metals

The PRB technology can be used for the removal of heavy metals from soils and groundwater. Although mobility of heavy metals may be limited in soils and groundwater with high content of clay and organic material, high alkalinity and low permeability [51], natural and anthropogenic complexing agents increase the solubility of heavy metals [52]. Precipitation, adsorption, precipitation subsequent to chemical reduction, or a combination of these processes can be used for the removal of heavy metals.

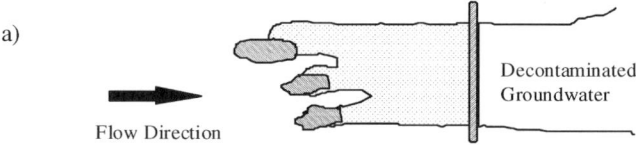

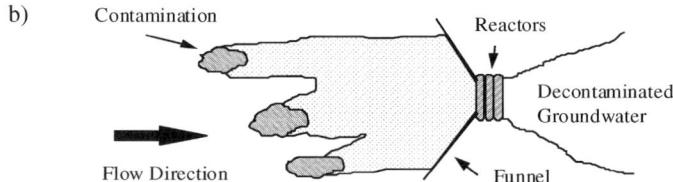

Fig. 7. Two principle arrangements used for PBR: (a) continuous permeable reactive wall and (b) funnel and gate system.

Chemical precipitation is a standard procedure in wastewater treatment, by which a soluble metal ion is transformed into an insoluble form by reacting with the precipitant. Such precipitants can be hydroxides, sulfides, phosphates, and carbonates. Their solubility is a function of pH and they exhibit amphoteric behaviour with high solubility both at low pH values (removal of hydroxide anions) and high pH values by formation of soluble hydroxo complexes with a minimum solubility for most heavy metals observed between pH 9 and 11 [53]. The precipitation of hydroxides allows a reduction in heavy metal content lower than 1mg/l, sometimes even lower than 0.1 mg/l.

$$Me^{2+} + 2OH^- \rightarrow Me(OH)_2 \text{ (s)} \tag{4}$$

$$Me(OH)_2 \text{ (s)} + 2H^+ \rightarrow Me^{2+} + 2H_2O \tag{5}$$

$$Me(OH)_2 + OH^- \rightarrow Me(OH)_2^- \tag{6}$$

In contrast to metal hydroxides, metal sulfides exhibit a much lower solubility and they are not amphoteric. At pH values below 8, H_2S is emitted due to hydrolysis of the sulfides, which results in increasing metal concentrations in solution.

$$Me^{2+} + S^{2-} \rightarrow MeS \text{ (s)} \tag{7}$$

$$MeS(s) + 2H^+ \rightarrow Me^{2+} + 2H_2S \tag{8}$$

A good example for the formation of precipitates with very low solubility is the reaction of lead ions with hydroxyapatite ($Ca_{10}(PO_4)_6(OH)_2$). After dissolution of the hydroxyapatite, Pb is precipitated as Pb hydroxypyromorphite, which has a solubility product of log K_{sp} = -62.79 [54] according to the following reactions:

$$Ca_{10}(PO_4)_6(OH)_2 + 14H^+ \rightarrow 10Ca^{2+} + 6H_2PO_4^- + 2H_2O \tag{9}$$

$$10Pb^{2+} + 6H_2PO_4^- + 2H_2O \rightarrow Pb_{10}(PO_4)_6(OH)_2 + 14H^+ \tag{10}$$

As the reaction between heavy metals and apatite is very fast [55], it is suitable for use in PRB systems. Another heavy metal, which can be removed from water by precipitation, is chromium, which exists in natural waters in two oxidation states +III and +VI. The toxic hexavalent chromium ion occurs as an oxyanion in the form of CrO_4^- or as $Cr_2O_7^{2-}$, which are more mobile than the Cr(III) compounds. Cr(VI) cannot be precipitated as a hydroxide and has to be reduced to Cr(III) prior to precipitation. According to Ref. 53, Cr(VI) can be removed with the help of sulfide in a single reaction step which combines reduction and precipitation:

$$Cr_2O_7^{2-} + 2FeS + 7H_2O \rightarrow 2Cr(OH)_3 + 2\,Fe(OH)_3 + 2S + 2OH^- \qquad (11)$$

Reduction of chromate with zero valent iron has been studied intensively [56]. The first reaction step occurs spontaneously and can be written as:

$$CrO_4^- + Fe^0 + 8H^+ \rightarrow Fe^{3+} + Cr^{3+} + 4H_2O \qquad (12)$$

In the next step, iron and chromium are precipitated as oxyhydrates:

$$_{(1-x)}Fe^{3+} + Cr^{3+} + 2H_2O \rightarrow Fe_{(1-x)}Cr_xOOH + 3H^+ \qquad (13)$$

Another element which can be precipitated is Uranium, which is the heaviest naturally occurring element. Uranium occurs in the oxidation states +IV and +VI, and similarly to Cr(VI) and Cr(III), the hexavalent uranium (the uranyl ion UO_2^{2+}) is more mobile than the U(IV) compounds. With phosphate, $(UO_2)_3(PO_4)_2$ is formed. This reaction product has a solubility product of log K_{sp} = -49.09 [51], with hydroxyapatite, autunite $(Ca(UO_2)_2(PO_4)_2)$ is precipitated (log K_{sp} = -47.28). Elemental iron can be used to reduce U(VI) to U(IV) according to the following reaction [57, 58]:

$$Fe + UO_2^{2+}(aq) \rightarrow Fe^{2+} + UO_2(s) \qquad (14)$$

The reduction of the uranyl ion UO_2^{2+} occurs spontaneously. Uranium becomes soluble below pH 4, and Uranitite (UO_2) can be transformed to the uranyl ion under oxidizing conditions. In the presence of complexing agents, solubility can increase considerably [52]. Uranium can also be removed by surface adsorption onto several materials such as lime, haematite, peat ferric oxyhydroxides, phosphates and TiO_2 [59, 60]. In a barrier system, precipitation and adsorption can be used simultaneously. The strong pH dependence of adsorption requires an optimum operation pH for the barrier system which can be difficult to achieve in the case of uranium. For example, molybdenum shows enhanced mobility at pH values above 8, while uranium shows only low mobility in these pH ranges [61]. Biosorption is also able to remove uranium and other heavy metals such as Pb, Cd, Cu, Zn, and Cr from solution [62]. Biosorption is an ion exchange process between the heavy metal ion and the protons introduced to the binding sites of certain types of inactive, dead, microbial biomass [63]. It is also possible to use microorganisms to reduce U(VI), which could also be used for field application [64].

Another possibility of removing heavy metals from solution is adsorption on inorganic sorbent materials like zeolites, ferric oxyhydroxides, and others [65]. Zeolites are alumosilicates with cage-like structures, which exhibit a high cation exchange capacity for heavy metal ions. They also offer the possibility of treating waters contaminated with heavy metals and non-ionic organic compounds simultaneously by modification of their surface with quaternary amines (organo-zeolites). Ref. 66 reports on the sorption of chromate, selenate, and sulfate on a clinoptilolite-dominated zeolite modified by the quaternary amine hexadecyltrimethylammonium (HDTMA). HDTMA is sorbed onto the zeolite`s external surface by Coulombic interactions. While adsorption of anions is not possible on untreated zeolite, HDTMA-zeolite showed strong adsorption of chromates, sulfate, and selenate (Fig. 8). Several potential adsorption mechanisms are possible. First, the formation of surfactant admicelles through tail-tail interaction of the HDTMA substituent chains could provide a positively charged interface on the organo-zeolite surface, which would attract anions. A general model of adsorption of ionic surfactants on a solid surface is the formation of a monolayer or "hemimicelle" at the solid-aqueous interface via strong Coulombic bonds. Hemimicelle formation will occur at surfactant concentrations at or below its critical micelle concentration (CMC). This mechanism is shown in Fig. 9a. If the surfactant concentration exceeds the CMC, the hydrophobic tails of the surfactant molecules associate to form a bilayer or "admicelle" on the surface (Fig. 9b) [67].

2.5.4. Engineering Methods for Execution of Permeable Reactive Barriers

Until recently, only little attention was given to construction methods for the execution of PBR in contrast to extensive investigations concerning physical, chemical and biological processes used in these barriers. In fact, execution is rather simple if shallow depths of up to 10 m are required for the barrier. Standard machinery can be used for trenching and backfilling the trench with reactive material.

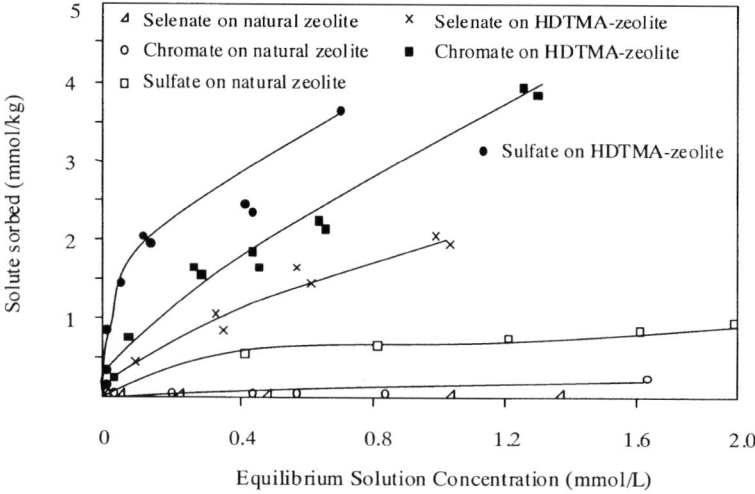

Fig. 8. Sorption of oxianions to natural zeolite and to HDTMA-zeolite (modified after Ref. 66).

If a funnel and gate system is used, cut-off wall construction methods can be easily adapted to the reactive barriers. However, if greater depths than 20 m have to be executed, more sophisticated methods are needed for execution. As for the impermeable funnel walls, the whole range of special deep construction methods can be used. These methods include technologies such as slurry trenching, jet grouting, and sheet pile walls, which have been discussed in detail in subchapter 2.2 (encapsulation). As for the permeable reactor, whether it may be an extended wall or a small gate used in a funnel and gate configuration, several requirements have to be fulfilled [68, 69]. First, the permeability of the reactor has to be higher than the permeability of the surrounding aquifer in order to ensure an undisturbed flow of groundwater through the reactor. Second, if the reactive material has to be replaced after some time of operation, the construction of the reactor must facilitate exchange of the reactive material (e.g. in form of filter cassettes). Fig. 10 shows a principle possibility for reactor execution. If required depths are not deeper than 10-15 m, the reactor can be executed in an open construction pit with reinforced walls in order to compensate earth pressure against the walls. Another possibility is the use of large-diameter drillings, which are then filled with reactive material.

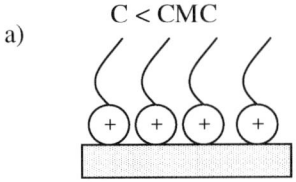

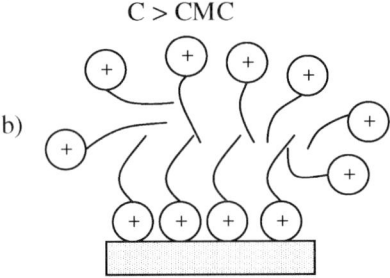

Fig. 9. Hemimicelle (a) and admicelle (b) formation by cationic surfactants on zeolite surface (modified after Ref. 66).

The reactor has to be stable against fines washed into the barrier from the surrounding soil, clogging the pores, and preventing water flow through the barrier. Therefore, gravel filters have to be added to the reactor. In greater depths, however, this technique has its limits both for constructional and economic reasons. Methods from special deep foundation can be adapted to the requirements of PRB. Other possibilities of execution are reactive thin walls, drilling, deep soil mixing, jet technology, injection technology, hydraulic fracturing, biobarriers, and biopolymer trenching. Reactive thin walls are a modification of the thin-wall technique developed for cut-off walls. They combine the advantages both of slurry trenching and of sheet piling [70]. Similar to conventional sheet piling, hollow steel beams are vibrated into the ground. Upon retracting, reactive material is then placed in the void space. By repeating this step, a number of panels can be constructed, which form a continuous reactive wall. This execution method offers a variety of advantages such as low demand on space, no extraction and disposal of contaminated water and soil, and a minimum impact on groundwater flow. The reactive material can be recovered by using the same steel beams. This method is time and cost effective and can be used for depths of up to 25 m.

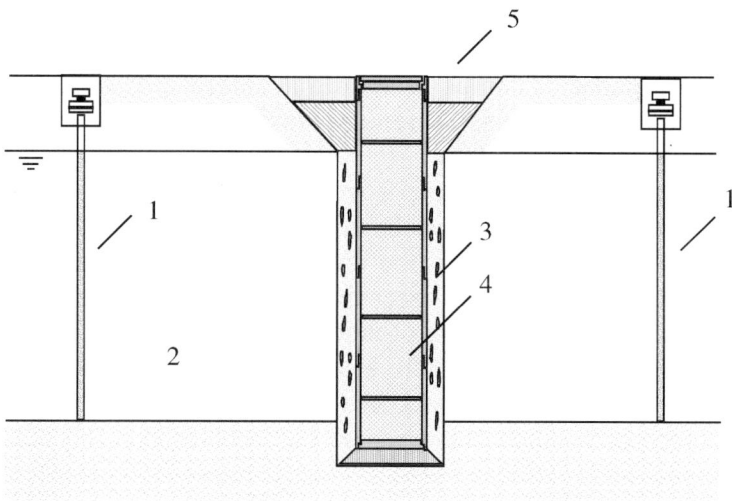

Fig. 10. Construction principle for permeable reactor construction. 1: monitoring wells, 2: aquifer, 3: gravel filter, 4: cassettes with reactive material, 5: impermeable clay covers.

Drilling and deep soil mixing methods can also be used for the installation of a PRB system. A barrier can be executed by drilled contiguous circular columns of diameters of 0.5 m up to 2.5 m containing reactive material [71] using caisson drilling. The casing is lowered into the ground down to the requested depth, then the soil is removed by augers and the hole is filled with reactive material. The disadvantage of this method is the handling and disposal of the contaminated soil recovered by the augers.

Casing can be substituted by using a supporting fluid, which can be biodegradable polymer slurry [72, 73] or a shear-thinning fluid [74]. Another possibility of avoiding excavating contaminated soil is deep soil mixing, which mixes the soil in situ with slurry [75]. A caisson is lowered into the ground, which is then penetrated by a set of multiple augers. The reactive material is then injected through the hollow kelly bar of the mixing tools. The reactive material is injected as slurry, which is then mixed with the soil by the augers. Jet technology offers another possibility for barrier execution. Several methods are possible. Ref. 76 proposes a jet technology, which uses high-pressure jets and jet pumps. Trench excavation can be executed by using a jet cutting head for low-strength soil or a milling head for rock. The slurry used for trench stabilization can either remain in the trench after the excavation is complete and form the barrier (single-phase technique) or can be replaced by another reactive material (twin-phase technique). Iron particles can be used both as the reactive material and the abrasive medium, which enables trench construction and reactive material placement simultaneously [51].

Injection technology is another attractive possibility of forming an underground barrier. Reactive material can be injected into the ground at high pressure and at great depths and the control of deep and large plumes is possible even if it is of irregular extension. As injection technology is used in petroleum engineering and deep foundation extensive experience can be used. As the recovery of the reactive material is not possible in injected systems, either reinjection of new material or degradation would be an appropriate approach. Injection may use the already existing pores in the ground or create new pathways (ground or hydraulic fracturing). The spacing of injection wells depends on the pores available, and the permeability limit below only pure liquids can be injected is as high as 10^{-5} m/s. As for the injected media, air microbubbles as oxygen suppliers [77] or cationic surfactants for enhancement of the retardation of organic contaminants [78] have also be suggested.

Hydraulic fracturing is a technology well used in petroleum engineering for decades to enhance permeability around oil wells by cracking the ground with water at high pressures and then pumping sand or other granular material into the cavities. It is obvious to use this approach for the barrier approach by filling in reactive material for contaminant treatment [79]. Another advantage is that a high-permeability zone is formed, which enhances groundwater flow. The fracturing process can be monitored by recording both pressure and deformation of the ground surface, which will lift up. Reactive materials which can be used for injection include materials that alter redox conditions, adsorb contaminants, or slowly release useful materials (e.g. oxygen, nutrients, porous ceramic granules, etc.).

Biobarriers are an innovative approach using subsurface biofilms comprising deposited cells of microorganisms and polymers together with captured organic and inorganic particles [80, 81]. Such biobarriers may reduce the hydraulic conductivity by five orders of magnitude, and in column tests 92-98% reduction in hydraulic conductivity was achieved [51]. Biobarriers may also be useful as a part of a funnel and gate system. To form a biobarrier, potential autochthonous barrier-forming bacteria must be identified and isolated from the site. Then these bacteria are reinjected to serve as an inoculum for biobarrier formation.

Bacteria producing extracellular polymer strains such as some species of *Pseudomonas* and *Klebsiella* are desirable candidates. These bacteria showed good growth on low cost nutrients such as molasses or distillery waste. Biobarriers can be constructed by using an array of well to introduce microorganisms, nutrients, and oxygen into the ground. The microbial activity then leads to the formation of biofilms, which form the barrier. Biopolymer slurry trenching is an increasingly popular and cost-effective construction method for PRB [82]. It is similar to constructing a conventional impermeable bentonite slurry wall (also see subchapter 2.2. Encapsulation). Biopolymers (e.g., guar gum) are used as additives to stabilize the trench walls during excavation. The reactive material can be placed into the trench using a tremie tube. To remove the biopolymer, a high-pH enzyme breaker is then added to the fluid which will break down the polymer, leaving the reactive material in place. This method prevents clogging of pores, which will occur when bentonite slurry is used and enables groundwater flow through the barrier.

There is also the possibility of applying the PRB technology to already existing containment systems. In the past many old landfills have been encapsulated by vertical barriers to separate them from the natural groundwater flow. Yet there is always a water inflow into the encapsulated area due to the permeability of the vertical barrier. As a consequence a hydraulic maintenance of the encapsulated site is necessary, which means that the costs for pumping and treating the contaminated groundwater are very high. It is possible to open the vertical barrier at distinct points and to fill the openings with a reactive material. The containment is thus transformed into a funnel-and-gate system for the treatment of the contaminated groundwater [83]. PRB is an innovative and very promising approach for removing heavy metals from soil and groundwater. Compared to traditional pump and treat techniques, it offers considerable cost savings and a wide range of in situ treatment reactor technologies such as redox reactions, precipitation, adsorption, and ion exchange.

3. CHEMICAL REMEDIATION TECHNIQUES

Chemical remediation techniques for the removal of heavy metals in contaminated groundwater and wastewater include precipitation of dissolved metals, ion exchange, flocculation, and membrane filter processes such as micro- and ultrafiltration and reverse osmosis. When dealing with contaminated soils, stabilization and solidification processes can be used. In fluid media, heavy metals occur as dissolved ions or bound to colloidal particles. A basic distinction must be made between precipitation and flocculation processes. Precipitation is a process, in which there is a phase transfer of soluble, ionic components into a non-soluble ionic phase. Flocculation is a process, in which small undissolved solids of colloidal size are aggregated into larger solid flocks. These flocks are then separated mechanically from the fluid by sedimentation, centrifugation, or flotation [84]. Solidification aims at the reduction of heavy metal mobility by trapping or immobilizing them within the soil by injecting or mixing immobilizing agents into the contaminated soil. Besides reducing chemical solubility, stabilization and solidification processes can be used to increase mechanical strength and reduction of soil permeability.

3.1. Precipitation

In precipitation, dissolved metal ions react with added precipitants by forming insoluble compounds. These solids sediment and can be removed from the supernatant liquid by different solid/liquid separation techniques. The main chemical parameters, which are of importance in the precipitation process, are pH and concentration. In general, heavy metals tend to be present in ionic form at low pH levels, while they tend to precipitate when pH is raised. Heavy metals can be precipitated as insoluble hydroxides, sulfides, carbonates, and others (see subchapter 3.5.2. for detailed discussion).

The pH of the wastewater is raced by addition of bases so that the solubility products are exceeded. Table 3 shows the pH range for removing heavy metals by precipitation from start of precipitation to point of redissolving [85]. It should be pointed out that high salt concentrations and the presence of organic complex formers influence the solubility products. The equilibrium conditions for heavy metals in solutions rich in these compounds cannot yet be satisfactorily presented [86].

There is a variety of precipitating agents such as digested sludge, Fe salts, calcium hydroxide, and Al salts. Table 4 lists the effect of these different agents on the elimination of heavy metals [84]. The precipitation of metal sulfides has been proven to be very effective. Fe salts are good for the removal for Ag, Cr, Pb, Cd, Hg, and Sn, yet show poor performance for Mn, Co, Sb, and Se. Also Al salts and calcium hydroxides are used. To ensure good performance, secondary devices such as ion exchange (for bivalent metals such as Cu, Ni, Zn, Cd, and Fe) or activated carbon filters (for Zn, Cu, and Hg) should be used.

3.2. Ion Exchange

Ion exchangers are commonly used for the removal of heavy metals, especially when dealing with wastewater from the metal processing industry. Surface treatment such as electroplating produces waters containing cyanide, copper, nickel, and cadmium. The basic principle of an ion exchanger is a matrix or resin laden with dissociable counter ions. The most common ion exchangers are made of interlaced polystyrene and polyacrylate or condensation resins made from phenol and formaldehyde [84]. In order to minimize metal emissions and wastewater, closed cycle systems are ideal [87, 88]. The recycled solutions can then be returned into the production process. A simple design for an ion exchanger is a solid bed device with a counter-current rinse process. The exchanger is charged from above, while the exchange proceeds from below. For regeneration, acids such as HCl and H_2SO_4 and alkaline solution (NaOH) are used. The substance cycle in an electroplating process is shown in Fig. 11. Two major approaches are made in order to minimize wastewater generation and waste generation. First, a primary treatment step or nonselective concentrating step is used, which concentrates electrolytes in the rinse water by using multi-stage rinsing processes SP_1 and SP_2. There is no need for adding further chemical compounds in the substance cycle, which prevents wastewater generation and additional salt concentration in the residual wastewater. Second, final or post-treatment processes are required, which remove heavy metals even when complex formers or high salt concentrations are present. There is a variety of process technologies available. The most important of these technologies are selective ion exchange using carbic acid and chelate exchangers and bonding of metals to activated and nonactivated biomass (biosorption).

Table 3
Effect of different precipitating agents on the elimination of heavy metals

Metal	pH at start of precipitation	pH at point of redissolving	Metal hydroxide	solubility product
Fe	2.8	-	$Fe(OH)_3$	$9 \cdot 10^{-38}$
Al	4.3	8.3	$Al(OH)_3$	$2 \cdot 10^{-32}$
Cr	5.8	9.2	$Cr(OH)_3$	$3 \cdot 10^{-28}$
Cu	5.8	-	$Cu(OH)_2$	$2 \cdot 10^{-19}$
Pb	6.5	-	$Pb(OH)_2$	$\sim 10^{-13}$
Zn	7.6	11.0	$Zn(OH)_2$	$4 \cdot 10^{-17}$
Ni	7.8	-	$Ni(OH)_2$	$6 \cdot 10^{-15}$
Cd	9.1	-	$Fe(OH)_3$	$1 \cdot 10^{-14}$

Modified after Ref. 84.

Heavy metal bonding to anaerobic biosludges, decomplexing and flocculation using substances such as lime, chalk, bentonites, and others (see subchapter 3.3. for detailed discussion), photo oxidative destruction of complex formers through UV, hydrogen peroxide, or ozone, and sand, micro- or nanofiltration as a pretreatment for ion exchange or as post treatment after precipitation are also used [89]. Dilute solutions can be treated using membrane filter processes (such as e.g. nanofiltration or reverse osmosis, see subchapter 3.4. for detailed discussion).

3.3. Flocculation

As it has been mentioned above, a clear distinction must be made between precipitation and flocculation. Flocculation is used to transform the suspended colloidal particles in a form so that they can be separated mechanically from the supernatant solution with the help of flocculants. Primary colloidal particles have a typical particle diameter in the range between 10^{-4} to 10^{-7} cm [90]. In order to understand the flocculation mechanisms, the different forces acting in a colloidal system must be taken into account.

3.3.1. Colloidal Systems

In general a colloidal system is defined as a two-phase system, in which larger kinetic units are dispersed in a homogeneous phase. The term colloid refers not only to the dispersed substance, but to the entire system. In principle there are two classes of colloidal systems: hydrophilic and hydrophobic colloids [90]. Hydrophobic colloids are dispersions of solid particles in a liquid. They are characterized by the large interface between the two components. The properties of the particle surfaces are of great importance for the system properties. Examples for this colloid class are dispersions of very fine inorganic particles in water like gold sols, hematite sols, and clay sols [91-93]. Hydrophilic colloids consist of macromolecular components and water. They are in fact genuine solutions of macromolecules in water and are called macromolecular colloids. Their colloid chemical properties are caused by the size of their molecules and ions. An important difference between the two classes consists in their reaction towards salt addition. Hydrophobic colloids flocculate immediately when salt is added, i.e. the particles, which had been distributed in the liquid form clumps and agglomerate, until a sediment of flocculated particles and a clear supernatant solution is formed [94].

Table 4
pH range for removing heavy metals by precipitation from start of precipitation to point of redissolving

Precipitating Agent	High Effectiveness	Low Effectiveness
Digested Sludge	Cu, Pb, Zn, Cd	Ni
Precipitation with Fe salts	Ag, Cr, Pb, Cd, Hg, Sn	Mn, Co, Sb, Se
Precipitation with CaOH	Ag, Co, Cr, Pb, Ni, Cd	Sb, Se, As
Precipitation with Al salts	Ag, Be, Hg, Cr, Cd, Pb	Zn, Mn, Ni

Modified after Ref. 85.

In contrast, macromolecular sols, which are genuine solutions in a thermodynamic sense, react relatively insensitive when salt is added. They do not flocculate like hydrophobic sols, but large salt addition influences the solubility of the macromolecules, which then precipitate.

The stability of a suspension against flocculation is determined by the interactions of the single particles. In stable state in a well dispersed system, attractive and repulsive forces between the particles are in equilibrium. If this equilibrium is disturbed by changing the electrolyte content of the medium, the suspension becomes unstable and will flocculate. The first theoretical treatment of the stability of colloidal systems was carried out by Derjaguin and Landau in Russia and, independently, Verwey and Overbeek [95] in the Netherlands. Therefore this theory is called the DLVO theory. The attractive force between the particles is based on the van der Waals interaction between all atoms of one particle with all atoms of another particle. The van der Waals force explains certain deviations from the behaviour of an ideal gas as they occur in real gases. The total attractive force between the particles is the sum of all forces between the atom pairs. At first sight it seems to be very unlikely that van der Waals forces can have a considerable effect on the total system, as the attractive effect between the atoms of one atom pair is very small and its range decreases inversely with the seventh power of their distance. Yet the van der Waals interaction is additive, i.e., the total attraction between particles consisting of a large number of atoms is equal to the sum of all attractive forces between each atom of the one and each atom of the other particle. This summation results in a higher total interaction and in a higher range. The suspension's stability against flocculation is determined by the interactions of its particles. In a stable suspension, attractive and repulsive forces are in an equilibrium state. The suspension becomes instable if the electrolyte content of the fluid medium is increased. The repulsive force results from the interaction of the two diffusive double layers [96]. The total net interaction energy results from the summation of attractive and repulsive energy for each particle distance. The attractive potential is counted negatively and the repulsive potential is counted positively [97]. Fig. 12 shows schematic potential curves for high, middle and low electrolyte concentrations. The curves for middle and low electrolyte content show a minimum with predominant attraction for small particle distances and a maximum with predominant repulsion for larger particle distances. This maximum does not exist in the curves for higher electrolyte content. The consequences for suspension stability are that the potential curve for high electrolyte content shows a minimum only for very small particle distances. Therefore, attraction predominates for each particle distance with the exception of very small distances. The net interaction energy results from the summation of attractive and repulsive forces for each particle distance. If two particles meet due to their Brownian motion, they will agglomerate if they reach a position where attraction is at minimum. The agglomeration rate for this process can be calculated according to the theory of diffusion [98]. For low and intermediate electrolyte content, there is an energy barrier for the particles. In order to pass this barrier, a certain amount of activation energy is necessary. The height of the barrier is a function of the amount of compression of the diffuse double layer, which is caused by the electrolyte [94].

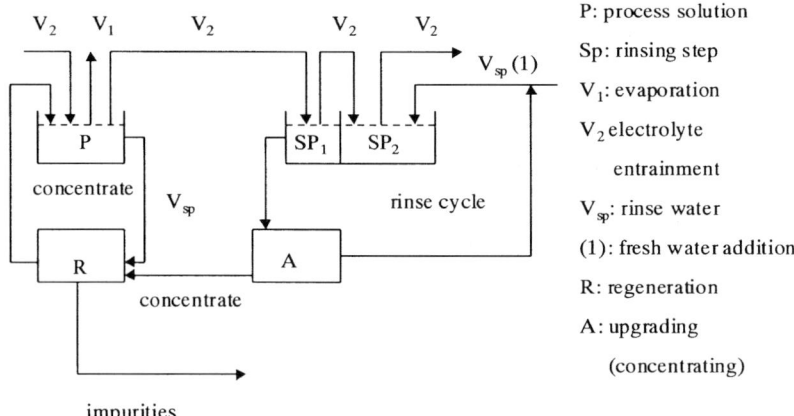

Fig. 11. Substance cycle in an electroplating process (redrawn after Ref. 84).

3.3.2. Flocculation Chemicals

The first step for successful flocculation is the destabilization of the colloid. As it has been discussed above, coagulation is enhanced by reducing the repulsive force between the particles. By adding multivalent counter-ions (e.g. trivalent Fe), the surface charge of the particles is reduced or neutralized. This process is called specific coagulation. The main inorganic flocculation chemicals are calcium hydroxide, Fe(II)- and Fe(III) salts, and Al salts. In order to enhance the agglomeration of positively and negatively charged particles, inorganic flocculation additives like activated alumina and kaolinitic and bentonitic clays are used. Al and Fe(III) facilitate coagulation at low pH, while Al and Fe hydroxo complexes are formed at higher pH. Usually a sequential dosing is preferred with the use of inorganic flocculants for the formation of micro-flocs, which are then transformed into macro-flocs by using polymer flocculants. There is an important quantitative relation between the socalled critical flocculation concentration CFC and the valence of the counter ion. The CFC is the smallest concentration, by which a colloid can be destabilized. The determination of the CFC is one of the easiest tests, which can be performed in colloid chemistry. The test dispersion is introduced into a series of test tubes and to each tube various proportions of water and electrolyte are added.

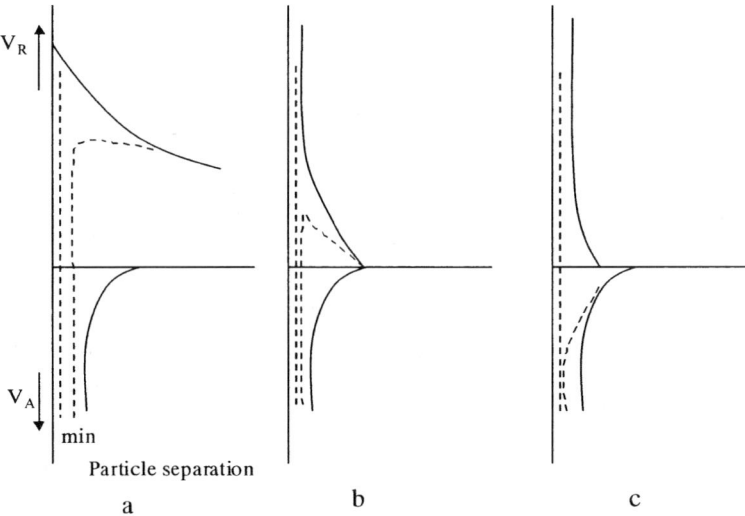

Fig. 12. Schematic potential curves for high (a), middle (b), and low (c) electrolyte concentrations.

After mixing and waiting, the tubes are visually inspected for flocculation. The settling of the dispersed phase is a clear evidence of flocculation. The highest concentration of salt, which leaves the colloid unchanged and the lowest concentration, which causes flocculation, brackets the CFC. Both colloids bearing negative and positive surface charges have been tested.

One of the earliest generalizations about the effect of added electrolyte is known as the Schulze-Hardy rule. This rule states that it is the valence of the ion of opposite charge to the colloid that has the principal effect on colloid stability. E.g., for a negative As_2S_3 sol, about 7×10^{-4} M/L of divalent cation is needed for flocculation, yet only 9×10^{-5} M/L are needed of trivalent cation. In short, the CFC value varies inversely with the sixth power of the valence of the ions in solution. The Schulze-Hardy rule can be theoretically derived from the DLVO theory by assuming that the maximum in the potential energy curve occurs at zero [90].

There is also a variety of organic flocculants like low and high molecular substances, highly absorbent and either positively or negatively charged polymers, as well as uncharged or nonionogenic polymers. In contrast to the specific coagulation mechanism occurring with multivalent counterions, the particles are not agglomerated by van der Waals forces. Fig. 13 shows the two steps of this flocculation mechanism. First, the polymer is initially adsorbed by the colloid. One end of the flocculant is bonded to the colloid, while the other end is immersed in the solution. This leads to destabilization of the colloid. In the second step, the other end of the polymer is bonded to a second colloidal particle. The polymers form bridge-like structures between the particles, and large flocs can be formed. These flocs settle down and can be separated mechanically. If necessary, the large flocs depicted in Fig. 13b can be broken up by strong shear forces, which lead to the restabilization of the colloid.

3.4. Membrane Filter Processes

The retention of particles by filtration is a widely used technique in wastewater treatment and is based upon physical interactions between the particles and the granular media. Filtration techniques are available for particle sizes between 1 mm and 0.1 μm. There are different filtration processes, which can be used such as pressure filtration, gravity filtration, single-layer, double-or triple-layer filters, downstream or upstream filtration, and surface filters [99]. All filters have to be controlled regularly with frequent backwashing. Membrane filters are used whenever a pressure gradient or an electrical field forces the solution through a membrane. Table 5 shows areas of use of membrane processes in wastewater treatment [100]. Depending on the pressure range, the following technologies can be applied: first, microfiltration in the 0.5 to 3 bar pressure range, second, ultrafiltration in the 1 to 10 bar pressure range, third, reverse osmosis in the 20 to 100 bar pressure range, and finally, electrodialysis, which uses an electrical field. From a practical point of view, membrane processes offer a variety of advantage such as continuous throughput, relatively low energy need, normal operating temperatures, simple maintenance, and flexible modular design [85]. Different materials are used for the membranes such as glass, polycarbonate, polyester, cellulose acetate, polyamide, polytetrafluoroethylene, or polysulfone. Also different shapes of the membranes are available such as spirally rolled membranes, hollow fibre membranes, tubular modules, disk modules, and others. In electrodialysis, anion or cation exchange resins are built into the polymer membranes to select positively or negatively charged ions. This method is used especially for heavy metal recovery.

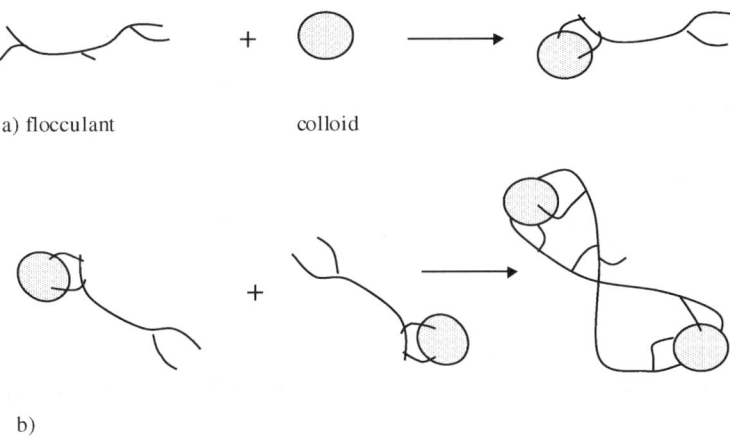

Fig. 13. Flocculation mechanisms occurring with polymer flocculants: a) initial adsorption with optimal flocculant addition and destabilization; b) floc formation by direct reaction (modified after Ref. 85).

3.5. Solidification and Stabilization (A. Xenidis)

Solidification and stabilization (S/S) are treatment processes that reduce the mobility of contaminants by trapping or immobilizing them within the soil through chemical and/or physical means. They were initially applied for the treatment of radioactive wastes and consolidation of industrial sludges; following changes on the legislation, they have been also applied to hazardous wastes and contaminated soils. In Solidification/Stabilization (S/S) treatment, agents are mixed or injected into the contaminated soil to accomplish one or more of the following objectives [101-103].

First, the handling and physical characteristics of a contaminated material has to be improved by the treatment. Second, the exposed surface area of the contaminated material, across which mass transfer or loss of contaminants may occur, must be decreased by the formation of a crystalline, glassy, or polymeric framework, which surrounds the material particles. Third, the contact between transport fluids and contaminants must be limited by reducing the material's permeability. Finally, the solubility of any hazardous constituents contained in the treated material must be reduced by the formation of sorbed species or low solubility precipitates (e.g., hydroxides, carbonates, silicates, phosphates, or sulphides).

3.5.1. Introduction

Solidification is described as the process, by which sufficient quantities of solidifying material are added to encapsulate wastes or contaminated soils in a solid of a high structural integrity. The encapsulation may consist of individual fine contaminated material particles (microencapsulation) or of a large block or container of contaminated material (macroencapsulation). The produced solid material, which is often referred to as "monolith" may present increased strength, decreased compressibility and decreased permeability [104].

Table 5
Areas of use of membrane processes in wastewater treatment

Pollution/Substance	Recommended Membrane
Particle > 0.1 mm	microfiltration, ultrafiltration
< 0.1 mm	ultrafiltration
Solution organic	ultrafiltration, reverse osmosis
inorganic	reverse osmosis
Microorganisms	microfiltration
	ultrafiltration

Modified after Ref. 100.

Solidification does not necessarily involve a chemical interaction between the contaminants and the solidifying reagents, but they may be mechanically locked within the solidified matrix. Therefore, contaminant migration is restricted by decreasing the surface area exposed to leaching and/or isolating the material within an impervious capsule [105].

Stabilization refers to those treatment techniques that employ additives to reduce the hazard potential of a contaminated material by converting the contaminants into their least soluble, mobile, or toxic form. Stabilization must be considered as a treatment process that reduces the movement of contaminants into the environment to an acceptable or geological slow rate [104]. The physical nature and handling characteristics of the treated material are not necessarily changed by stabilization [105, 106].

Many treatment processes are designed to accomplish both solidification and stabilization, although stabilization normally occurs prior to solidification [104]. Hence, the term solidification/stabilization (S/S) is frequently used to describe both techniques. Solidification/stabilization is a proven technology for the treatment of liquids, soils, and sludges contaminated with heavy metals. It can be accomplished primarily through the use of inorganic binders such as cement and fly ash, organic binders such as bitumen, and other additives that aim to convert the metals to a less mobile form or to counteract adverse effects of the contaminants on the solidified/stabilized mixture (e.g., accelerated or retarded setting times, and low physical strength). Biodegradable stabilizers are not recommended because their long-term effectiveness is unproven [101]. The form of the final product from S/S treatment can range from a crumbly, soil-like mixture to a monolithic block [102]. The contaminated soil can be excavated and mixed with the S/S agents aboveground (ex situ S/S), in either continuous feed or batch operations, or it can be mixed in situ by using appropriate equipment.

Another term, which is widely used in the past for either stabilization or solidification or even the combination of the two processes, is chemical fixation. Soils usually do not require to be converted into solid, hardened mass by solidification, although some soils may be processed to suppress dust [105, 107]. For metal contaminated soils, stabilization is also referred to as chemical immobilization or in situ or in place inactivation, and is considered as a very promising low cost remediation alternative.

S/S processes can be broadly classified into two main categories: the chemical and physical processes. Chemical processes require that a chemical reaction takes place between the added reagents and the contaminated soil, whereas physical processes do not involve such a reaction.

For the assessment of stabilization performance, it is necessary to measure the physical, mechanical and chemical properties of the stabilized material. Since S/S techniques primarily aim at the reduction of the rate of contaminants migration into the environment, chemical extraction or leaching tests are the main tool for the evaluation of solidification/stabilization performance. Furthermore, a number of tests mainly adapted from the civil engineering area to evaluate the mechanical properties of the stabilized materials (strength, compressibility, and permeability) are also used.

3.5.2. Solidification/Stabilisation Mechanisms

S/S primarily aims at increasing the resistance of the treated material to leaching and preventing subsequent migration of contaminants to the environment. Therefore, it is essential to understand the pertinent S/S mechanisms in order to achieve the maximum degree of chemical immobilization. There are a variety of basic mechanisms for the immobilisation of metals in contaminated soils [106, 108]. One basic mechanism is pH control, which can be achieved by using alkaline additives such as lime, hydrated lime, limestone, soda ash, magnesium hydroxide, and others. Heavy metals can also be precipitated as low-solubility species, i.e. hydroxides, sulphides, silicates, and phosphates, or can be immobilised by organic complexation by using materials with many oxygen-containing functional groups, particularly –COOH and –OH. Another mechanism is sorption, which is based on the use of natural materials such as clays, peat moss, and sawdust, wastes or synthetic materials such as fly ash, zeolite, activated carbon, and metal oxides. Ion exchange is involved at some degree in almost all stabilization methods. Encapsulation involves the complete coating of fine particles or large blocks of contaminated soil with material such as cement, polymer, asphalt, or similar material.

In many systems it is difficult to clearly distinguish the main mechanism responsible for the immobilisation of heavy metals in soils, since more than one mechanism may take place and the mechanism, which applies for one metal, may be different for another.

pH Control

In general, metal leachability strongly depends on pH. In order to control pH in stabilized soil and prevent dissolution of metals, acidic or alkaline materials may be used. The most common additives for pH control are lime (either CaO or Ca(OH)$_2$), soda ash (sodium carbonate, Na$_2$CO$_3$), sodium hydroxide, and to a lesser extent, magnesium hydroxide. The latter, although rarely used in commercial S/S systems due to the availability of other alkalies at lower costs, it is considered as an interesting reagent where more precise pH control is required, since it presents low solubility in water, and consequently does not immediately raise the pH. Most of the solidification reagents are alkaline and can substitute in part or entirely for traditional alkalies, acting both as pH controls and as binding agents. Alkaline binders include Portland cement, cement and lime kiln dusts, type C fly ash, and sodium silicate [106].

The presence of pH buffering materials in stabilized soil is desirable to maintain pH at target values for long periods, and thus to promote long term stability. Buffers provide resistance to rapid pH changes upon exposure of the contaminated soil to acid or base. Limestone may be used to buffer soil acidity [109].

Precipitation

Precipitation as low solubility and low availability compounds is considered by far the most important fixation mechanism for metals in contaminated soils. Dissolved metals may be re-precipitated as hydroxides, sulphides, carbonates, silicates, and phosphates and remain in the stabilized soil. Furthermore, metals present in soils in water leachable forms may be dissolved and re-precipitated as low solubility compounds. Usually, a combination of mechanisms is active and the products of treatment are frequently not simple compounds [106]. In practice, sufficient quantities of additives that are as specific as possible to the metal to be immobilized are used. These additives may remain in the system until used up in the desired reactions, unless the stabilization reactions are very rapid [107]. It is well known that the solubility of the most common metal hydroxides is a function of pH. The precipitation reaction of metal hydroxides can be represented by the equilibrium [110]:

$$Me(OH)_{x\,(s)} + zH^+ = Me^{z+} + zH_2O \qquad (15)$$

from which a convenient graphical representation of $\log[Me^{z+}]$ vs. pH can be derived:

$$\log[Me^{z+}] = \log K_s - z\,pH \qquad (16)$$

where K_s is the equilibrium constant for the reaction (15). This expression is plotted in Fig. 14 for a number of metal ions. In order to evaluate the complete solubility of oxides and hydroxides, the hydroxo–metal complexes in equilibrium with the solid must be taken into consideration [110]. In this case, the total metal ion concentration in equilibrium with the solid oxide or hydroxide is expressed as:

$$Me_T(aq) = [Me^{z+}] + \sum_{1}^{n} [Me(OH)_i^{z-i}] \tag{17}$$

For instance, the existence of zinc hydroxyl species will modify the solubility of ZnO. Fig. 14 shows the modified solubilities of Fe^{3+}, Zn^{2+} and Cu^{2+}, taking into consideration the hydroxo complexes. As illustrated in Fig. 15, solid oxides and hydroxides have amphoteric characteristics; they can react with protons or hydroxyl ions. There is a pH value, for which a particular metal presents minimum solubility. It is well known that most metals present minimum solubility within the approximate pH range of 7.5 to 11 (Fig. 16). In more alkaline or acidic pH regions, the solubility increases.

If the contaminated soil under consideration for S/S contains a number of different metals, it is possible that their solubility minima may not entirely overlap. In cases that pH values at these solubility minima are not too different, it may be sufficient to choose an average pH to precipitate metals. However, in many cases, minimum solubility is not required since metal leachability values are below target levels. In cases that the solubility of metals are not low enough for effective stabilization, even at optimum pH values or the optimum pH values are very different, other forms (e.g. sulphides, phosphates etc) should be selected for metals precipitation [109]. It must be borne in mind that theoretical metal solubility values (given in Fig. 16), may differ significantly from actual ones in real systems, since the former were calculated using stability constants of species at low total ionic strength, and without taking into account various effects such as common ion, complexation, and redox potential. Furthermore, coprecipitation and surface adsorption may lead to lower than the predicted residual concentrations.

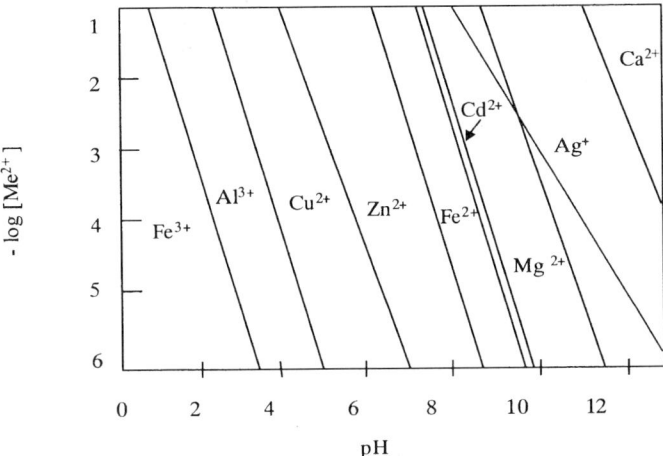

Fig. 14. Free metal ion concentrations in equilibrium with solid oxides and hydroxides. The occurrence of hydroxo–metal complexes must be considered for evaluation of the complete solubility.

Stabilization of metals in contaminated soils by precipitation as sulphides is an alternative to hydroxide precipitation. The solubilities of sulphides are several orders of magnitude lower than those of the corresponding hydroxides (Fig. 16). Therefore, sulphide precipitation has the potential to reduce metals leachability to extremely low levels. However, metal sulphides are prone to oxidation and resolubilisation under the action of water and oxygen. There are three categories of sulphide reagents that have been investigated and used in S/S systems. First, there are the soluble inorganic sulphides, second, there are the "insoluble" inorganic sulphides and third, there are the organosulphur compounds [108]. Soluble sulphides include sodium sulphide (Na_2S), hydrogen sulphide (H_2S), sodium hydrosulphide (NaHS), and calcium sulphide (CaS). By using soluble sulphides, it is necessary to maintain pH equal or higher than 8 to prevent formation and evolution of H_2S. The soluble sulphides are usually added as solutions before the addition of solidification reagents to avoid consumption of sulphide species by calcium, magnesium, iron, and other metals that are contained in these reagents. Excess sulphide ions are necessary to be added for metals precipitation; however, this excess should be kept to a minimum to avoid treatment of the final free sulphides.

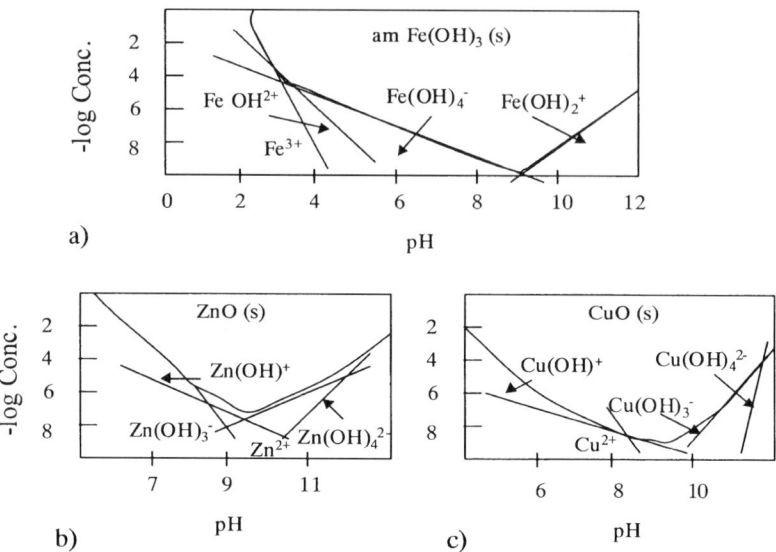

Fig. 15. Total metal ion concentrations in equilibrium with Fe(OH)₃ (a), ZnO (b) and CuO (c).

Hydrogen sulphide is not usually used for precipitation in S/S systems, since it is very dangerous to handle [106]. Low solubility or "insoluble" inorganic sulphides used in S/S systems involve ferrous sulphide (FeS) and elemental sulphur (S). Ferrous sulphide, which has low solubility in water, is dissolved at very low rates and provides the S^{-2} ions that are used for metal precipitation. Thus, there is a continuous dissolution of FeS until it has been consumed, or all the toxic metals have been precipitated. The advantage of slow dissolution rate is that very little, if any excess sulphide is present in the system at any time, so the problem of H_2S evolution is eliminated. If metals in the soil are present in the form of hydroxides and sufficient time is available, respeciation of metal hydroxides to more stable metal sulphides is possible. There are a number of organosulphur compounds proposed for metals stabilisation including dithiocarbamates ([R-NH-CS-S]⁻), thiourea (H_2N-CS-NH_2), thioamides (R-CS-NH_2) and xanthates ([RO-CS-S]⁻) [108]. However, all of the applications using organosulphur compounds are aimed at mercury stabilisation. This is due to the extremely low regulatory leaching levels for mercury and/or the very low solubility of mercury sulphides [106].

It is well known that phosphorus reacts with many heavy metals, metalloids, and radionuclides to form secondary phosphate precipitates that are considered as stable over a wide range of environmental conditions. As seen in Table 6, Pb phosphates present low solubility, generally several orders of magnitude lower than the analogous oxides, hydroxides, carbonates, and sulphates. Therefore, the application of phosphate amendments has been identified as a potentially efficient in situ chemical immobilization technique for heavy metal contaminated soils. Many natural or industrial materials including natural or synthetic apatites, phosphate rocks, phosphoric acid, or phosphoric salts (mainly of calcium and sodium) were used as sources of phosphates for immobilizing heavy metals in soils. Among all inorganic P sources, apatites are the most economical to be used for the stabilization of Pb contaminated soils because of their availability and low cost. The reaction of metal species in solution and the soluble silicates result in the production of "insoluble" precipitates that are usually noncrystalline and therefore difficult to be characterized structurally. They are most often described as hydrated metal ions associated with silica or silica gel. Metal silicates are nonstoichiometric compounds, in which the metal is coordinated to a silanol group, $\equiv$SiOH, in an amorphous silica matrix [106]. The precipitation of metal silicates strongly depends on pH. Since metals can be also precipitated as metal hydroxides, pH should be kept lower than the value required for hydroxide precipitation. The order of metals precipitation as metal silicates is copper, zinc, manganese, cadmium, lead, nickel, silver, magnesium, and calcium [106].

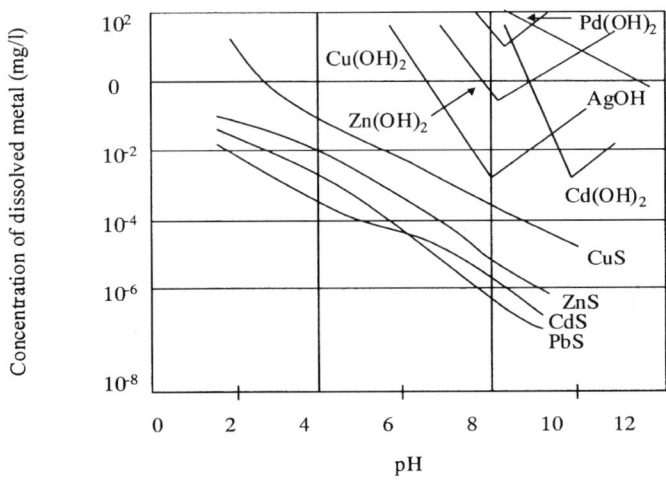

Fig. 16. Solubility of heavy metal hydroxides and sulphides.

Encapsulation

Encapsulation is primarily based on solidification and aims at preventing the contact of contaminated soil with potential leaching agents such as water. It involves the complete coating or jacketing of a contaminated material with binders such as cement, polymer, asphalt, or similar material. As it is already stated, there is a distinction between the microencapsulation that refers to the encapsulation of individual particles, and the macroencapsulation that is the encapsulation of a large block or container of contaminated soil or microencapsulated material.

Microencapsulation is the mechanism, in which contaminated materials are physically entrapped within the crystalline structure of the solidified matrix at a microscopic level. The microencapsulation process involves mixing the contaminated soil with the encasing agent before solidification occurs. Since the contaminants are physically entrapped into the solidified structure and are not chemically altered or bound, the rate of contaminants release from the stabilized mass may increase by increasing the exposed surface area. However, opposite to macroencapsulation, even if the treated material breaks down, the contaminants still remain entrapped [104]. Portland cement, pozzolans, or lime/hydrated lime and organic polymers may be used for microencapsulation of contaminants in contaminated soils.

Table 6
Solubility products (K_{sp}) of selected Pb compounds.

Name	Reaction		$\log K_{sp}$
Anglesite	$PbSO_4$	$= Pb^{2+} + SO_4^{2-}$	-7.79^a
Galena	PbS	$= Pb^{2+} + S^{2-}$	-27.51^b
Litharge	PbO	$= Pb^{2+} + H_2O - 2H^+$	12.89^a
Cerrusite	$PbCO_3$	$= Pb^{2+} + CO_3^{2-}$	-13.1^b
Lead oxyphosphate	$PbHPO_4$	$= Pb^{2+} + HPO_4^{2-}$	-11.45^a
Lead phosphate	$Pb_3(PO_4)_2$	$= 3Pb^{2+} + 2PO_4^{3-}$	-44.36^a
Hydroxypyromorphite	$Pb_5(PO_4)_3OH$	$= 5Pb^{2+} + 3PO_4^{3-} + OH^-$	-76.79^a
Chloropyromorphite	$Pb_5(PO_4)_3Cl$	$= 5Pb^{2+} + 3PO_4^{3-} + Cl^-$	-83.70^a
Fluoropyromorphite	$Pb_5(PO_4)_3F$	$= 5Pb^{2+} + 3PO_4^{3-} + F^-$	-71.63^a
Bromopyromorphite	$Pb_5(PO_4)_3Br$	$= 5Pb^{2+} + 3PO_4^{3-} + Br^-$	-78.14^a
Corkite	$PbFe_3(PO_4)(SO_4)(OH)_6$	$= Pb^{2+} + 3Fe^{3+} + PO_4^{3-} + SO_4^{2-} + 6OH^-$	-112.6^c
Hinsdalite	$PbAl_3(PO_4)(SO_4)(OH)_6$	$= Pb^{2+} + 3Al^{3+} + PO_4^{3-} + SO_4^{2-} + 6OH^-$	-99.1^c
Plumbogummite	$PbAl_3(PO_4)_2(OH)_5 \cdot H_2O$	$= Pb^{2+} + 3Al^{3+} + 2PO_4^{3-} + 5OH^-$	-99.3^c

Modified after: [a]Ref. 111, [b]Ref.110, [c] Ref. 112

Organic polymers have one major advantage over most of the inorganic systems: once cured, they have lower permeability and tend to remain in monolithic form due to the strength and elasticity properties of the polymer [107]. Like thermoplastic systems, most of these products are hydrophobic after curing and tend to resist leaching even if crushed to a small particle size. Thus, they are good at retaining highly toxic metals and organic compounds. Another characteristic of organic polymer systems is that very quick gelation and development of physical strength can be obtained if desired [107]. Macroencapsulation is the mechanism, which allows the physical entrapment of hazardous material constituents in a larger structural matrix; the hazardous material is held in discontinuous pores within the stabilizing material. A number of agents were proposed and applied for encapsulation of contaminated materials including concrete, organic materials (polyethylene, polyester etc), sulphur cement, and others. The macroencapsulation process involves pouring the encasing agent over and around a large mass of contaminated material, thereby enclosing it in a solidified block. The encapsulated materials usually exhibit high retention of contaminants under leaching by a broad spectrum of aqueous solutions, and they can withstand considerable mechanical stress. However, several kinds of physical or environmental stress, such as wet/dry or freeze/thaw cycles, high mechanical stresses, or even contact with reactive fluids, can result in physical degradation (breakdown) of the solidified material [104]. Therefore, the contaminated material may be exposed to leaching and the contaminants can be mobilized.

3.5.3. Evaluation of S/S processes

As it has been already stated, the primary objective of soil treatment by S/S is to decrease the mobility of contaminants. Depending on the specific process applied for S/S, several types of tests can be performed for the determination of S/S efficiency. They include physical-geotechnical tests to determine properties like particle size distribution, moisture, hydraulic conductivity, etc., and leaching tests that provide information about the leachability and bioavailability of contaminants under certain environmental conditions.

Physical-geotechnical tests

A number of physical-geotechnical tests, which are mainly adapted from the civil engineering area, are conducted to evaluate the performance of a S/S process. These tests can be applied to both the contaminated soil in order to provide basic information on the material's treatability and the stabilized soil. The most common physical-geotechnical tests used for the evaluation of S/S processes are index property tests, density tests, compaction tests, permeability (hydraulic conductivity) tests, strength tests, and durability tests [101]. The index property tests provide information about the general physical characteristics of the soil (particle size distribution, Atterberg limits, moisture content). These tests are most frequently performed on the untreated soil to determine the feasibility of various S/S processes. Density tests (bulk density) determine the in-place density of soils. These data are required to determine weight to volume relationships of materials and to estimate the required reagents quantity. Compaction tests determine the relationship between the moisture content and density of soil. Permeability (hydraulic conductivity) tests are used to estimate the quantity and flow rates of water through treated and non-treated soil under saturated conditions. Strength tests (unconfined compressive strength, flexural strength, etc.) provide a means for judging the performance of a S/S process under mechanical stresses. Durability tests evaluate the resistance of a S/S treated soil to degradation due to external environmental stresses such as repeated freezing and thawing, or wetting and drying cycles.

Leaching tests

When contaminated soil comes into contact with ground or surface water, some constituents will be dissolved at a greater or lesser extent. The degree and rate of dissolution of individual elements from contaminated soil depend on a number of physical, chemical and biological factors. The process of element dissolution is called leaching, the fluid, to which the contaminants are leached (usually water or aquatic solution) is called leachant, the contaminated water that has passed through or around the solid material is termed leachate, and the capacity of any solid material to leach is called leachability. The main physical factors that influence leaching include the contact time, temperature during leaching, particle size, porosity, and permeability of the solid material. These parameters greatly affect the surface exposed to leaching. The main chemical factors affecting leaching are pH, oxidation-reduction potential, presence of inorganic or organic complexing agents in the leachant, and form of contaminants in the solid material [113]. Leaching tests are often employed to simulate field leaching scenarios or to assess specific properties of a material, such as the contaminants mobility under certain environmental conditions, the release of constituents that are relevant for the fertility of soil (nutrient availability), etc. Numerous leaching procedures have been developed worldwide addressing various material types as well as their management scenarios and exposure conditions. There are a number of test variables that may affect leachability of contaminants and are taken into consideration, when a leaching test is designed and implemented.

The most important variables are type of leachant, pH of the leachant, Liquid to Solid ratio (L/S), particle size of material, contact time, temperature, extent of agitation, and number of contacts with fresh solution. Several solutions are used in standard or modified leaching tests, including deionized water, acetic acid, nitric acid, simulated acid rain, etc. In many leaching tests, pH is not controlled and therefore the final pH value is dictated by the characteristics of material subjected to leaching. In other tests, a buffered solution is used to keep pH at around a certain value, or pH is adjusted using acids and/or bases. The volume of leachant (L), which has been in contact with a certain amount of material tested, is related to the dry mass (S) of material prior to leaching. The most common L/S ratios are 10 or 20 l/kg. If the material is fine then it presents higher surface areas exposed to leaching and consequently higher contaminant dissolution rates will occur. When element dissolution is kinetically controlled, the contact time significantly affects the leachability results. Leaching is generally conducted at room temperature.

The leaching tests used in a given situation depend upon the nature of the questions to be answered. The major aspects of leaching to be considered in this context are potential leaching, actual or expected leaching as a function of time or liquid to solid ratio, and the influence of various factors on the leaching behaviour. Moreover, consideration has to be given to whether the leaching is controlled primarily by equilibrium or steady - state like conditions, or by diffusion from a solid body into the aqueous phase.

Leaching tests are particularly useful when S/S is selected as a remediation scheme for contaminated soil. Apart from their use to characterize the soil, they can also be applied to compare the effectiveness of one stabilization mixture or process with another, or to select the proper application rate of stabilizing agents. Leaching test methods can be divided into three general categories. First, there are single extraction tests, in which leaching takes place with a single specified volume of leaching fluid. Second, there are dynamic extraction tests, in which the leaching fluid is renewed throughout the test, and third, there are specific tests focused on chemical speciation or bioavailability issues. A summary of single and dynamic extraction test procedures used in Europe and North America is given in Table 7. Single extraction tests involve the equilibration of solid material under examination (that may be crushed up to a desired grain size) with leaching solution at fixed liquid to solid ratio, temperature, and contact time. Then, the slurry is filtered and the leachate quality (pH, heavy metals concentration, etc.) is determined.

Examples of single extraction tests used for regulatory purposes are the US EPA TCLP, the German standard DIN 38414-S4, and the European Standards EN 12457 1-4 developed for the evaluation of leaching properties of granular wastes and sludges. The US EPA Toxicity Characteristics Leaching Procedure (TCLP) is a regulatory test that was adopted by the US EPA as a replacement for the EP toxicity test to classify materials as hazardous or non-hazardous. This test was initially developed for the characterization of industrial wastes, based on the scenario of co-disposal with municipal wastes producing organic acids during decomposition. The TCLP is also used to evaluate the effectiveness of stabilization.

Table 7
Single extraction and dynamic tests

Standard Test	Application	Leaching Medium
Single Extraction Tests		
U.S EPA Ep Tox	Classification in terms of toxicity	Acetic acid 0.04 M, pH 5.0
U.S EPA TCLP	Classification in terms of toxicity	Acetic acid pH 2.88 or pH 4.93
U.S EPA SPLP	Assess impact of materials	Synthetic acid rain
German DIN 38414 S4	Sludges and sediments	Deionised water
European EN 12457 1-4	Granular material and sludges	Deionised water
Dynamic Tests		
US EPA MEP Serial batch test	Granular material	a) acetic acid b) synthetic acid rain
Dutch NEN 7341	Dutch waste management maximum leachability	Deionised water a) pH 7.0 and b) pH 4.0
Dutch NEN 7343 Column test	Simulate leaching in the short and medium term (<50 years)	Deionised water with HNO_3 at pH 4.0
Dutch NEN 7349 Serial batch test	Long term leaching behaviour of wastes	Deionised water with HNO_3 at pH 4.0
US EPA MWEP	Granular materials/ monoliths	Deionised water
Dutch NEN 7345	Tank leaching test for monoliths and stabilised wastes	Deionised water
ANS-16.1	Low level/ hazardous wastes	Deionised water

Table 7 (continued)
Single extraction and dynamic tests

Standard Test	Max. Particle Size	L/S Ratio (l/kg)	No. of extractions	Test duration
Single Extraction Tests				
U.S EPA Ep Tox	9.5 mm	16:1	1	24 h
U.S EPA TCLP	9.5 mm	20:1	1	18 h
U.S EPA SPLP	9.5 mm	20:1	1	18 h
German DIN 38414 S4	10 mm	10:1	1 or more	24 h
European EN 12457 1-4	90% > 4 mm	2:1 up to 10:1	1	24 h
Dynamic Tests				
US EPA MEP	9.5 mm	16:1	1	24 h
Serial batch test		20:1	≥ 9	24 h
Dutch NEN 7341	125 µm	50:1	2	3 h per extraction
Dutch NEN 7343 Column test	4 mm	0.1:1 up to 10:1	7	21 days
Dutch NEN 7349 Serial batch test	4 mm	20:1 up to 100:1	5	23 h per extraction
US EPA MWEP	9.5 mm or monolith	10:1	4	18 h per extraction
Dutch NEN 7345	0.1 m x 0.1 m x 0.1 m > 4 mm	5:1	8	6 h to 64 days
ANS-16.1	Monoliths	Vol:Surf. Area: 10 cm	11	2 h to 90 days

Modified after Ref. 101 and Ref. 113.

In general, this test is widely used and recognized by many environmental authorities in North America and Europe for all the categories of waste materials and for stabilised materials [114]. Since the test simulates leaching in sanitary landfills, it involves leaching of a sample with either an acetic acid solution (pH = 2.88) or a buffer solution of acetic acid / NaOH (pH = 5) for 18 hours at L/S equal to 20 l/kg. The choice between the two leachants depends on the alkalinity of the material: very alkaline materials are leached with the first solution, whereas other materials are leached with the second. In general, the resulting pH in the leachate will be approximately 5, but for strongly alkaline materials the pH may be anywhere between 5 and 12, which may result in variable test results for components with highly pH dependent solubilities [113]. If the dissolved metals exceed the specified limits (Table 8), the spoil is characterized as hazardous [114].

These limits were based on the potable water limits using a dilution factor of 100 to account for attenuation, dilution, and degradation between the point of leachate generation and the point of water use. The validity of the TCLP for the evaluation of stabilization effectiveness has often been criticized for several reasons. First, the test protocol requires particle size reduction of the solidified monolithic mass to pass the 9.5 mm sieve. As a consequence, the beneficial effects of macroencapsulation and microencapsulation are reduced. In addition, the application of low pH values in the TCLP extraction test may overestimate the potential leachability of contaminants, since in actual field conditions the pH values are usually higher. However, in cases that stabilization is performed using highly alkaline material such as fly ash or cement, pH in the TCLP leaching solution increases rapidly, therefore, extraction is performed under basic rather than acidic conditions. Despite these criticisms, TCLP is particularly useful in comparing the performance of alternative treatment technologies and stabilization mixtures [104].

In dynamic extraction protocols, the leaching fluid is renewed, either continuously or intermittently, to drive the leaching process. Because the physical integrity of the material is usually maintained during the test, and the information is generated as a function of time, dynamic extraction tests provide information about the kinetics of contaminant mobilization. In general, dynamic extraction tests can be categorized as serial batch tests, flow-around tests, flow-through tests, and Soxhlet tests. In a serial batch test, a portion of a crushed, granular sample is mixed with leachant and agitated for a specified time period. At the end of the time period, the leachate is separated, fresh leachant is added, and the process is repeated until the desired number of leaching periods has been completed.

Table 8
Toxicity limits specified for TCLP leaching tests (modified after Ref. 114)

Element	TCLP (mg/l)
As	5
Cd	1
Pb	5
Cr	5
Se	1
Ag	5
Ba	100
Hg	0.2

The concentrations of contaminants measured in the serial leachates can provide kinetic information about contaminant dissolution. Examples include the Multiple Extraction Procedure (US EPA Method 1320), the Availability Test (NEN 7341), and Serial Batch Test (NEN 7349) from the Netherlands.

Flow-around tests use either monolithic samples or samples that are somehow contained. The sample is placed in a test vessel, with space around the sample, and leachant is added so that it flows around the sample. The leachant may be renewed continuously and sampled periodically, or it may be replaced intermittently. In either case, the liquid to solid ratio is expressed as the ratio of volume of leachant to surface area of sample. Examples of flow-around tests include the ANSI 16-1 and the Monolithic Diffusion test (NEN 7345) from Netherlands. Flow-through tests differ from flow-around tests in that the leachant flows through the sample rather than around it, conditions simulating the disposal of wastes or contaminated soils. Flow-through tests are usually conducted in columns or lysimeters, and can be set up to mimic site-specific conditions. These tests, however, pose particular experimental challenges such as channelling, flow variations caused by the hydraulic conductivity of the material, clogging of the system by fine particulates, and biological growth in the system. An example of a flow-through test is the Dutch standard column test (NEN 7343).

There are some special extractions that are used to investigate specific aspects of soil leaching such as the partitioning of heavy metals into different chemical forms, the phytoavailability of elements in soil, or bioavailability of contaminants. Sequential chemical extraction procedures (SEP) have been used for the chemical speciation of metals in soils and sediments providing useful information about the mode of occurrence and mobilization of elements under various environmental conditions. Several sequential extraction procedures were developed usually involving 5 or 7 extraction steps. The most known and widely applied procedure involves the application of five sequential leaching stages, using $MgCl_2$, CH_3COOH/Na, $NH_2OH \cdot HCl$, $HNO_3 + H_2O_2$ and $HF + HClO_4$ solutions [115]. With this method, the total concentration of metals is separated into the following five operationally defined fractions: exchangeable, carbonate, bound to iron and manganese oxides or reducible, bound to organic matter or oxidizable, and residual. The exchangeable and carbonate fractions represent the mobile metal species, which are easily released in neutral (exchangeable) or slightly acidic (carbonate) waters, and are considered to provide indirect information about the bioavailable portion of metals in soils. The reducible and oxidizable fractions describe the potential mobilization under reducing or oxidizing conditions. Finally, the residual fraction represents the inert forms of metals, which are not expected to be released under the conditions normally encountered in nature.

Several reagents have been used to determine the phytoavailable fraction of heavy metals in soils including acids, chelating agents and salts of ammonium, calcium, barium, etc. However, EDTA (ethylenediaminetetraacetic acid) and DTPA (diethylenetriamine pentaacetic acid) are the most generally acceptable extractants, in relation to the extractable metal content and plant available forms [113]. In general, EDTA is preferred as it extracts greater amounts of metal and is simple to prepare and use. DTPA is applied only to evaluate the available metal fractions from calcareous soils [116]. There are various EDTA leaching schemes in terms of EDTA concentration (0.01-0.05 M), leaching pH, and the presence of other salts such as CH_3COONH_4. The fraction of metals reported to the EDTA leachate is characterised as phytoavailable. When assessing risks associated with heavy metals in contaminated soil, one exposure pathway typically evaluated is soil ingestion by children. Furthermore, it has been indicated that only a fraction of heavy metals in ingested soil, which is called bioavailable or bioaccessible fraction, reaches the central (blood) compartment from the gastrointestinal tract [117, 118]. Animal dosing studies (in vivo studies) used for measuring bioavailability of metals in soil have limited value since they are expensive, complex, time consuming, and may create ethical concerns. The Physiologically-Based Extraction Test (PBET) is a quick chemical extraction that is being developed to serve as an alternative to animal studies. It has been extensively used to determine the bioavailability of Pb, As, and other contaminants in both untreated and stabilized soils [119]. The PBET is an in vitro test simulates conditions in the gastrointestinal tract to assess the human bioaccessibility of potentially harmful elements by ingestion [117].

3.5.4. Technology Description

There is a number of S/S processes developed for the treatment of wastes and soils. Of these, the following three technologies applicable to contaminated soils are described: cement and pozzolan-based technologies, polymer microencapsulation, and phosphate-based technologies. Among these techniques, emphasis was given to phosphate-based technologies applied on metal contaminated soils, which are considered as a very promising low cost remediation alternative.

Cement and pozzolan-based technologies

Out of all the inorganic binders, cement, which can be used either alone or combined with other materials such as soluble silicates and fly ash, is the most frequently used material for solidification/stabilisation of contaminated soils. The major objectives of S/S treatment are to reduce the mobility of contaminants, minimize free liquids, and, occasionally, increase the strength of the material. Cement-based processes accomplish these objectives by forming a granular or monolithic solid that incorporates the contaminated soil and immobilizes contaminants.

Portland cement is the most common type of cement with annual world production over 1.5 billion tons. The main chemical components of Portland cement are calcium, silica, alumina, and iron. Calcium is derived from limestone, whereas sands, clays, and iron ore are used as sources of silica, alumina, and iron, respectively. These raw materials are finely ground and mixed, then fed into a rotary cement kiln and heated progressively up to about 1400 to 1600°C. The first important reaction that occurs is the calcining of limestone (calcium carbonate) into lime (calcium oxide) and carbon dioxide, which occurs in the lower-temperature portions of the kiln, up to about 900°C. The second reaction is the bonding of calcium oxide and silicates to form dicalcium ($2CaO \cdot SiO_2$) and tricalcium ($3CaO \cdot SiO_2$) silicates. Small amounts of tricalcium aluminate ($3CaO \cdot Al_2O_3$) and tetracalcium aluminoferrite ($4CaO \cdot Al_2O_3 \cdot Fe_2O_3$) are also formed. The relative proportions of these four principal compounds determine the key properties of the Portland cement and the type classification (Type I, Type II, etc.). Under the presence of water, calcium silicates and aluminates are able to react together. The heated substance, which is called clinker, usually occurs in the form of small grey-black pellets about 12.5 mm in diameter. The clinker is then cooled, pulverized into a fine powder that almost completely passes through a 0.075 mm sieve, and fortified with a small amount of gypsum ($CaSO_4 \cdot 2H_2O$) to give the final product known as Portland cement. The most common type of cement kiln today (accounting for 70% of plants in the U.S.) is a dry process kiln, in which the ingredients are mixed dryly. Many older kilns use the wet process.

Tricalcium and dicalcium silicate (also notated as C_3S and C_2S, respectively) are the main constituents amounting to 40-60 % and 20-30 % respectively of the total Portland cement weight. Tricalcium aluminate (C_3A) and tetracalcium aluminoferrite (C_4AF) content in Portland cement is relatively lower (around 12 % and 8 %, respectively). By mixing Portland cement with water, reactions take place resulting in the chemical alterations of cement constituents that cause it to harden (or set). All these reactions involve the addition of water to the basic chemical compounds of cement; this chemical reaction with water is called "hydration". Hydration reactions involve hydrolysis of the calcium silicates followed by dissolution of calcium hydroxide (lime) into solution, and reaction of calcium hydroxide with the other oxides leading to the formation of hydrated calcium aluminosilicates. In the simplest form, the reaction of dicalcium and tricalcium silicate can be described by the following equations [104]:

$$2(3CaO \cdot SiO_2) + 6H_2O \rightarrow 3CaO \cdot 2SiO_2 \cdot 3H_2O + 3Ca(OH)_2 \qquad (18)$$

$$2(2CaO \cdot SiO_2) + 4H_2O \rightarrow 3CaO \cdot 2SiO_2 \cdot 3H_2O + 4Ca(OH)_2 \qquad (19)$$

Each one of these reactions occurs at a different time and rate. Tricalcium silicate reacts and hardens rapidly and is largely responsible for initial set and early strength of cement, whereas dicalcium silicate hydrates and hardens slowly and is responsible for strength increases beyond one week. The hydration reaction of tricalcium aluminate is very rapid and highly exothermic, liberating a large amount of heat according to the following equation:

$$3CaO \cdot Al_2O_3 + 6H_2O \rightarrow 3CaO \cdot Al_2O_3 \cdot 6H_2O + heat \qquad (20)$$

Sudden and rapid crystallization takes place until the alumina is consumed. The setting rate is controlled by the amount of gypsum added to the cement. If sufficient gypsum is present, sulphates combine with tricalcium aluminate to form calcium aluminate sulphate, which coats the cement particles and retards hydration reactions. Tetracalcium aluminoferrite also hydrates rapidly but it contributes very little to strength. However, the presence of tetracalcium aluminate allows lower kiln temperatures in Portland cement manufacturing.

Concrete is produced by mixing cement with fine aggregate (sand), coarse aggregate (gravel or crushed stone), water, and often small amounts of various chemicals called additives that control such properties as setting time and plasticity. The properties of concrete are determined by the type of cement used, the additives, and the overall proportions of cement, aggregate, and water. Pozzolan is a finely-divided material that reacts with calcium hydroxide and alkalies to form compounds possessing cementitious properties. A more formal definition is given by ASTM C618 as "a siliceous or siliceous and aluminous material which, in itself, possesses little or no cementitious value but which will, in finely divided form in the presence of moisture, react chemically with calcium hydroxide at ordinary temperature to form compounds possessing cementitious properties". The reaction of aluminosilicious material, calcium hydroxide, and water results in the formation of a concrete-like product termed pozzolanic concrete.

Pozzolans are present on earth's surface such as diatomaceous earth, volcanic ash, opaline shale, pumicite, and tuff. Artificial pozzolanic materials include fly ash, ground blast furnace slag, and cement kiln dust. Fly ash is an artificial pozzolan produced when pulverized coal and lignite is burned in electric power-generating plants. Fly ash is composed of predominantly silt-sized, spherical, amorphous ferroaluminosilicate minerals [120] and is generally characterised as having low permeability, low bulk density, and high specific surface area [121]. The physical, chemical, and mineralogical characteristics of fly ash depend on the parent coal source, the method of combustion, and the efficiency and type of emission control device. Two major classes of fly ash are specified in ASTM C618 on the basis of their chemical composition resulting from the type of coal burned. These are designated Class F and Class C. Class F is fly ash normally produced from the burning of anthracite or bituminous coal ($SiO_2 + Al_2O_3 + Fe_2O_3 \geq 70\%$). Class C is normally produced from the burning of subbituminous coal and lignite ($SiO_2 + Al_2O_3 + Fe_2O_3 \geq 50\%$). Class C fly ash has cementitious properties in addition to pozzolanic properties due to free lime, whereas Class F is rarely cementitious when mixed with water alone.

Both ex-situ and in-situ S/S methods can be applied for the treatment of contaminated soils using the cement-based process. This process involves mixing of contaminated soil with Portland cement or pozzolanic materials, or blends of both, water, and other additives. The composition of the cement and pozzolan, as well as the amount of water, aggregate, and other additives, determines the set time, cure time, pour characteristics, and material properties (e.g., pore size, compressive strength) of the stabilized soil.

Cement-based S/S reduces the mobility of inorganic compounds by several mechanisms including formation of insoluble hydroxides, carbonates, or silicates, physical encapsulation, substitution of the metal into a mineral structure, and sorption. It involves a complex series of reactions, and there are many potential interferences that can prevent attainment of S/S treatment objectives for physical strength and leachability. Soils with free calcium hydroxide can contribute to the strength-forming reactions, but excess hydroxide will increase pH and the solubility of amphoteric metals. Salts of some metals such as Mn, Sn, Zn, Cu, and Pb can increase set time and reduce strength. Fine particulates such as silt, clay, or coal dust can coat cement particles and prevent the growth of calcium-silicate-hydrate crystals from the cement grain. To address these adverse effects, several approaches using appropriate additives or other changes in formulation can be identified and confirmed via treatability studies [102].

Fly ash provides many desirable properties to concretes and significantly reduces the total S/S cost because it can replace 25 to 35% of the Portland cement normally used. The greatest disadvantage of using fly ash is the volume increase associated with large additions of fly ash that replace some of the cement otherwise used. Fly ash combined with lime can also be used for S/S, since fly ash mixed with lime and water presents cementitious properties. The reaction product formed is initially a non-crystalline gel, but eventually turns to calcium silicate hydrate, a compound found in hydrated Portland cements. However, these reactions are slower than those of cement and do not produce exactly the same products in terms of chemical and physical properties [107].

If a single metal is the predominant contaminant in soil, then Cd and Pb are the most amenable to cement-based S/S. These metals form insoluble precipitates in the pH ranges found in cured cement. However, if pH is not carefully controlled, the metals may resolubilize. For example, Pb in aqueous solutions may be resolubilized at pH around 10 and above, since lead hydroxyl anions are predominant at these pH values resulting in increased Pb solubility. Several problems exist for the implementation of cement-based processes on soils contaminated with Hg, As, and Cr(VI). Mercury and Arsenic do not form low solubility precipitates in the high pH cement environment and therefore, cement-based solidification is generally not expected to be successful for these elements. On the other hand, although Cr(VI) forms anions that are soluble at high pH and therefore it is difficult to be stabilized in cement, it can be reduced to Cr(III) that does form insoluble hydroxides. Although Hg and As are particularly difficult candidates for cement-based S/S, this should not necessarily eliminate S/S (even cement-based) from consideration for several reasons. First, as it is the case with Cr(VI), it may be possible to devise a multistep process that will produce an acceptable product for cement-based S/S. Second, a non-cement based S/S process (e.g., lime and sulfide for Hg, oxidation to As(V), and coprecipitation with iron) may be applicable. Third, the leachable concentration of the contaminant may be sufficiently low so that a highly efficient S/S process may not be required to meet treatment goals [102].

Polymer microencapsulation

S/S by polymer microencapsulation can include application of thermoplastic materials or thermosetting resins. Thermoplastic materials are the most commonly used in organic-based S/S treatment techniques. Soils can be stabilized by blending molten thermoplastic materials including asphalt, paraffin, bitumen, polyethylene, propylene, and sulphur cement [102]. Of these, asphalt is the least expensive and by far the most commonly used. Polyethylene, which is widely applied for S/S, is very resistant to most chemicals, even very aggressive ones and also to many acids and alkalies.

The process of thermoplastic encapsulation involves heating and mixing the contaminated material and the binders at elevated temperature, typically 130°C to 230°C, in an extrusion machine. The extruder consists of a heated cylinder containing one or two mixing and transport screws where the mixture is homogenized as the polymer melts [102]. Each individual soil particle is embedded within the polymer block and is surrounded by a durable, leach-resistant coating. Any water or volatile organics in the soil boil off during extrusion and are collected for treatment or disposal. The heated, homogenous mixture exits the cylinder through an output die into a mould, where it cools and solidifies.

Unlike conventional hydraulic cement binders, thermoplastic polymers such as polyethylene do not require chemical reaction for solidification. Polymer microencapsulation has been mainly applied for the treatment of low-level radioactive wastes. However, organic binders have been tested or applied to materials containing chemical contaminants such as As, metals, inorganic salts, polychlorinated biphenyls, and dioxins [102]. Polymer microencapsulation is especially effective with high concentrations of soluble salts such as nitrates, chlorides, or sulphates that are not effectively stabilized with chemical S/S processes.

Phosphate-based techniques

Unlike most S/S processes, the addition of phosphates in contaminated soil either in dry form or as a solution of phosphate salts does not convert soils into a solid, hardened mass. For this reason the treated material is different from those in most S/S processes as it still behaves as a normal soil. Furthermore, it has been indicated that phosphates exhibit very low solubility over a wide range of pH values. In some cases, cementitious materials can be also added to improve the physical characteristics of the treated soil. However, in these cases the process is usually considered as a cement or pozzolan-based process with phosphates as additives. The primary advantages of phosphate processes are the low cost and ease of application. In the basic form, phosphates are mixed with the contaminated soil by spraying or combining in a mixer. No other physical or chemical operations are required [107].

Apatite is a general name for four minerals, of which the most common is fluoroapatite $(Ca_5(PO_4)_3F)$. More rare are chloroapatite, $(Ca_5(PO_4)_3Cl)$, hydroxylapatite $(Ca_5(PO_4)_3OH)$, and bromoapatite $(Ca_5(PO_4)_3Br)$. Apart of halide substitution for hydroxyl anions to form the aforementioned apatite minerals, calcium cations present in the apatite minerals can also be substituted by several cations such as Ba^{2+}, Pb^{2+}, Zn^{2+}, Fe^{3+}, Na^+, K^+, Sr^{2+}, Mn^{2+}, Cd^{2+}, Mg^{2+}, Fe^{2+}, and Al^{3+}, whereas the oxyanions AsO_4^{3-}, VO_4^{3-}, CO_3^{2-}, HPO_4^{2-}, $P_2O_7^{4-}$ SO_4^{2-}, and SiO_4^{4-} can replace structural PO_4^{3-}. Therefore, the general formula for the apatite group is $C_5(AO_4)_3(OH, F, Cl, Br)$, where C and AO_4 are the aforementioned cationic and anionic species, respectively. These substitutions affect the crystallinity, morphology, and lattice parameters of the apatites and, as a consequence, they alter their reactivity and stability [122, 123]. Naturally occurring apatites can be divided into the following two major groups: first, the apatite series, where Ca is the dominant cation (Table 9) and second, the pyromorphite series, in which Pb is the dominant cation [122].

Apatites occur commonly in soils and sediments and are the principal mineral components of phosphate rock, which serves as the raw material in all phosphate fertilizers. Sedimentary phosphorite is one of the primary sources of phosphate rock (>80% worldwide) and can be obtained readily and inexpensively. The principal phosphate mineral in sedimentary phosphorites is carbonate fluoroapatite, which is microcrystalline and differs in composition from pure synthetic apatites because of extensive and complex ion substitutions in apatite structure [124]. Geochemical studies have indicated that a variety of trace elements are enriched in apatites. It was found that the metals adsorbed remain in the apatite structure indefinitely with little subsequent desorption, rendering the apatite minerals as natural collectors for metals and radionuclides [125].

Table 9
Solubility products (K_{sp}) of main apatite minerals

Mineral	Dissolution reaction	log K_{sp}	Source
Hydroxyapatite	$Ca_5(PO_4)_3OH = 5Ca^{2+} + 3PO_4^{3-} + OH^-$	-58.19	[111]
Chloroapatite	$Ca_5(PO_4)_3Cl = 5Ca^{2+} + 3PO_4^{3-} + Cl^-$	-53.08	[112]
Fluoroapatite	$Ca_5(PO_4)_3F = 5Ca^{2+} + 3PO_4^{3-} + F^-$	-58.86	[111]

Thus, apatites could control the concentrations of calcium, lead, cadmium, zinc, and other heavy metals [111] in natural environments and may provide a cost-effective technology to remediate metal-contaminated soils and water [124]. Lead phosphates have been demonstrated to present very low solubility values (Table 6) and to be the most stable form of Pb under a wide variety of environmental conditions. Furthermore, they can form rapidly in the presence of adequate Pb and phosphate.

A detailed thermodynamic basis for the reaction of Pb and phosphates in aqueous solutions has been established by Nriagu [126 – 128]. Among all the Pb-P minerals, it was found that chloropyromorphite, $Pb_5(PO_4)_3Cl$ and plumbogummite, $PbAl_3(PO_4)_2(OH)_5 \cdot H_2O$ have the lowest solubility. They are the most stable compounds under favourable environmental conditions [126]. Since pyromorphites are the most stable Pb phosphate minerals found in nature, thermodynamics predict that other solid phase species would be converted to pyromorphite by a dissolution-precipitation mechanism. Several researchers identified chloropyromorphite as a common weathering product of Pb bearing compounds in mine-waste, urban, and motorway roadside soils having normal soil phosphorus concentrations [129]. Based on the solubility products of the pyromorphite mineral family with the general formula $Pb_5(PO_4)_3X$ (X = halide or hydroxide), the thermodynamic stability sequence of lead pyromorphites is $Pb_5(PO_4)_3Cl > Pb_5(PO_4)_3Br > Pb_5(PO_4)_3OH > Pb_5(PO_4)_3F$ [130]. Due to the ubiquity of chloride in nature, it is expected that chloropyromorphite is the dominant environmental form of pyromorphites [131]. It was found that lead phosphates are naturally formed in contaminated soils. For instance, the presence of lead phosphates in soils was reported at a port facility historically used for shipment of ore concentrates [130]. It has been demonstrated that the weathering of galena and the presence of elevated P concentrations in soil from an adjacent phosphoric plant has led to the formation of low solubility lead phosphate compounds. Pyromorphite was found to be the predominant lead phosphate phase formed at the site. It was also identified as a weathering product of Pb compounds in mine waste contaminated soils, motorway roadside soils, and highly contaminated garden soils [129, 132].

Effective Pb immobilization by the formation of pyromorphite has been reported upon addition of apatite [133], phosphate rock [134], or commercially available phosphates, mainly in the form of phosphate fertilizers on Pb contaminated soils at laboratory and field scale. The use of natural or synthetic apatites for the treatment of lead contaminated soils and water has been extensively investigated by many authors. These studies suggest that the reaction of apatites with aqueous Pb can result in the formation of pyromorphites (or elsewhere lead apatites). However, the mechanisms involved are not agreed upon in the literature. In general, polyvalent metal phosphates are capable to interact with metal ions in solution via two possible mechanisms: first, ion-exchange mechanism, involving a partial replacement of ions in the initial phosphate structure without destruction of the latter, and second, chemical reaction, involving the destruction of the lattice of the initial compound and the formation of new crystalline or amorphous products.

The first of the interaction mechanisms has been intensively studied [135-139]. It was proposed that Pb is initially adsorbed on the surface of apatite particles, followed by a cation exchange reaction between apatite Ca^{2+} and Pb^{2+} ions in solution according to the following reaction:

$$Ca_{10}(PO_4)_6(OH)_2 + x\ Pb^{2+} \rightarrow Ca_{10-x}Pb_x(PO_4)_6(OH)_2 + x\ Ca^{2+} \tag{21}$$

According to the second mechanism, hydroxyapatite is initially dissolved providing phosphate ions in the solution followed by precipitation of hydroxypyromorphite according to the following reactions [132]:

$$Ca_{10}(PO_4)_6(OH)_2 + 14\ H^+ \rightarrow 10\ Ca^{2+} + 6\ H_2PO_4^- + 2\ H_2O \tag{22}$$

log K = 28.92

$$10\ Pb^{2+} + 6\ H_2PO_4^- + 2\ H_2O \rightarrow Pb_{10}(PO_4)_6(OH)_2 + 14\ H^+ \tag{23}$$

log K = 8.28

The same mechanism of dissolution and precipitation was proposed for chloroapatite and fluoroapatite amendments to contaminated soil or water leading to the formation of chloro- and fluoropyromorphites, respectively

[123, 134]. Similarly, it was proposed that dissolution of phosphate rock, which composition is close to $Ca_{10}(PO_4)_6F_2$ with substantial CO_3^{2-} substitution in the structure, and precipitation of carbonated fluoropyromorphite-like minerals is the primary mechanism for Pb removal by phosphate rocks. However, it was proved that the new solid phases formed from the interaction of carbonate fluoroapatite with Pb solutions depend on the final pH of solution. Assuming $Ca_{10}(PO_4)_{6-x}(CO_3)_{2-x}F_{2+x}$ as the general formula for carbonate fluoroapatite, the following chemical reactions involved at circum neutral equilibrium pH values (6.6-6.8) were proposed [124, 134]:

$$Ca_{10}(PO_4)_{6-x}(CO_3)_{2-x}F_{2+x}(c) + (12-x)\ H^+ \rightarrow 10\ Ca^{2+} + (6-x)\ H_2PO_4^- +$$

$$x\ HCO_3^- + (2+x)\ F^- \tag{24}$$

$$10\ Pb^{2+} + 6\ H_2PO_4^- + 2\ (F^-, OH^-) \rightarrow Pb_{10}(PO_4)_6(F,OH)_2(c) + 12\ H^+ \tag{25}$$

At acidic pH, fluoropyromorphite ($Pb_{10}(PO_4)_6F_2$) is formed since the activities of both carbonates and OH ions are extremely low, resulting in the difficulty of incorporating them into the pyromorphite structure. At pH 7.1 - 10.6 due to the increased activities of OH^-, HCO_3^- and CO_3^{2-} ions, hydrocerussite ($Pb_3(CO_3)_2(OH)_2$) and minor amounts of carbonate hydroxyl fluoropyromorphite ($Pb_{10}(PO_4,CO_3)_6(F,OH)_2$) are formed. Finally, at pH above 10 the activity of both OH^- and CO_3^{2-} ions is extremely high leading to their increased incorporation in the pyromorphite sites. At the same pH values lead oxide fluoride is also formed according to the following reaction:

$$2\ Pb(OH)_3^- + 2\ F^- \rightarrow Pb_2OF_2(c) + H_2O + 4\ OH^- \tag{26}$$

In most cases, stabilization of Pb by apatite application is not determined by the decrease of Pb concentrations in the soil solution but rather on the conversion of pre-existing high-solubility Pb compounds to more geochemically stable pyromorphites. Effective soil remediation may not require the conversion of total Pb content to pyromorphite but only the most chemically and/or biologically reactive forms of Pb [140]. The application of hydroxyapatite or phosphate rocks to calcareous soil may restrict the transformation of Pb to pyromorphite because dissolution of Pb bearing compounds and availability of soluble P are limited. Therefore, addition of phosphoric acid or oxyphosphate salts to calcareous soil were applied in order to lower soil pH, facilitate Pb dissolution, and increase soluble P concentration that lead to pyromorphite formation. Applications of either phosphoric acid or oxyphosphate salts to various industrial wastes [141, 142] or contaminated soils [143] have been shown to stabilize and reduce leaching of Pb and other heavy metals.

Although phosphates have been mainly applied to remediate soils contaminated with Pb, the technique is also applicable to soils contaminated with other metals such as Zn, Cu, and Cd. Studies have shown that the effectiveness of hydroxyapatite in removing other heavy metals from solution was in the order Al > Zn > Fe(II) > Cd > Cu > Ni [122]. However, the presence of certain ions such as carbonates in natural apatites and consequently in the aqueous phase may change this order dramatically. The proposed removal mechanisms of metallic ions by apatite differ from those of Pb and include, first, ion exchange at the apatite surface, second, surface complexation, third, precipitation of amorphous to poorly crystalline mixed metal phosphates, and fourth, substitution for Ca in apatite by other metals during recrystallization (coprecipitation). Some researchers indicated that surface complexation with hydroxyapatite surface functional groups such as $\equiv$POH and $\equiv$CaOH and coprecipitation are the most important mechanisms of Zn and Cd removal from aqueous solutions by hydroxyapatite [144]. However, Cd^{2+} coprecipitation is more significant than Zn^{2+} precipitation, whereas Zn^{2+} is held more strongly than Cd^{2+} on the apatite surfaces. The surface protonation on apatite surfaces was also studied and the following reactions were proposed [145]:

$\equiv$PO$^-$ + H$^+$ = $\equiv$POH $\log\beta(int) = 6.6 \pm 0.1$ (27)

$\equiv$CaOH$_2^+$ = $\equiv$CaOH + H$^+$ $\log\beta(int) = -9.7 \pm 0.1$ (28)

Under the presence of Zn^{2+} (or Cd^{2+}) in the soil solution, the following reactions may occur [144]:

$\equiv$POH + Zn^{2+} = $\equiv$POZn$^+$ + H$^+$ (29)

$\equiv$PO$^-$ + Zn^{2+} = $\equiv$POZn$^+$ (30)

$$\equiv CaOH + Zn^{2+} = \equiv CaOZn^+ + H^+ \tag{31}$$

The coprecipitation of Cd^{2+} with Ca^{2+} can be represented by the following reaction:

$$x\ Cd^{2+} + (5-x)\ Ca^{2+} + 3\ H_2PO_4^- \rightarrow (Cd_x,Ca_{5-x})(PO_4)_3OH + H^+ \tag{32}$$

The performance of soil stabilization tests using phosphates was evaluated by applying leaching tests. In a study [133], batch leaching tests at a liquid to solid ratio equal to 5 l/kg were performed to investigate the immobilization of dissolved Pb in a Pb-contaminated soil by hydroxyapatite. The initial aqueous Pb in the untreated soil sample was 2273 μg/l that was reduced to 36 μg/l after reacting with hydroxyapatite at the higher addition rate. The same researchers [134] mixed a contaminated soil containing Pb in the form of $PbHAsO_4$ insecticide with phosphate rock and performed leaching experiments in small columns. They reported significant reduction of the water-soluble Pb due to the phosphate rock addition ranging from 56.8 to 100 % depending on the phosphates addition rate and the incubation time.

In another study [140], particle size and density separation techniques were applied to obtain Pb-enriched fractions from contaminated soil, which were reacted with synthetic hydroxyapatite at ambient pH (7.7) or at pH 5. It was found that the concentration of the Pb in the suspension of enriched soil at pH 5 was 38.42 mg/l after 14 days, whereas it was reduced to 0.22 mg/l after reacting with hydroxyapatite.

Phosphate rock and diammonium phosphate (DAP) were evaluated as amendments to soil using solute transport experiments with repacked columns [146]. DAP application at 1% rate was effective in reducing Cd, Pb, and Zn dissolution nearly more than 95%. However, a negative effect of As mobilization was reported. This mobilization, which has also been reported by other researchers, can be restricted mainly by applying iron salts. As it can be seen in Fig. 17, the addition of phosphates in contaminated soil resulted in a significant increase of concentration of As in the TCLP leachate. However, upon addition of ferrous sulphate, this adverse effect was addressed [147]. In another case, batch and field experiments were conducted to stabilize soil contaminated with Cd, Cu, Pb, and Zn using $CaHPO_4$, $CaCO_3$, and $Ca(H_2PO_4)_2$ as stabilizing agents, respectively.

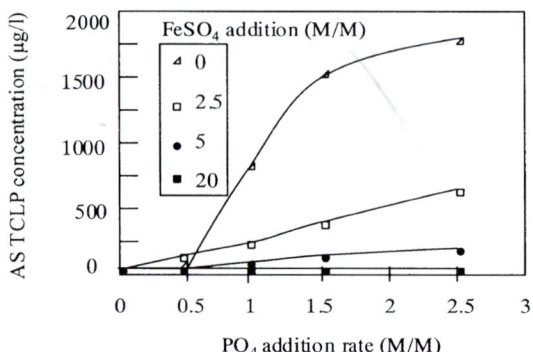

Fig. 17. Mobilisation and stabilization of As in contaminated soil as a function of PO_4 and $FeSO_4$ addition rates (redrawn after Ref. 147).

Based on the TCLP results, calcium oxyphosphate addition in soils at rates 0.5%- 6.25% resulted in a decrease of Cd, Cu, Pb and Zn concentration in the TCLP leachate by 98 %, 97 %, 99 %, and 96%, respectively. The chemical and physical evaluation of the efficiency of phosphate amendment as a remediation technique for heavy metals polluted soils is not sufficient. Biological evaluation methods are required to assure that immobilization results in lower soil to plant transfer of heavy metals and phytotoxicity [148]. Moreover, these methods can give indications about possible adverse effects (toxicity, deficiency of other elements, and dissolution of phosphates) of phosphate application.

P is an essential element for plants and is one of the most limiting nutrients for plant growth in soils. Plant roots can only absorb free phosphate and several mechanisms exist for increasing the availability of soil-P to roots [129]. Fungi associated with roots (mycorrhizae) are thought to play an important role in the liberation of phosphate ions from organic and inorganic compounds. Roots exudates containing phosphatase enzymes can convert organic-P to phosphate in the rhizosphere. Rhizosphere microorganisms (bacteria and fungi), which are not infecting the root, are also important in the mineralization of soil organic P. Phosphate available to plant roots is also available to heavy metal compounds to form heavy metal phosphates. This may also lead to the decrease of essential micronutrients that affects plant growth.

Several plants have been used to determine changes in the bioavailability of heavy metals in soils by applying phosphates, such as maize (*Zea Mays*), barley (*Hordeum vulgare*), radish (*Raphanus sativus*) [148,149], Sudax grass (*Sorghum bicolor* L. Moench) [140], oat (*Avena sativa*) [125], *Agrostis capillaries* [129], *Phaseolus vulgaris* [148, 150, 151], St. Augustine grass (*Stenotaphrum secundatum*) [152], and others. Almost for all of these cases, it was proved that the Pb content in shoot tissue decreased as the quantity of added phosphates increased. Furthermore it was found that the Pb and P contents in the plant roots increase and pyromorphite is formed at the root zone.

A recent study [140] examined the effect of apatite amendments on the bioavailability of Pb in soil containing 37026 mg/kg Pb as well as high levels of Zn, Cr, Cu, and Cd, and concluded that apatite amendments lower the bioavailability and increase the geochemical stability of soil Pb. Natural and synthetic apatites were applied as soil amendments and the bioavailability of Pb was determined in plant uptake studies. Sudax grass (*Sorghum bicolor* L. Moench), a hybrid of sorghum (*Sorghum vulgare* L. Moench) and Sudan grass (*Sorghum vulgare* var. *sudanense*), was chosen for greenhouse pot experiments because it does not only accumulate metals readily but also grows well under greenhouse conditions. Although it was found that the Pb content in shoot tissue decreased as the quantity of added apatite increased, the Pb and P contents in the plant roots increased. XRD and SEM analysis indicated that apatite reacted with Pb in the contaminated soil to form pyromorphite in situ. However, accumulation of Pb in the roots and formation of pyromorphite on root surfaces was noted. In the absence of other phosphates sources, plants can induce the dissolution of pyromorphite to facilitate uptake of P, which is an essential macroelement for plants.

However, it has been indicated that plant uptake of P does not result in destabilization of pyromorphite, if the quantity of apatite P present in soil exceeds the amount required for conversion of soil Pb to pyromorphites. Therefore, it seems prudent to maintain an excess quantity of unreacted phosphate in Pb contaminated soils, even after pyromorphite formation has occurred in order to inhibit the uptake of pyromorphite P, and thus to dissolve pyromorphite [153]. Some other studies [139, 148] showed that the addition of apatite (0.4% by weight) to a Zn, Cd, and Pb polluted soil led to an increasing yield and lowered the Zn, Cd, and Pb content in three week-old maize plants, in mature maize tissues (roots, young leaves, old leaves, stems, grains), and in barley.

Maize (*Zea Mays*) and *Phaseolus vulgaris* were also used to evaluate the effectiveness of hydroxyapatite for stabilization of heavy metal (Zn, Pb, Cu, Cd) and As polluted soils [148]. It was found that plant growth was partly restored on the 0.5 and 1% hydroxyapatite treated soils, but at 5% hydroxyapatite addition rate, growth was inhibited again. It was found that the metal concentration in the leaves of test plants decreased after hydroxyapatite addition. However, the uptake of some essential trace elements also decreased and probably led to Mn-deficiency in maize. Another adverse effect was the increase of As uptake after treatment, which is attributed to the increased concentration of phosphate ions.

At a pilot-scale field experiment in central Florida that was conducted to assess the efficiency of P-induced metal immobilization in soils, the following phosphate treatments were applied to contaminated soils: H_3PO_4, H_3PO_4, $Ca(H_2PO_4)_2$, and H_3PO_4, and phosphate rock [152, 154]. These P amendments reduced the TCLP leachability of elements below the respective regulatory limits. Furthermore, they enhanced metal uptake in the roots of St. Augustine grass that was planted in the treated soil, and significantly reduced metal translocation from root to shoot, especially Pb via formation of a pyromorphite-like mineral on the membrane surface of the root. The mixture of H_3PO_4 and phosphate rock was proved to be more effective in metal immobilization, with less soil pH reduction and less soluble P.

3.5.5. Field Applications

Solidification and stabilization processes can be performed either ex situ or in situ. In ex situ applications, the contaminated soil is excavated from its previous location and processed through a treatment system before being redeposited in the excavation or at another location. In situ treatment of contaminated soil is generally performed by introducing reagents (additives and/or binders) into the ground and mixing them with the contaminated soil by using augers. Almost all present commercial S/S processes are relatively simple using standard or slightly modified mechanical equipment, which depends on the type of the method applied (ex situ or in situ). In case that both stabilization and solidification is required, stabilization reagents are added prior to solidification reagents to allow sufficient time for the stabilizing agents to react with the contaminants. Otherwise, the presence of solidifying agents may inhibit the stabilization reactions.

Ex situ applications

Ex situ methods involve excavation and staging of the contaminated soils, screening and crushing materials too large in diameter to be effectively treated, blending the binding reagents and additives with contaminated soil, and stockpiling treated material prior to shipment off site or placement back in the excavation. A typical ex situ S/S process flow diagram is given in Fig. 18. Apart from the main mixing equipment, feeding devices, pumps, conveyors, chemical storage facilities, and other ancillary equipment are also required. Mixing can be performed via in-drum, in-plant, or area mixing processes. In-drum mixing is preferred for the treatment of small volumes of contaminated material. In-plant processes utilize mixers such as rotary drum or pug mill mixers and may be set up in either batch or continuous mode, depending on the expected use. As described in previous paragraphs, specific extrusion equipment is required to properly mix the contaminated soil with thermoplastic binders. Larger volumes of contaminated soil may be excavated and moved into a contained area for mixing. This process involves layering the contaminated material with the stabilizer/binder, and subsequent mixing with a backhoe or similar equipment [155]. Mobile and fixed treatment plants are available for ex situ S/S treatment.

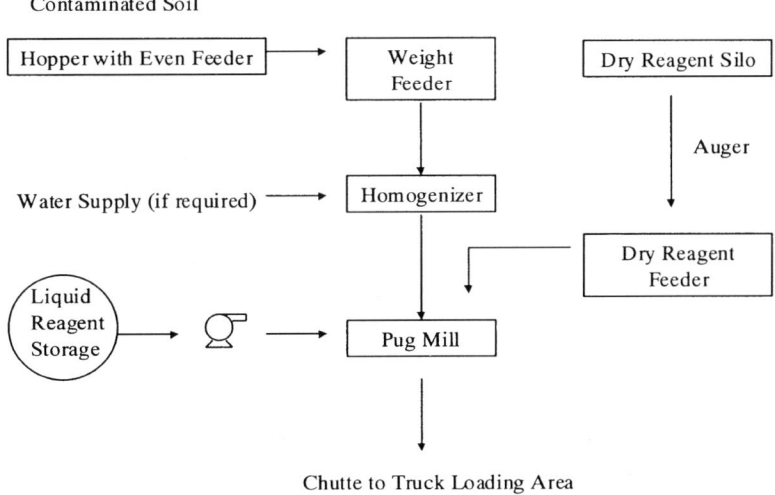

Fig. 18. Ex situ S/S process flow diagram.

At a batch process, a predetermined amount of contaminated soil, reagents, and water are added in mixers (usually pug mills) and mixed for a certain time. The mixing process normally takes 1 to 15 min, depending on the mechanical system used, the size of the batch, the type of contaminated soil, and the amount and types of reagents used. In continuous systems, residence time in the mixer itself generally depends on the design of the feed and mixing equipment and is limited to 0.5 to 2.0 min. Therefore intensive mixing is required [107]. Small scale plants can treat up to 100 tons of contaminated soil per day, whereas 500-1000 tons of soil per day can be stabilized in large plants [156]. The mixer-based systems provide significantly improved control over the mix proportions and blending efficiency. Following mixing, the blended contaminated soil is allowed to cure either in temporary containers or impoundments, or at its final disposal area.

In situ applications

There are several techniques available for the in situ mixing of stabilizing agents with contaminated soil. In situ S/S is less labour and energy intensive than ex situ processes, which require excavation, transport, and disposal of the treated soil. Therefore, in situ application is the most widely applied method for implementing stabilization and solidification of contaminated soils. In situ mixing, which uses commonly applied construction equipment such as backhoe, is the oldest and most frequently performed mixing technique. Reagents are placed on the surface of the contaminated soil to be treated and mixed up to the required depth with the backhoe or tilled in with a harrow for treatment of shallow layers. Although these techniques are cost effective under certain conditions, the working depths are limited by the size of the backhoe and it is difficult to provide uniform and complete mixing.

Most recently, earth drilling and foundation construction equipment properly modified to allow well controlled reagent injection and mixing to great depths is used. It utilizes a crane-mounted mixing system with modified augers and blades (Fig. 19). Depending on the depth of the mixing, large-diameter single or multiple augers can be used. For the treatment of soils with shallow contamination, the Shallow Soil Mixing (SSM, GeoCon) can be applied [157]. In this technique, a single shaft large diameter auger consisting of two or more cutting edges and mixing blades (Fig. 20) is used. The crane-mounted auger with diameter from 1 to 3.7 m may proceed up to a maximum depth of 10 m. As it proceeds downwards into the contaminated soil, the treatment chemicals are pumped through hollow drill shaft and are then injected into the soil.

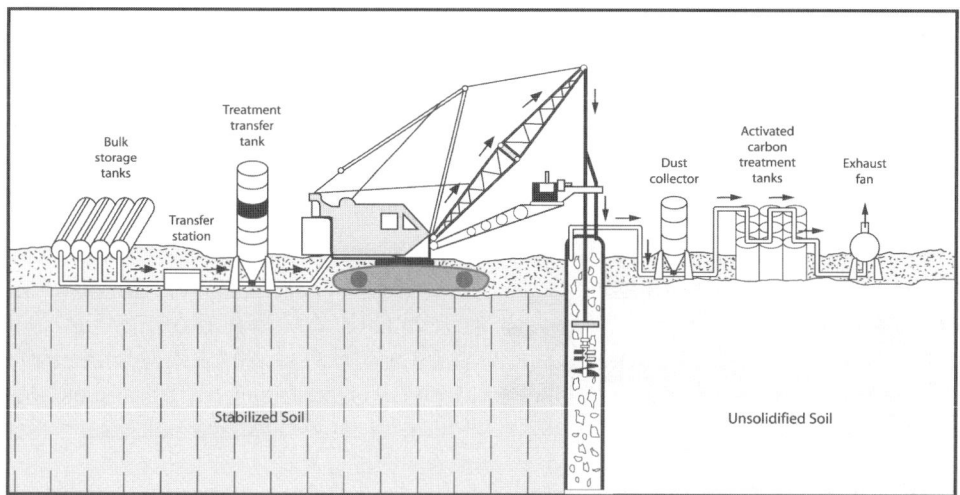

Fig. 19. In situ S/S with crane-mounted augers (GeoCon process).

Then due to the rotation of the cutting edges and mixing blades, the soil and the treatment chemicals are mixed. When the auger head reaches the maximum depth, it is raised to expose the mixing blades at the surface and then it is allowed to advance again to the bottom. If the contaminated soil is in a slurry state or if there are emissions from the treatment process, a semi-closed system using a bottom-opened cylinder as presented in Fig. 20 can be applied to obtain a well controlled mixing system and reduce the risks associated with emissions. Dry treatment chemicals are transferred pneumatically and fluid chemicals are pumped at a constant predetermined rate to obtain the proper mixing proportions. Since a boring creates a column of contaminated soil mixed with additives, the columns must overlap to assure full treatment.

A cross section of the columns describing the drilling pattern sequence for a single auger is given in Fig. 21. In the deep soil mixing (DSM) systems, two to four overlapping mixing paddles and auger flights that are guided by a crane supported set of leads are used. The size of the augers ranges from 0.6 to 0.9 m in diameter. Similarly to the case of the shallow soil mixing systems, as the augers advance into the soil the treatment, agents are injected into the soil in a slurry form through the tip of the hollow-stemmed augers. When the augers reach the desired depth, they are withdrawn and the mixing process is repeated on the way to the surface.

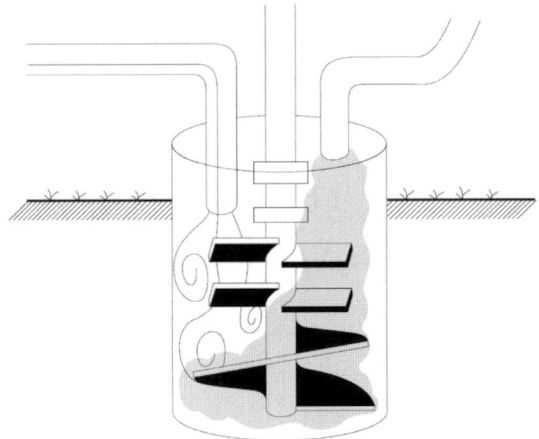

Fig. 20. Shallow soil mixing using modified auger and bottom-opened cylinder (Geo-Con process).

Each boring, depending on the number of augers, creates 2 to 4 circular columns of treated material, and thus, a sequence of partially overlapping strokes is followed to cover the entire contaminated area. It is reported that depending on soil conditions, deep soil mixing can extend down to 46 m. Another important in situ mixing method is known as jet grouting or jet mixing (see also subchapter 2.2.4. Jet Grouting). This technique involves forcing a binder containing dissolved or suspended treatment agents into the subsurface, allowing it to permeate the soil. Grout injection can be applied to contaminated soil lying well below the ground surface. The injected grout cures in place to produce an in situ treated mass. A high-pressure jet is used to both excavate and mix the soil with the injection fluid. First, a jetting pipe is positioned at the bottom of the treatment zone. The pipe can be the drill pipe itself, or in case of hard or rock-bearing soils, holes may need to be drilled first, and the jetting pipe is then placed in a separate step [158]. The jetting pipe rotates slowly and lifts upwards, while it injects grout under high pressure to finally form a column of treated soil.

The main advantage of jet mixing is the small size and the relative quiet, low vibration equipment rendering the technique applicable to residential areas. In cases that solidification is not required and stabilization aims at immobilization of heavy metals in soil, direct injection of treatment solution into the contaminated soil can be applied.

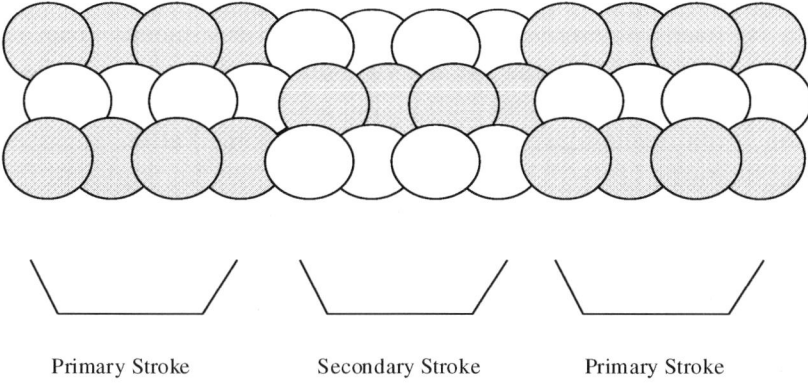

Fig. 21. Drilling pattern sequence using single auger for in situ S/S of soils.

An example of such possibility is the treatment of Pb contaminated soil with soluble phosphates. In many cases contamination of soils is restricted to the upper layers. In these situations, agricultural soil conditioning equipment is cost-effective for in situ mixing of contaminated soil with additives.

4. PHYTOREMEDIATION OF HEAVY METALS (H.B. Bradl)

In the past several years, the use of plants for the removal of heavy metals in various media such as water, groundwater, industrial wastewaters, soils, sediments, and sludges has attracted the interest of the environmental engineering community as a potentially useful remediation tool [159]. This chapter introduces various aspects of phytoremediation, beginning with basic physiological aspects, which allow plants to contain, sequester, or remove heavy metals, explains several mechanisms of phytoremediation such as phytoextraction, phyto/rhizofiltration, phytovolatilization, and phytostabilization, and finally assesses advantages and disadvantages of phytoremediation including the economical aspects in comparison to other remediation techniques available at the market.

4.1. Introduction

Plants can be used to remove potentially toxic heavy metals from soils, sediments, sludges, and waters. The use of plants for the removal of heavy metals is referred to as phytoremediation, but sometimes also the terms bioremediation, botanical bioremediation, or green remediation are used in various publications [160-163]. Also the term phytotechnology can be encountered, which is defined as "the use of vegetation to contain, sequester, remove, or degrade inorganic and organic contaminants in soils, sediments, surface waters, and groundwater" [164]. Here the term "phytoremediation" is used.

It has been reported as early as 1855 that some plants have adapted to increased heavy metal concentration such as outcropping ore bodies and may contain large amounts of heavy metals [165, 166]. Thus they can be used as indicators for prospecting for these metals [167]. Therefore, the investigation of metal tolerance and accumulation in plants and the establishment of vegetation on contaminated sites have been of some interest [168-170]. To understand how phytoremediation works, it is necessary to get some deeper insights into several complex processes, which comprise interactions between soil organisms and plants including protein and enzyme production, bio-catalysis, and symbiosis of these organisms with plant roots. In addition, processes within the plant system such as inorganic nutrition, water uptake, evapotranspiration, photosynthetic production of phytochemicals, root exudation, root turnover, and bio-metabolism play an important role.

4.2. Basic Physiological Processes

The following chapter gives an overview on the basic physiological processes, which take place in the vicinity of plants in the subsurface environment. During evolution, plants and soil microorganisms have developed symbiotic and synergistic relationships in order to maintain protection, nutrition, and enhanced water uptake capacities for the plants on the one side and enhanced nutrition for the soil microorganisms on the other side. A multiplicity of interactions takes place in the root-soil interface known as the rhizosphere. The term "rhizosphere" has been introduced in 1904 for the description of the specific interaction between bacteria and the roots of legumes [171]. The rhizosphere is a region of soil of approximately 1-3 mm thickness surrounding individual roots and is characterized by high bioactivity [172]. Some of the mechanisms vital for phytoremediation focus on the physiological and microbiological processes occurring in the rhizosphere.

4.2.1. Processes involving Microorganisms

Microorganisms such as bacteria, actinomycetes, molds, fungi, algae, viruses, and protozoa inhabit the subsurface environment ubiquitously in extremely large numbers. There are more than 400 named genera and estimated 10^4 species of soil bacteria [173]. Typical values for 1 gram dry-weight soil are 10^8 for bacteria, 10^6 for actinomycetes, and 10^5 for fungi. There is a significant variation of the microbial population density in the soil profile both in vertical and in lateral directions, which is caused by the availability of oxygen and organic material as potential food sources [174]. Bacteria in the subsurface environment tend to attach to solid particle surfaces in the soil and thus form a biofilm, which is rich in substrates desirable for microorganisms [175]. Biofilms build up rapidly within 50 μm from the root surface and show an increase in microbial population size which can be 50-100 times greater than that in non-vegetated soil. Therefore, microbial activity and as a consequence biodegradation rates increase significantly within the rhizosphere.

Table 10 shows the number of bacteria at increasing distance from the root surface. The rhizosphere microbial community is characterized by proportionally more denitrifiers and gram negative bacteria species such as *Pseudomonas sp.* and *Achromobacter sp.* and differs from that of the bulk soil. Rhizospheric soils have been reported to contain microbial populations that are 1 to 2 orders of magnitude higher than bulk surface soils [176]. Their major source of carbon stems from plant roots and organic residues following plant growth such as amino acids, simple sugars, carbohydrates, and enzymes. Analyses of organic compounds found in, on, or in the vicinity of roots comprise of a large variety of aliphatic, amino, and aromatic acids as well as amines, sugars, and amino sugars. Additionally, insoluble compounds such as cellulose, lignin, and proteins are found. Microorganisms need inorganic nutrients such as nitrogen and phosphorus, which are essential for their active growth. They are able to deplete a soil of available carbon and other essential sources rapidly. Therefore, the relationship between microbial population and available nutrient source is of utmost importance for biodegradation processes in soils.

The microorganisms present in the subsurface environment can be classified into three different types: aerobic, anaerobic, and facultative microorganisms. Aerobic microorganisms use oxygen as the only terminal electron acceptor, while anaerobic organisms cannot use oxygen but utilize other electron acceptors to cause the preferential donation of electrons from the contaminant to nitrate, manganese oxide, iron(III), and sulphate. In the absence of sufficient oxygen, aerobic microorganisms are unable to metabolize, and therefore, oxygen content becomes an important parameter in aerobic growth. Facultative microorganisms are able to use oxygen if it is available and then switch to other electron acceptors if the level of oxygen becomes too low [177]. The soil microorganisms live in a symbiotic community with the plant roots. They provide a protective sheath around each root in the rhizosphere by restricting contact with the potentially toxic compounds. The plant in return provides the microorganisms with nutrients such as oxygen, carbohydrates, and inorganic nutrients. Additionally, it has been found that vesicular-arbuscular mycorrhizal (VAM) fungi limit the uptake of non-essential inorganic elements such as Cd, Ni, Pb, and others thus protecting the plants. There is no clear understanding of how the VAM fungi achieve this protection. They could sequester the metals into their tissues, restricting their transfer into the plant, or they could enhance the water uptake capacity of the plant, which leads to a dilution of the extracted elements in the plant root cells [177]. Fungi also provide plants with resistances to drought, salinity, and acidic soils. Their hyphae also attract other microorganisms to form water-stable aggregates, which enhance soil tilth. VAM fungi in general enhance plant growth and nutrition by extending their hyphae out into the soil beyond the reaches of the root system, thus providing additional water uptake capacities by increasing the total surface area. Also nutrient uptake is enhanced by up to five times compared to non-infected roots [178].

Table 10
Number of bacteria at increasing distance from the root surface

Distance from Surface (mm)	Estimated frequency 10^9 cells cm^{-3}	Morphological types distinguished
0-1	120	11
1-5	96	12
5-10	41	5
10-15	34	2
15-20	13	2

Reprinted with permission from: R. Karthikeyan, and P.A. Kulakow, Soil Plant Microbe Interactions in Phytoremediation, in "Phytoremediation" (D.T. Tsao, ed.), Springer, Berlin, 2003, p. 61.

4.2.2. Plant Processes

Typical plant growth processes include many different but interrelated processes such as water uptake and transpiration of water vapour into the atmosphere, absorption of inorganic nutrients and the photosynthetic production of photochemicals. Each of these processes is vital for contaminant remediation. The plant requires 13 essential inorganic nutrient ions, which include nitrate or ammonium nitrogen (NO_3^-, NH_4^+), phosphate ($H_2PO_4^-$, HPO_4^{2-}, PO_4^{3-}), potassium (K^+), calcium (Ca^{2+}), magnesium (Mg^{2+}), sulfate (SO_4^{2-}), iron (Fe^{2+}, Fe^{3+}), chloride (Cl^-), zinc (Zn^{2+}), manganese (Mn^{2+}), copper (Cu^+, Cu^{2+}), borate (BO_3^{3-}, $B_4O_7^{2-}$), and molybdate (MoO_4^{2-})[178]. These nutrients are dissolved in water and are taken up by the root system. Once inside the root system, the inorganic nutrients can be transported throughout the whole plant through its vascular system, the xylem. The transport can be achieved either passively in the transpirational stream or actively by transport proteins.

The plant takes up other non-essential elements such as As, Cd, Na, Se, and Pb, which are of concern as well either by transporting them through the passive channels or by substituting onto transport proteins. Substituting is typical for multivalent cations. Also elements normally considered essential for the plant such as Cu, Zn, and Mn, are of concern when occurring in high concentrations.

Plants use several mechanisms to sequester or stabilize these elements and prevent translocation into the more sensitive terrestrial portion. One mechanism is to bind these inorganic elements in the soil, which is achieved by the exudation of biochemicals by the plant which facilitate precipitation or adsorption onto the soil particles. Also, the plants exude proteins, which bind the inorganics irreversibly onto the root surface thus preventing them from entering the plant system. Another mechanism is the sequestration of the inorganics into the vacuoles of the plant cells, which serve as a storage receptacle and prevent further transport into the plant system. Finally, to reduce the toxicity of the inorganic, its speciation is changed by incorporation into an organometallic compound [179]. Based upon these mechanisms, phytotechnologies for heavy metal remediation such as immobilization in the rhizosphere or extraction and subsequent removal into the terrestrial plant tissue can be developed for contaminated sites.

All nutrients and contaminants alike must be dissolved in water in order to be taken up into the plant system. The process of water uptake on the one hand and transpiration of water vapour on the other hand is vital for the plant. Water uptake takes place through the root system when the root takes up water from the soil aqueous phase. The aqueous phase is then transported throughout the plant into the leaves. This process is called translocation. Translocation occurs primarily by the equilibrium driving force between the liquid water phase in the leaves and the water vapour phase present in the atmosphere, the air humidity. This process is regulated by the plant with the help of the production of abscisic acid, which signals to the microscopic leave pores, the stomata, to alter their aperture size [180]. The size of the stomatal aperture is reduced during stress in order to decrease the amount of water vapour exiting from the plant. The processes of water uptake and transpiration are the primary mechanisms used to control hydraulics and contain contaminants. Certain water soluble contaminants can be taken up by the plant and subsequently are transported through the stomata. This process, which is called phytovolatilization, is a useful tool of removing contaminants such as volatile organic compounds (VOC) and certain metalloids.

All plants are able to use the carbon dioxide present in the atmosphere for the photosynthetic production of phytochemicals. The CO_2 enters into the plant through the stomata in a counter current gas exchange with water vapour. Then it is incorporated into photosynthetic products (phytochemicals), which exist in two primary biochemical pathways. The C3 pathway refers to the 3-carbon compound, 3-phosphoglycerate, while the C4 pathway refers to the 4-carbon compounds, malate, or aspartate, which represent the first stable compounds of CO_2 sequestration in the plant. After incorporation, the carbon is subjected to the photosynthetic carbon reduction cycle, which produces a variety of compounds such as cellulose, lignin, starch, carbohydrates, sugars, amino acids, proteins, enzymes, and others. These photosynthates can then be incorporated into the biomass, metabolized during cell respiration for energy production, or exuded into the root zone [178].

Plants secrete a variety of carbonaceous materials such as mucigel, which is a secretion from the growing root sections in order to protect the root apex from desiccation, to promote nutrient uptake, and to aid in root penetration into the soil. Other secreted compounds include metabolites, which protect the plants from soil pathogens such as viruses, bacteria, fungi, insects, nematodes, etc. These phytoalexins include terpenoids, flavonoids, alkaloids, phenylpropanoids, anthraquinoids, polyphenolics, and others [181, 178]. Plants also produce chemicals, which prevent other plant species from growing in order to minimize competition for water and nutrients. This process is called allelopathy. Another important process is root turnover, where a portion of the root system is sloughed off prior to winter dormancy. The root turnover leads to the deposition of additional organic material, which provides a feeding source for the soil microorganisms and thus helps to maintain the bioactivity of these microorganisms during the winter period [182]. Additionally, this process leads to the release of VAM fungal hyphae, which are able to inoculate other plants during the next growing season.

These mechanisms are vital for the cleaning up of organically impacted sites. Rhizodegradation can be used to treat several recalcitrant contaminants such as pentachlorophenol (PCB), polychlorinated biphenyls (PCB), polycyclic aromatic hydrocarbons, and others [183, 184]. Plants are also capable of producing compounds which have bioactivity for certain contaminants such as explosives, perchloroethylene (PCE), phenolic compounds, herbicides, and pesticides [185-187].

4.3. Mechanisms of Phytoremediation

There are several mechanisms used by plants, which allow cleaning up water, groundwater, soils, sediments, and sludges contaminated by heavy metals. These physiological processes which occur in the root zone, within the vegetation, and through the transpiration stream, are utilized in phytoremediation.

4.3.1. Phytoextraction

Phytoextraction is defined as "the use of metal-accumulating plants that can transport and concentrate metals from the soil to the roots and aboveground shoots" [188]. This ability is used to extract toxic metals from the soil and provides an interesting tool for the remediation of metal contaminated sites [189, 190]. Some plants are able to accumulate and tolerate high levels of heavy metals in their tissues. Plants accumulating nickel and zinc have been reported to contain as much as 5% of these metals on a dry-weight basis which results in a 10-fold bioaccumulation factor given a soil with a total metal concentration of 5000 mg/kg [191, 192]. Table 11 gives examples of some plant species, which hyperaccumulate over 1% of a heavy metal in their shoot dry matter. These concentrations are about 100 times higher than concentrations tolerated by normal crop plants. If the plant produces high amounts of biomass during the accumulation process, significant quantities of metal can be transferred from the soil and removed by harvesting the plant material. The harvested biomass can then be treated in different ways to reclaim the metals or to reduce volume and weight prior to disposal (e.g. composting, compaction, thermal treatment). Three main parameters are vital for the successful use of phytoextraction when remediating a site. First, the selection of a site conducive to phytoextraction is of utmost importance. Second, the solubility and availability of metals in the soils are crucial for metal uptake by the plant, and third, the ability of the plant to accumulate metals in the harvestable plant tissues plays an important role [193]. As metal concentrations in soils may vary in a wide range from < 1mg/kg to > 100 000 mg/kg depending on parent rock material and deposition it is necessary to define particular values, to which a level of risk has been assigned. Risks vary with the different metals and associated exposure pathways, and regulating limits for heavy metals concentrations in soils vary considerably from country to country. Most limits established in legislation are based upon human health impacts from direct contact; some are based on ecological risks or secondary exposure pathways. Additionally, soil cleanup criteria may differ for residential and non-residential areas. Most limits established in legislation are based upon human health impacts from direct contact; some are based on ecological risks or secondary exposure pathways. Additionally, soil cleanup criteria may differ for residential and non-residential areas. Metal solubility depends on soil characteristics and is mainly influenced by soil pH [194] and the degree of complexation with soluble ligands [195]. Metals may exist in various forms in soils, such as free ions and soluble metal compounds in the soil solution, exchangeable ions adsorbed onto inorganic solid phase surfaces, nonexchangeable cations and precipitated or insoluble inorganic metal compounds such as oxides, hydroxides, phosphates, and carbonates.

Table 11
Examples of some hyperaccumulator plant species

Metal (ppm)	Species	Maximum level in leaves (ppm)
Zn	*Thlaspi calaminare*	39.600
Cu	*Aoellanthus biformifolius*	13.700
Ni	*Phyllanthus serpentinus*	38.100
Co	*Haumaniastrum roberti*	10.200
Se	*Astralagus racemosus*	14.900
Mn	*Alyxia rubricaulis*	11.500
Cd	*Thlaspi caerulescens*	1.800

Modified after Ref. 160.

Usually, indigenous background concentrations are indicated by the latter, while the other forms indicate anthropogenic contamination [196]. Metals have to be soluble in the soil solution in order to be taken up by the plant. The effectivity of phytoextraction is therefore dependent on the amount of soluble metal. To increase this soluble metal amount in the soil, sometimes soil conditions may need to be altered.

An example for predominantly insoluble phases of heavy metals in soils is given by some Pb minerals such as Pb phosphates, Pb carbonates, and Pb (hydr)oxides commonly present in soils. Lead is generally insoluble and is therefore not available for plants unless soil pH decreases below 5.5 [197]. Decrease in soil pH, however, means increase in Al solubility and subsequent Al toxicity for plants. Acidification treatment of soils may result in enhanced metal solubility, but the different plant tolerances to acid conditions have to be taken into account. To increase solubility, chelating or complexing agents such as EDTA may be used. These agents complex the free metal ion in solution, thus allowing further dissolution of adsorbed or precipitated phases until equilibrium between the complexed metal, free metal ion, and insoluble phases is reached [198]. The amount of heavy metal available in the soil solution is a function of the concentration of the complexing agents, affinity of the agent for the respective heavy metal and competing ions, and the solubility of heavy metal compounds in the soil. Predictions regarding the ability of a complexing agent to solubilize heavy metals in a soil can be made theoretically by using equilibrium constants and assuming that heavy metal solubility is controlled by the solid phases of these metals. Nevertheless, the prediction of competing reaction is difficult due to the extreme variation of both chemical conditions and soil properties frequently encountered when dealing with contaminated sites [199, 200]. Therefore, empirical data evaluation is important to determine whether a complexing agent will enhance metal solubility for a specific site.

To evaluate the suitability of a specific soil type for phytoextraction treatability studies are conducted which use sequential extraction techniques to determine the respective heavy metal solubility, chemical relationships, and metal speciation for a defined soil fraction. Successive extraction steps are carried out and the metal concentration in the resulting solution is then determined. The standard procedure is reported in Ref. 115. The first step comprises extraction of the soluble-exchangeable fraction with a neutral salt followed by carbonate extraction using a buffered solution at pH 5, then Fe and Mn oxides are extracted with a reducing agent, organic matter is oxidized by acidic hydrogen peroxide, and finally a total digestion of the remaining material to constitute the residue is carried out.

Plant availability is then estimated by the distribution of the metals between the different fractions. It is expected to be the greatest for the soluble-exchangeable fraction and decreases with each succeeding extraction step. Soil tests have been developed, which use chelating agents such as diethylenetrinitrilopentaacetic acid (DTPA) to estimate plant availability [201]. EDTA also proved to be a successful agent for providing heavy metals such as Pb from the fractions considered bioavailable [192].

The efficiency of phytoextraction is controlled by several parameters such as plant species, metal availability to plant roots, metal uptake by roots, metal translocation from roots to shoots, and plant tolerance to toxic metals. Some plant species endemic to metalliferous soils are able to tolerate and accumulate high levels of toxic metals [193]. These plant species are called metal hyperaccumulators. There are about 400 known metal hyperaccumulators in the world [202], which are reported to accumulate more than 0.1 % of Pb, Co, Cr, or more than 1% of Mn, Ni, or Zn in plant shoots in their natural habitat [203]. Species such as *Thlaspi caerulescens*, *Alyssum murale*, *Alyssum lesbiacum*, *Alyssum tenium*, and *Silene vulgaris* have been tested as metal hyperaccumulators for Zn and Cd [189], but their potential as a possible remediation tool for phytoextraction is limited by their slow growth rate and low biomass production. There has been extensive research to find metal hyperaccumulators with high biomass production, which yielded good results when combining selected high biomass agronomic crops with soil amendments [204-207].

Heavy metals such as essential plant elements (Zn, Ni, Mn, and Cu) and toxic elements (Pb, Cd, U, and ^{137}Cs) enter the plant primarily by root absorption of the soluble cation from the soil solution. These elements are transported into the root cell via voltage-gated Ca channels in the plasma membrane of those cells [208], which are produced by a large chemical potential gradient across the plasma membrane [209]. The negative membrane potentials and the low intracellular metal activity result in an influx of these metals into the root cells. The heavy metals are then translocated from the roots to the plant shoots. A number of physiological processes such as metal unloading into root xylem cells, long distance transport within the xylem to the shoots, and metal reabsorption from the xylem stream by leaf mesophyll cells are involved in this transport. In general, the metals are taken up by the root symplasm by absorbing them from the soil solution, and then the absorbed metals are unloaded from the xylem parenchyma into mature xylem vessel and are finally transported to the shoots by the transpirational stream. It has been found that the rate of metal translocation from roots to shoots is much lower compared to the rate of uptake. Most heavy metals are rapidly accumulated in the plant roots but only a small portion of the absorbed metals is transported to the shoots [210]. Therefore, the two major limitations to the use of phytoextraction of heavy metals are the metal bioavailability in the soil solution and the poor metal translocation from roots to shoots. Bioavailability can be increased by the use of chelating or complexing agents.

4.3.2. Phytostabilization

Another possibility of using plants for the remediation of impacted sites is phytostabilization, also called inplace inactivation or phytorestoration. It is a site stabilization technique, which uses soil amendments that induce the formation of insoluble contaminant species, thus reducing hazards to human health and the environment. The plants are used to cover the soil surface to prevent erosion and direct contact with the contaminated ground and to reduce percolation [211]. Phytostabilization uses soil amendments such as lime, gypsum, and fertilizers, which are known from traditional agriculture to improve soil properties and soil fertility. Phytostabilization techniques have also been applied to the management of sewage-sludge amended soils and mine spoils [212]. Sewage sludges are often used for agricultural purposes because they contain valuable plant nutrient. Nevertheless, they may also contain high amounts of potentially toxic heavy metals, which are of environmental concern. To reduce the mobility of these metals, liming agents are added in order to maintain a near neutral or slightly alkaline soil pH so that the metals become less soluble and may even precipitate [213].

Various soil amendments are used for phytostabilization in order to inactivate metals rapidly, prevent leaching and plant uptake, and reduce bioavailability. They should also be inexpensive, easy to handle and apply, non-toxic, readily available or easy to produce, and environmentally friendly. Some soil amendments such as phosphate fertilizers and organic material have secondary advantageous effects on the treated soil such as enhancing soil fertility and soil moisture holding capacity. Waste products or byproducts, which have to be disposed off or recycled can be of interest. Examples of such materials are sludges from wastewater treatment plants, animal and plant manures, and byproducts from industrial processes. Widely used are soil amendments such as phosphate fertilizers, organic matter, Fe/Mn oxyhydroxides, clay minerals, or mixtures of these materials [211]. Table 12 gives an overview on the most common soil amendments used in phytostabilization and their suggested mechanisms of inactivating heavy metals. Phosphate amendments such as phosphoric acid, phosphate fertilizers, or byproducts containing high amounts of phosphates may promote formation of highly insoluble forms of heavy metals. As it has been shown in the case of Pb, Pb phosphate minerals have been formed which are extremely insoluble even under highly acidic conditions [214]. The ability of hydrous Fe oxides and hydroxides to strongly absorb heavy metals such as As, Cd, Pb, Ni, and Zn and reduce their availability in soils is well known [215]. Organic materials have also been shown to absorb or complex heavy metals such as As, Cd, Cu. Pb, and Zn. The adsorption ability of alumosilicates such as clays and zeolites for heavy metals is also well known [216-218]. Liming agents are widely used to maintain a neutral soil pH but their ability to inactivate heavy metals is limited. Liming alone is therefore not a suitable long-term treatment for phytostabilization [211]. When selecting plant species for phytostabilization, several factors have to be taken into account.

Table 12
Most Common Soil Amendments in Phytostabilization

Amendment Type	Possible Target Metals	Suggested Mode of Inactivation
Phosphate Materials H_3PO_4, apatite, calcium orthophosphates, Na_2HPO_4, KH_2PO_4, other phosphate fertilizers, high-phosphate byproducts	Pb	Formation of insoluble Pb phosphate minerals e.g. Pb Pyromorphites
Hydrous Fe oxides Iron rich or other byproducts containing hydrous Fe oxides, isolated hydrous Fe oxides	As, Cd, Cu, Ni, Pb, Zn	Sorption of contaminants on oxide surface exchange sites, coprecipitation, or formation of contaminant-Fe compounds
Organic Materials Manures, composts, sludges, and other biosolids	As, Cd, Cu, Pb	Sorption of contaminants on exchange sites, or incorporation into organic material
Inorganic Clay Minerals Synthetic zeolites, natural alumosilicates, or alumosilicate byproducts from burning of coal refuse	As, Cd, Cu, Mn, Ni, Pb, Zn	Sorption of contaminants on mineral surface exchange sites, or incorporation into the mineral structure

Modified after Ref. 211.

Unlike plants chosen for phytoextraction, those plants used in phytostabilization should be poor translocators for heavy metals to the aboveground plant tissue in order to prevent uptake by humans or animals. These plants should also be tolerant to the soil metal levels and other site specific conditions such as soil pH, salinity, soil structure and water content. They should grow fast, have dense rooting systems and canopies, have high transpiration rates, must be easy to establish and to care for, and have a relatively long life. Plants, which can be found growing on contaminated soils, however, do often not grow quickly enough or cannot provide sufficient ground cover. Plants most suited for the purposes of phytostabilization are plants with well-known agronomic characteristics. Phytostabilization can be adapted to a variety of sites and may also serve as an interim strategy until the most appropriate remediation technique is chosen for a specific site.

Phytostabilization can be adapted to a variety of sites and may also serve as an interim strategy until the most appropriate remediation technique is chosen for a specific site. At least it can be used to protect barren contaminated areas from erosion and leaching. Nevertheless, phytostabilization may have some limitations, which require further investigations. Research, which has been conducted on a small scale in laboratories and greenhouses must be transferred to larger scales. Soil amendments and their rates of application, which have been chosen empirically must be more thoroughly examined as well as their possible environmental impacts such as nutrient imbalances and eutrophication. Phytostabilization can also be included with other remediation technologies as part of a comprehensive remediation strategy for a specific site.

4.3.3. Phytovolatilization

Certain heavy metals such as Se and Hg are capable of being volatilized into the atmosphere while present in a dissolved form in the transpirational stream of plants. This mechanism is known as phytovolatilization [164, 219, 220]. In the biogeochemical cycling of Hg in the environment, Hg(II) is the predominant form of mercury in soils, which usually occurs as stable HgS. Hg(II) may then be reduced to volatile Hg(0) abiotically or by bacteria using *mer* genes, or methylated by anaerobic bacteria. In the case of Hg, it has been noticed that some bacteria colonize mercury-contaminated habitats and are able to enzymatically detoxify Hg(II) and organomercurial compounds [221]. This process is conferred by the *mer* operon, which comprises of a series of genes that encodes two regulatory proteins, two mercury transport peptides, and two catalytic peptides [222]. Resistant bacteria have been found to express the enzyme organomercurial lyase (MerB), which catalyzes the protonolytic cleavage of the carbon-mercury bonds of several organic mercury compounds [223, 224]. These genes have been transferred to a laboratory model plant *Arabidopsis thaliana*, and to tobacco *Nicotiana tabacum* in order to express the *merA* and *merB* genes and enable the plants to take up dissolved methylmercury, convert it to Hg(II), and then volatilize it as elemental mercury [225]. Elemental mercury is roughly six orders of magnitude less toxic than methylmercury. Nevertheless, the introduction of volatilized Hg into the atmosphere has yet to be accepted.

The removal of Se from polluted soil and water and the transfer into plant tissue depends on the chemical species of this element [226]. The more oxidized forms of Se, e.g. selenate or selenite, are in general well soluble in water and can therefore be taken up easily while reduced organic forms, such as selenide or elemental Se, are much less bioavailable. Selenate uptake from the soil into the roots is mediated actively by a cotransporter protein in the root plasma membrane [227] and is also enhanced by rhizosphere bacteria. Selenite uptake, on the other hand, seems to be passive [228]. Selenate and selenite are then assimilated to selenocystein in a series of reductive steps [229] and are finally transformed into a dimethylated organometallic, which can be volatilized from the leaves. The dimethyl selenite is less toxic than the selenate [230]. Recently, transgenic plants and plant-microbe combinations with enhanced abilities to accumulate and remove Se have been used for phytoremediation [220].

4.3.4. Phytofiltration

The use of plants to remove heavy metals from polluted ground waters and surface waters is another tool in phytoremediation. This procedure is referred to as phytofiltration [231]. Aquatic water plants such as the water hyacinth (*Eichhornia crassipes*), pennywort (*Hydrocotyle umbellata*), and duckweed (*Lemna minor*) are known to be able to remove heavy metals from waters [232-234]. As these plants are only of small size and their roots show only slow growing, the efficiency of heavy metal removal is limited. Also the high water content of aquatic plants is a disadvantage, which complicates them to be dried, composted or incinerated. Terrestrial plants, however, develop longer, fibrous root systems with root hairs, which provide an extremely high surface and easily dry out in the open air [235]. Hydroponically cultivated roots of some terrestrial plans show remarkable abilities of removing toxic metals from polluted waters. This process is called rhizofiltration [236]. It has recently reported that young plant seedlings of Indian mustard may be as effective as roots in removing heavy metals from water when there are grown as aquaculture in aerated water [237]. This process has been termed blastofiltration [238]. Blastofiltration may turn out to be an alternative plant-based water treatment technology, which takes advantage of the fast increase in surface:volume ratio after germination and the ability of the seedlings to adsorb large quantities of heavy metals, which makes it suitable for water remediation.

An ideal plant for rhizofiltration should exhibit characteristics such as maximum heavy metal removal from the contaminated water together with easy handling, low maintenance costs, and a minimum of secondary waste, which has to be disposed off. The plant should also produce hydroponically significant large amounts of roots and should be able to tolerate high amounts of heavy metals in solution, have a high root:shoot ratio and grow safely. Sunflower plants showed production rates of up to 50 g dry weight/m^2 per day [231].

4.4. Advantages and Limitations of Phytoremediation

The use of plants for the removal of heavy metals from soil and water offers a wide range of advantages. Phytoremediation is a technology which can be applied in situ without moving or excavating large amounts of contaminated soil and leaves the topsoil in an undisturbed and usable condition [239]. It uses solar energy and is in general easy to apply. A variety of metals and radionuclides can be treated. As for the contaminated sites, phytoremediation provides a useful tool for sites, which cannot be readily remediated by other methods, e.g., sites of large extension with only low contaminant concentrations at shallow depths. Several plant species used for phytoremediation belong to well-studied crop plants so there is a wide knowledge available for application and management of those plants [240].

Another advantage of phytoremediation is the reduction or elimination of secondary air or water borne wastes as the plants provide ground cover, which stabilizes the soil and reduces wind or water erosion [241]. Wind blown dust and erosion result in exposure pathways of direct inhalation of contaminated air and ingestion of food contaminated by deposition of suspended matter onto food plants [242]. If hyperaccumulators are used, their biomass can be disposed of by incineration thus reducing the mass and volume of waste, which has to be deposited at landfills. Ratios as high as 200:1 have been reported for the comparison of conventional remediation methods (soil excavation and landfill disposal) with plant ash from incineration [243].

Phytoremediation is also very cost-effective compared to other remediation methods. The remediation of a contaminated site may include such processes as excavation, removal, isolation, or incineration of the contaminated soil, the transport and return of the residue to the site. Prices for this kind of remediation are reported to range from 200 € up to 600 € depending on site characteristics, transport, and landfill costs. In comparison, costs for the phytoremediation including site preparation, planting, and harvesting are as low as 5000 € for a one acre site [244]. In general, cost estimate for soil remediation is about 25 € to 100 € per ton and 0.2 € to 1.5 € per m^3 water [159] compared to approximately 20 € per m^3 water when standard microfiltration is used [240]. Moreover, phytoremediation makes contaminated sites more aesthetically appealing and helps turning brownfields into greenfields [245].

Finally, public acceptance for phytoremediation is very high. The relationships that humans have to plants are rooted deep in our evolutionary past and *Homo sapiens* has a deeply held innate appreciation for their value as we all depend on plants giving food, medicine, and shelter [246]. Table 13 lists some general advantages and disadvantages of phytoremediation. Although phytoremediation has many advantages when compared to conventional remediation technologies it is also necessary to mention some limitations. Hyperaccumulators often accumulate only one specific element, which excludes their use to sites with multiple contaminations. The amount of hyperaccumulators available is limited and for some heavy metals, plants have yet to be found. Often these plants show slow growth rates and small production of biomass. A lot of research has still to be conducted as for the use of genetic engineering to introduce genes into fast growing plants, to regulate root growth, or to increase production of selected plant enzymes. Another serious limitation of phytoremediation is the long time required for the clean up of a site, which will take several growing seasons. In some cases, 18-60 months may be needed for site closure [159]. It has been estimated that natural hyperaccumulators might take 13-16 years to clean up a typical site [247]. Therefore, it cannot be used when there is an imminent danger to human health and the environment. Investors and property developer may not wish to wait years until a site is cleaned up by phytoremediation. The use of plants does not result in a 100% removal of contaminants.

Table 13
General advantages and disadvantages of phytoremediation of heavy metals

Advantages	Disadvantages
Cost:	Time:
Low capital and operating costs	Slower than some alternatives
Metal recycling provides further economic advantages	Seasonally dependent
Performance:	Performance:
Permanent treatment solution	May not be applicable to all mixed wastes
In situ application avoids excavation	Soil phytoremediation applicable only to surface soils
Other:	Other:
High public acceptance	Need to displace existing facilities
Aesthetically pleasing	Regulatory problems
Turns brownfields into greenfields	Lack of recognized economic performance data
Can be used during site investigation or after closure	

Modified after Ref. 159.

High heavy metal concentrations on some sites may cause toxic effects on plants. Only the topsoil (e.g. in general the top 1 m of soil) is available for phytoremediation. The effectiveness is controlled by the bioavailability of the heavy metals. Parameters such as soil texture and pH, contaminant concentration, salinity, and toxicity must be within the limit of plant tolerance. Costs may rise when the soil has to be pretreated with complexing or chelating agents, with soil amendments or fertilizers and insecticides in order to enhance bioavailability and plant growth. Although phytoremediation appears to be very cost effective, further validation of actual clean-up projects is needed as well as studies of economic performance and comparison to other competing technologies. Finally, regulatory problems have to be mentioned. Regulators are often unfamiliar with phytotechnology and its capabilities. Phytoremediation is a technology with the potential to remediate many sites contaminated with heavy metals. In many sites conducive to this technology it may fill the role of a cost-effective and less invasive form of treatment.

REFERENCES

[1] D.C. Adriano, Trace Elements in Terrestrial Environments, Springer, New York, Berlin, Heidelberg, 2001.
[2] M. Wilichowski, in" Treatment of Contaminated Soil" (R. Stegmann, G. Brunner, W. Calmano and G. Matz, eds.), p. 417-433, Springer, Berlin, Heidelberg, 2001.

[3] P.C. Hiemenz, Principles of Colloid and Surface Science, Marcel Dekker, New York, 1986.
[4] J. Werther, O. Malerius and J. Schmidt, in" Treatment of Contaminated Soil" (R. Stegmann, G. Brunner, W. Calmano and G. Matz, eds.), p. 435-459, Springer, Berlin, Heidelberg, 2001.
[5] T. Venghaus and J. Werther, Adv. Environ. Res., 2 (1998) 77.
[6] B. Yarar, in "Ullmann`s Encyclopedia of Industrial Chemisty" (W. Gerhartz, ed.), p. 23-1-23-30, VCH, Weinheim, 1988.
[7] M. Pearl and P. Wood, in "Altlastensanierung `93" (F. Arendt, R. Bosman and W.J. van den Brink, eds.), p. 1325-1333, Kluwer Academic Publishers, Dordrecht, Boston, London, 1993.
[8] M.S.E. Abdo and A.M. Darwish, Chem. Eng. Tech., 14 (1991) 119.
[9] L.K. Philip, Engineering Geology, 60 (2001) 209.
[10] P. Arz, Bauwirtschaft, 110 (1988) 831.
[11] T. Meggyes and N. Pye, in "Landfill Liner Systems" (U. Holzlöhner, H. August, T. Meggyes and M. Brune, eds.), p. N1-28, Penshaw Press, Sunderland, 1995.
[12] H.B. Bradl, Geotechnik, Special Edition (1995) 175.
[13] H.B. Bradl, in "Proceedings of the International Containment Technology Conference" St. Petersburg, FL, p. 645-651, 1997.
[14] H.G. Schmidt and J.M. Seitz, Grundbau, Ernst und Sohn, Berlin, 1998.
[15] H.L. Jessberger and M. Geil, Geotechnik, 4 (1992) 237.
[16] R.G.H. Boyes, Civil Engineering 6 (1986) 42.
[17] H. Meseck, in "Dichtwände und Dichtsohlen" (H. Meseck, ed.), p. 155-170, Technische Universität Braunschweig, Braunschweig, 1987.
[18] G. Ghezzi, P. Ghezzi and M. Pellegrini, in "Sardinia 99, 7th Waste Management and Landfill Symposium" (T.H. Christensen, R. Cossu and R. Stegmann, eds.), p. 549-554, CISA, Cagliari, 1999.
[19] H.H. Weber, W. Fresenius, G. Matthess, H. Müller-Kirchenbauer, K. Storp and E. Wessling, Altlasten: Erkennen, Bewerten, Sanieren, Springer, Berlin, 1990.
[20] S. Tóth, Geotechnik, 1 (1989) 1.
[21] C. Kutzner, Injektionen im Baugrund, Enke, Stuttgart, 1991.
[22] German Geotechnical Society, Geotechnics of Landfills and Contaminated Land: Technical Recommendations for the International Society of Soil Mechanics and Foundation Engineering, Ernst und Sohn, Berlin, 1993.
[23] L.N. Reddi and H.I. Inyang, Geoenvironmental Engineering – Principles and Applications, Marcel Dekker, New York, 2000.
[24] M.B. Jones, Tunnels and Tunnelling, 14 (1982) 31.
[25] J.G. Dash, H.Y. Fu and R. Leger, in "Proceedings of the International Containment Technology Conference" St. Petersburg, FL, p. 607-613, 1997.
[26] D. Lesmes, D. Cist and D. Morgan, in "Proceedings of the International Containment Technology Conference" St. Petersburg, FL, p. 1074-1077, 1997.
[27] A.P. Annan and J.L. Davis, Radio Science, 11 (1976) 383.
[28] M. Orfeuil, Electric Process Heating, Battelle Press, Columbus, Richland, WA, 1987.
[29] M.B.Volf, Chemical Approach to Glass, Elsevier, New York, 1984.

[30] K. Czurda, P. Huttenloch, G. Gregolec and K.E. Roehl, in "Advanced Groundwater Remediation: Active and Passive Technologies (F.G. Simon, T. Meggyes and C. McDonald, eds.), p. 173-192, Thomas Telford, London, 2002.
[31] Y.B. Acar and A.N. Alshawabkeh, Environ. Sci. Technol., 27 (1993) 2638.
[32] C.J. Bruell, B.A. Segall and M.T. Walsh, J. Environ. Eng., 122 (1992) 68.
[33] R.F. Probstein and R.E. Hicks, Science, 260 (1993) 498.
[34] R.F. Probstein, Physicochemical Hydrodynamics – An Introduction, John Wiley & Sons, New York, 1994.
[35] J. Voss, M. Altrogge, D. Golinske, O. Kranz, D. Nünnecke, D. Petersen and E. Waller, in" Treatment of Contaminated Soil" (R. Stegmann, G. Brunner, W. Calmano and G. Matz, eds.), p. 547-562, Springer, Berlin, Heidelberg, 2001.
[36] D. Rahner, G. Ludwig and J. Roehr, Electrochim. Acta, 47 (2002) 1405.
[37] S.A. Jackman, G. Maini, K.A. Sharman and C.J. Knowles, Enzyme and Microbial Technology, A24 (1999) 316.
[38] S.V. Ho, P.W. Sheridan, C. J. Athmer, M.A. Heitkamp, J.M. Brackin, D. Weber and P.H. Brodsky, Environ. Sci. Technol., A29 (1995) 2528.
[39] S.V. Ho, C. J. Athmer, P.W. Sheridan, B.M. Hughes, R. Orth, D. McKenzie, P.H. Brodsky, A. Shapiro, R. Thornton, J. Salvo, D. Schultz, R. Landis, R. Griffiths and S. Shoemaker, Environ. Sci. Technol., 33 (1999a) 1086.
[40] S.V. Ho, C. J. Athmer, P.W. Sheridan, B.M. Hughes, R. Orth, D. McKenzie, P.H. Brodsky, A. Shapiro, T.M. Sivavec, J. Salvo, D. Schultz, R. Landis, R. Griffiths and S. Shoemaker, Environ. Sci. Technol., 33 (1999b) 1092.
[41] D.M. Mackay and J.A. Cherry, Environ. Sci. Technol., 23 (1989) 630.
[42] G. Teutsch, P. Grathwohl, H. Schad and P. Werner, Grundwasser, 1 (1996) 12.
[43] R.C. Starr and J.A. Cherry, Ground Water, 32 (1994) 465.
[44] P. Grathwohl, Zeitschr. Umweltchem. Ökotox., 4 (1992) 231.
[45] H.B. Bradl, Bautechnik, 12 (1996) 832.
[46] C. Reeter, S. Chao and A. Gavaskar, Permeable Reactive Wall Remediation of chlorinated Hydrocarbons in Groundwater, US Department of Defense, ESTCP Cost and Performance Report, Washington, D.C., 1999.
[47] H.D. Stupp, Terratech, 8 (2000) 34.
[48] H.G. Edel, and T. Voigt, Terratech, 1 (2001) 40.
[49] P. Bayer, C. Bürger, M. Finkel and G. Teutsch, in "Advanced Groundwater Remediation" (F.-G. Simon, T. Meggyes and C. McDonald, eds.), p. 267-282, Thomas Telford, London, 2002.
[50] R.W. Gillham and S.F. O`Hannesin, Ground Water, 32 (1991) 958.
[51] F.-G. Simon, T. Meggyes and T. Tünnermeier, in "Advanced Groundwater Remediation" (F.-G. Simon, T. Meggyes and C. McDonald, eds.), p. 3-34, Thomas Telford, London, 2002.
[52] D. Langmuir, Geochim. Cosmochim. Acta, 42 (1978) 547.
[53] N.K. Chung, in "Standard Handbook of Hazardous Waste Treatment and Disposal" (H.M. Freeman, ed.) p. 7.21-7.32, McGraw-Hill, New York, 1989.
[54] Q.Y. Ma, S.J. Traina and T.J. Logan, Environ. Sci. Technol., 27 (1993) 1803.
[55] X.B. Chen and J.V. Wright, Water, Air, Soil Poll., 98 (1997) 57.
[56] R.M. Powell, R.W. Puls, S.K. Hightower and D.A. Sabatini, Environ. Sci. Technol., 29 (1995) 1913.

[57] B. Gu, L. Liang, M. J. Dickey, X. Yin and S. Dai, Environ. Sci. Technol., 32 (1998) 3366.
[58] J. N. Fiedor, W.D. Bostick, R.J. Jarabek and J. Farrell, Environ. Sci. Technol., 32 (1998) 1466.
[59] S.J. Morrison and R.R. Spangler, Environ. Sci. Technol., 26 (1992) 1922.
[60] S.J. Morrison, R.R. Spangler and V.S. Tripathi, J. Contaminant Hydrol.,176 (1995) 333.
[61] S.J. Morrison and R.R. Spangler, Environ. Progress, 12 (1993) 175.
[62] J. Yang and B. Volesky, in „Conference Proceedings International Biohydrometallurgy Symposium" (A. Ballester and R. Amils, eds.), p. 483-492, San Lorenzo de El Escorial, Elsevier, Amsterdam, 1999.
[63] B. Volesky, Biosorption of Heavy Metals, CRC Press, Boca Raton, FL, 1990.
[64] Y.A. Gorby and D.R. Lovley, Environ. Sci. Technol., 26 (1992) 205.
[65] M. Lehmann, A.I. Zouboulis and K.A. Matis, Chemosphere, 39 (1999) 881-892.
[66] G.M. Haggerty and R.S. Bowman, Environ. Sci. Technol., 28 (1994) 452-458.
[67] Y.L. Chen, S. Chen, C. Frank and J. Israelachvili, J. Colloid Interface Sci., 153 (1992) 244.
[68] E. Beitinger and E. Bütow, in "Grundwassersanierung 1997" (H.P. Lühr, ed.), p. 342-356, Erich Schmidt, Berlin, 1997.
[69] D.A. Smyth, S.G. Shikaze and J.A. Cherry, Land Contamination and Reclamation, 5 (1997) 131.
[70] T. Jansen and A. Grooterhorst, Terratech, 3 (199) 46.
[71] S.R. Day, S.F. O`Hannesin and L. Marsden, J. Hazardous Materials, 67 (1999) 285.
[72] D.W. Hubble, R.W. Gillham and J.A. Cherry, in "Proceedings of the International Containment Technology Conference" St. Petersburg, FL, p. 872-878, 1997.
[73] H.B. Bradl, Verfahren zur Herstellung eines permeablen Reaktorraumes im Boden, German Patent No. DE 197 29 303 (1997).
[74] K.J. Cantrell, D.I. Kaplan and T.J. Gilmore, in "Proceedings of the International Containment Technology Conference" St. Petersburg, FL, p. 774-780, 1997.
[75] A.R. Gavaskar, J. Hazardous Materials, 68 (1999) 41.
[76] E. Debreczeni and T. Meggyes, in "Sardinia 99, Proceedings of the 7th International Waste Management and Landfill Symposium" (T.H. Christensen, R. Cossu and R. Stegmann, eds.), p. 533-540, CISA, Cagliari, 1999.
[77] A.G. Duba, K.L. Jackson, M.C. Jovanovich, R.B. Knapp and R.T. Taylor, Environ. Sci. Technol., 30 (1996) 1982.
[78] D.R. Burris and C.P. Antworth, J. Contaminant Hydrol., 10 (1992) 325.
[79] L. Murdoch, B. Slack, B. Siegrist, S. Vesper and T. Meiggs, in "Proceedings of the International Containment Technology Conference" St. Petersburg, FL, p. 445-451., 1997.
[80] A.B. Cunningham, W.G. Characklis, F. Aberdeen and D. Crawford, Environ. Sci. Technol., 25 (1991) 1305.
[81] A.B. Cunningham, B. Warwood, P. Sturmann, K. Horrigan, G. James, J. W. Costerton and R. Hiebert, in "The Microbiology of the Terrestrial Deep Surface" (P.S. Amy and D.L. Haldeman, eds.), p. 325-344, Lewis, Boca Raton, FL, 1997.

[82] T. Sivavec, T. Krug, T. Berry-Spark and R. Focht, in "Advanced Groundwater Remediation" (F.-G. Simon, T. Meggyes and C. McDonald, eds.), p. 87-100, Thomas Telford, London, 2002.
[83] H.B. Bradl and U. Bartl, in "Sardinia 99, Proceedings of the 7th International Waste Management and Landfill Symposium" (T.H. Christensen, R. Cossu and R. Stegmann, eds.), p. 525-540, CISA, Cagliari, 1999.
[84] U. Förstner, Integrated Pollution Control, Springer, Berlin, Heidelberg, New York, 1998.
[85] H.H. Hahn, Wassertechnologie, Springer, Berlin, 1987.
[86] B.D. Honeyman and P.H. Santschi, Environ. Sci. Technol., 8 (1988) 862.
[87] J.W. Patterson, Industrial Wastewater Treatment Technology, Butterworth Publishers, Boston, 1985.
[88] G.C. Cushnie, Jr., Electroplating Wastewater Pollution Control Technology, Noyes Publication, Park Ridge, 1985.
[89] K. Fischwasser and H. Schilling, Wasser, Luft und Boden, 4 (1992) 34.
[90] G. Lagaly, in „Ullmann`s Encyclopedia of Industrial Chemistry"(F.T. Campbell, R. Pfefferkorn and J.R. Rounsaville, eds.), A7, 341-367, Verlag Chemie, Weinheim, 1986.
[91] P.A. Thiessen, Z. Elektrochemie, 48 (1942) 675.
[92] E. Matijevic, R.J. Kuo and H. Kolmy, J. Colloid Interface Sci., 80 (1981) 94.
[93] U. Hoffmann, Angew. Chemie, 18 (1968) 736.
[94] H. Van Olphen, An introduction to clay colloid chemistry, John Wiley and Sons, New York, 1977.
[95] E.J.W. Verwey and J.Th.G. Overbeek, Theory of the stability of lyophobic colloids, Elsevier, Amsterdam, 1948.
[96] E. Barouch, E. Matijevic, T.A. Ring and J.M. Finlan, J. Colloid Interface Sci., 67 (1978) 1.
[97] B.V. Derjaguin and L.V. Landau, Acta Physicochim. U.R.S.S., 14 (1941) 635.
[98] E.P. Honig, G. J. Roebersen and P. H. Wiersema, J. Colloid Interface Sci., 36 (1971) 97.
[99] P. Kunz, Behandlung von Abwasser, Vogel Verlag, Würzburg, 1990.
[100] P.S. Cartwright, in „Pretreatment in Chemical Water and Wastewater Treatment"(H.H. Hahn and R. Klute, eds.), p. 189-200, Springer, Berlin, 1988.
[101] US Environmental Protection Agency, Stabilization/Solidification of CERCLA and RCRA Wastes, Physical Tests, Chemical Testing Procedures, Technology Screening and Field Activities, EPA/625/6-89/022, Office of Research and Development, Cincinnati, OH, 1989.
[102] US Environmental Protection Agency, Engineering Bulletin: Technology Alternatives for the Remediation of Soils Contaminated with As, Cd, Cr, Hg, and Pb, EPA/540/S-97/500, Office of Research and Development, Cincinnati, OH, 1997.
[103] C.C. Wiles, in "Standard Handbook of Hazardous Waste Treatment and Disposal" (H. M. Freeman, ed.), p. 7.31-7.59, McGraw-Hill, New York, 1997.
[104] M.D. LaGrega, P.L. Buckingham and J.C. Evans, Hazardous Waste Management, McGraw-Hill, New York, 2001.
[105] J.R. Conner and S.L. Hoeffner, Critical Reviews in Environmental Science and Technology, 28 (1998) 325.

[106] J.R. Conner, Chemical Fixation and Solidification of Hazardous Wastes, Van Nostrand Reinhold, New York, 1990.
[107] J.R. Conner and S.L. Hoeffner, Critical Reviews in Environmental Science and Technology, 28 (1998) 397.
[108] D. Grasso, Hazardous waste site remediation - Source control, Lewis Publishers, Boca Raton, FL, 1993.
[109] US, Environmental Protection Agency, Solidification/Stabilization and its application to waste materials, Technical Resource Document, EPA/530/R-93/012, Office of Research and Development, Washington DC, 1993.
[110] W. Stumm and J.J. Morgan, Aquatic Chemistry, John Wiley and Sons, New York, 1981.
[111] W.L. Lindsay, Chemical equilibria in soils, John Wiley & Sons, New York, 1979.
[112] J.O. Nriagu and P.B. Moore, Phosphate Minerals, Springer, Berlin, 1984.
[113] H.A. Van der Sloot, L. Heasman and P. Quevauviller, Harmonization of leaching/extraction tests, Elsevier, Amsterdam, 1997.
[114] US Environmental Protection Agency, Test Methods for Evaluating Solid Waste Physical/ Chemical Methods, Office of Research and Development, Washington DC, 1996.
[115] A. Tessier, P.G.C. Campbell and M. Bisen, Anal. Chem., 51 (1979) 844.
[116] P.Quevauviller, M. Lachia, E. Barahona, G. Rauret, A. Ure, A. Gomez, A. and H. Muntau, Sci. Total Environ., 178 (1996) 127.
[117] M.V. Ruby, A. Davis, S. Eberle and C.M. Sellstone, Environ. Sci. Technol., 30 (1996) 422.
[118] M.V. Ruby, R. Schoof, W. Brattin, G. Goldack, P. Post, M. Harnois, D.E. Mosby, S.W. Casteel, W. Berti, M. Carpenter, D. Edwards, D. Cragin and W. Chappell, Environ. Sci. Technol., 33 (1999) 3697.
[119] Interstate Technology Regulatory Cooperation (ITRC) Work Group, Emerging technologies for the remediation of metals in soils - insitu stabilization / inplace inactivation, Final report, www.itrcweb.org., 1997.
[120] D.C. Adriano, A.L. Page, A.A. Elseewi, A.C. Chang and I. Straughan, J. Environ. Qual., 9 (1980) 333.
[121] L.Y. Sale, D.S. Chanasyk and M.A. Naeth 1997, Can. J. Soil Sci., 77 (1997) 677.
[122] Q.Y. Ma, S.J. Traina, T.J. Logan and J.A. Ryan, J.A., Environ. Sci. Technol. 28 (1994a) 1219.
[123] Q.Y. Ma, T.J. Logan, S.J. Traina and J.A. Ryan, J.A., Environ. Sci. Technol., 28 (1994b) 408.
[124] X. Chen, J.V. Wright, J.L. Conca and L.M. Peurrung, L.M., Environ. Sci. Technol., 31 (1997) 624.
[125] A.S. Knox, J.C. Seaman, M.J. Mench and J. Vangronsveld, in" Environmental Restoration of Metals-contaminated Soils" (I.K. Iskandar, ed.), p. 21-60, Lewis Publishers, London, 2001.
[126] J.O. Nriagu, Geochim. Cosmochim. Acta, 37 (1973) 367.
[127] J.O. Nriagu, J.O., Geochim. Cosmochim. Acta, 37 (1973) 1735.
[128] J.O. Nriagu, J.O., Geochim. Cosmochim. Acta, 38 (1974) 887.
[129] J. Cotter-Howells and S. Capron, S., Appl. Geochem., 11 (1996) 335.

[130] M.V. Ruby, A. Davis and A. Nicholson, A., Environ. Sci. Technol., 28 (1994) 646.
[131] J.A. Ryan, P. Zhang, D. Hesterberg, J. Chou and D.E. Sayers, D.E., Environ. Sci. Technol., 35 (2001) 3798.
[132] A. Davis, J.W. Drexler, M.V. Ruby and A. Nicholson, Environ. Sci. Technol., 27 (1993) 1415.
[133] Q.Y. Ma, S.J. Traina and T.J. Logan, Environ. Sci. Technol., 27, 1803–1810.
[134] Q.Y. Ma, T.J. Logan and S.J. Traina,. Environ. Sci. Technol., 29 (1995) 1118.
[135] T. Suzuki, K. Ishigaki and M. Miyake, J. Chem. Soc. Faraday Trans., 80 (1984), 3157.
[136] Y. Takeuchi, T. Suzuki and H. Arai, J. Chem. Eng. Jap., 21 (1988) 98.
[137] I.L. Shashkova, A.I. Rat'Ko and N.V. Kitikova, Colloids and Surfaces A: Physiochemical and Engeneering Aspects, 160 (1999) 207.
[138] S. Sugyiama, N. Fukuda, A. Matsumoto, H. Hayashi, N. Shigemoto, Y. Hiraga and J.B. Moffat, J. Colloid Interface Sci., 220 (1999), 324.
[139] A. Chlopecka and D.C. Adriano, D.C., Sci. Total Environ., 207 (1997)195.
[140] V. Laperche, S.J. Traina, P. Gaddam and T.J. Logan, Environ. Sci. Technol., 30 (1996) 3321.
[141] T.T. Eighmy, B.S. Crannell, L.G. Butler, F.K. Cartledge, E.F. Emery, D. Oblas, J.E. Krzanowski, J.D. Eusden Jr, E.L. Shaw and C.A. Francis, Environ. Sci. Technol., 31 (1997) 3330.
[142] Eighmy, T.T., B.S. Crannell and J.F. Krzanowski, Waste Management, 18 (1998) 513.
[143] J. Yang, D.E. Mosby, S.W. Casteel and R.W. Blanchar, Environ. Sci. Technol. 35 (2001) 3553.
[144] Y. Xu, F.W. Schwartz and S.J. Traina, Environ. Sci. Technol., 28 (1994)1472.
[145] L. Wu, W. Forsling and P.W. Schindler, J. Colloid Interface Sci., 147 (1991) 178.
[146] N.T. Basta S.L. and McGowen, Environ. Poll., 127 (2004) 73.
[147] A. Xenidis, G. Bouboukioti and P. Theodoratos, P., In situ immobilization of Pb and As in contaminated soils, 1st International Conference-Advances in Mineral Resources Management and Environmental Geotechnology, Chania, Crete, 7-9 June, 2004.
[148] J. Boisson, A. Ruttens, M. Mench and J. Vangronsveld, Environ. Poll., 104 (1999) 225.
[149] A. Chlopecka and D.C. Adriano, Environ. Sci. Technol., 30 (1996) 3294.
[150] A. Xenidis, C. Stouraiti and I. Paspaliaris, J. Soil Contamin., 8 (1999) 681.
[151] P. Theodoratos, N. Papassiopi and A. Xenidis, J. Hazard. Mat., 94 (2002) 135.
[152] R.X. Cao, L.Q. Ma, M. Chen, S.P. Singh and W.G. and Harris, Environ. Poll., 122 (2003) 19.
[153] V. Laperche, V., 2001, in "Environmental Restoration of Metals-contaminated Soils" (I.K. Iskandar, ed.), p. 61-76, Lewis Publishers, London, 2001.
[154] M. Chen, L.Q. Ma, S.P. Singh, R.X. Cao and R. Melamed R., Advan. Environ. Res., 8 (2003) 93.
[155] C.R. Evanko and D.A. Dzombak, Remediation of metals-contaminated soils and groundwater, Ground-Water Remediation Technologies Analysis Center (GWRTAC), Technology Evaluation Report TE-97-01, Pittsburgh, PA, 1997.
[156] C.N. Mulligan, R. N. Yong and B.F. Gibbs, Eng. Geol., 60 (2001) 193.

[157] US Environmental Protection Agency, International Waste Technologies/Geo-Con In Situ Stabilization/Solidification - Applications Analysis Report, EPA/540/A5-89/004, Office of Research and Development, Cincinnati, OH, 1989.
[158] S.S. Suthersan, Remediation Engineering – Design Concepts, Lewis Publishers, New York, 1997.
[159] D.J. Glass, in "Phytoremediation of Toxic Metals: Using Plants to Clean Up the Environment" (J. Raskin and B.D. Ensley, eds.), p. 15-31, John Wiley & Sons, New York, 2000.
[160] F.R. Siegel, Environmental Geochemistry of Potentially Toxic Metals, Springer, Berlin, Heidelberg, 2002.
[161] R.L. Chaney, M. Malik, Y.M. Li, S.L. Brown, E.D. Brewer, J.S. Amgle and A.J.M. Baker, Current Opin. Biotechnol., 8 (1997) 279.
[162] J. Raskin and B.D. Ensley, Phytoremediation of Toxic metals – Using Plants to clean up the Environment, John Wiley & Sons, New York, 2000.
[163] W.H.O. Ernst, in "Heavy metals" (W. Salomons, U. Förstner and P. Mader, eds.), p. 141-149, Springer, Berlin, Heidelberg, 1995.
[164] D.T. Tsao, in "Phytoremediation" (D.T. Tsao, ed.), p. 1-50, Springer, Berlin, Heidelberg, 2003.
[165] J.G. Forchhammer, Poggendorf's Annal. Phys. Chem., 91 (1855) 60.
[166] R.R. Brooks, Plants that Hyperaccumulate Heavy Metals – Their Role in Phytoremediation, Microbiology, Archaeology, Mineral Exploration and Phytomining, CAB International, New York, 1998.
[167] A.S. Moffat, Science, 269 (1995) 302.
[168] C.F. Harrington, D.J. Roberts and G. Nickless, Can. J. Botany, 74 (1996) 1742.
[169] K. Winterhalder, Environ. Rev., 4 (1996) 185.
[170] R.D. Reeves, R.M. McFarlane and R.R. Brooks, Am. J. Botany, 70 (1983) 1297.
[171] R. Karthikeyan and P.A. Kulakow, in "Phytoremediation" (D.T. Tsao, ed.), p. 51-74, Springer, Berlin, Heidelberg, 2003.
[172] J.F. Shimp, J.C. Tracy, L.C. Davis, E. Lee, W. Huang and L.E. Erickson, Environ. Sci. Technol., 23 (1993) 41.
[173] E.A. Paul and F.E. Clark, Soil Microbiology and Biochemistry, Academic Press, San Diego, 1996.
[174] R. Hissett and T.R.G. Gray, in"The Role of Terrestrial and Aquatic Organisms in Decomposition Processes" (J.M. Anderson and A. Mac Fadden, eds.), p. 23-39, Blackwell, Oxford, 1976.
[175] B.E. Rittmann, Water Resources Res., 29 (1993) 2630.
[176] T.A. Anderson, E.A. Guthrie and B.T. Walton, Environ. Sci. Technol., 27 (1993) 2630.
[177] D.M. Sylvia and S.E. Williams, in "Mycorrhizae in Sustainable Agriculture" (G.J. Bethlenfalvay and R.G. Linderman, eds.), p. 101-124, ASA Special Publication Number 54, Madison, WI, 1992.
[178] D.T. Tsao, The Elicited and Enhanced Production of Phytoalexins in Cotton Suspension Cultures, MS Thesis, Purdue University, IN, 1990.
[179] J. Koch, L. Wang, C.A. Ollson, W.R. Cullen and K.J. Reimer, Environ. Sci. Technol., 34 (2000) 22.
[180] C.L. Trejo, A.L. Clephan and W.J. Davies, Plant Physiology, 109 (1995) 803.

[181] R.S. Hedge and J.S. Fletcher, Chemosphere, 32 (1996) 2471.
[182] P.E. Olsen and J.S. Fletcher, Bioremediation J., 3(1999) 27.
[183] A.M. Ferro, R.C. Sims and B. Bugbee, J. Environ. Qual., 23 (1994) 272.
[184] W. Aprill and R.C. Sims, Chemosphere, 20 (1990) 253.
[185] L.E. Newman, S.E. Strand, N. Choe, J. Duffy, G. Ekuan, M. Ruszay, B.B. Shurtleff, J. Wilmoth, P. Heilman and M.P. Gordon, Environ. Sci. Technol., 31 (1997) 1062.
[186] V.A. Nzengung, L.N. Wolfe, D.E. Rennels, S.C. McCutcheon and C. Wang, Internat. J. Phytoremediation, 1 (1999) 203.
[187] C. Flanders, J. Dec and J.-M. Bollag, Bioremediation J., 3 (1999) 315.
[188] B.D. Ensley, in " Phytoremediation of Toxic metals – Using Plants to clean up the Environment" (J. Raskin and B.D. Ensley, eds.), p. 3-31, John Wiley & Sons, New York, 2000.
[189] A.J.M. Baker, S.P. McGrath, C.M.D. Sidoli and R.D. Reeves, Res. Conserv. Recycl., 11 (1994) 41.
[190] J. Raskin, P.B.A. Nanda Kumar, S. Dushenkov and D.E. Salt, Curr. Opin. Biotechnol., 5 (1994) 285.
[191] S.L. Brown, J. Environ. Qual., 23 (1994) 1151.
[192] M.J. Blaylock and J. W. Huang, in " Phytoremediation of Toxic metals – Using Plants to clean up the Environment" (J. Raskin and B.D. Ensley, eds.), p. 53-70, John Wiley & Sons, New York, 2000.
[193] R.D. Reeves and A.J. Baker, in " Phytoremediation of Toxic metals – Using Plants to clean up the Environment" (J. Raskin and B.D. Ensley, eds.), p. 193-229, John Wiley & Sons, New York, 2000.
[194] R.D. Harter, Soil Sci. Soc. Am. J., 47 (1983) 47.
[195] W.A. Norvell, Soil Sci. Soc. Am. J., 48 (1984) 1285.
[196] L. Ramos, L.M. Hernandez and M.J. Gonzalez, J. Environ. Qual., 23 (1994) 50.
[197] M.B. McBride, Environmental Chemistry of Soils, Oxford University Press, New York, 1994.
[198] W.A. Norwell, in "Micronutrients in Agriculture" (J.J. Mortvedt, ed.), p. 187-227, Soil Science Society of America, Madison, 1991.
[199] N. Albasel and A. Cottenie, Soil Sci. Soc. Am. J., 49 (1985) 86.
[200] R. Prost, Contaminated Soils, INRA, Paris, 1997.
[201] W.L. Lindsay and W.A. Norvell, Soil Sci. Soc. Am. J., 42 (1978) 421.
[202] A.J.M. Baker and P.L. Walker, in"Heavy Metal Tolerance in Plants: Evolutionary Aspects" (A.J. Shaw, ed.), p. 155-176, CRC Press, Boca Raton, 1989.
[203] A.J.M. Baker and R.R. Brooks, Biorecovery, 1 (1989) 81.
[204] P.B.A. Kumar, V. Dushenkov, H. Motto and I. Raskin, Environ. Sci. Technol., 29 (1995) 1232.
[205] M.J. Blaylock, D.E. Salt, S. Dushenkov, C. Gussman, Y. Kapulnik, B.D. Ensley and I. Raskin, Environ. Sci. Technol., 31 (1997) 860.
[206] J.W. Huang and S.D. Cunningham, New Phytol., 134 (1996) 75.
[207] J.W. Huang, J. Chen, W.R. Berti and S.C. Cunningham, Environ. Sci. Technol., 31 (1997) 800.
[208] J.W. Huang, D.L. Grunes and L.V. Kochian, Proc. Natl. Acad. Sci. USA, 91 (1994) 3473.

[209] J.W. Huang, J.E. Shaff, D.L. Grunes and L.V. Kochian, Plant Phys., 98 (1992) 230.
[210] S.D. Ebbs and L.V. Kochian, J. Environ. Qual., 26 (1997) 776.
[211] W.R.Berti and S.D. Cunningham, in "Phytoremediation of Toxic metals – Using Plants to clean up the Environment" (J. Raskin and B.D. Ensley, eds.), p. 71-88, John Wiley & Sons, New York, 2000.
[212] J.R. Pitchel, W.A. Dick and P. Sutton, J. Environ. Qual., 23 (1994) 766.
[213] B.J. Alloway, Heavy Metals in Soils, Blackie Academic and Professional, New York, 1995.
[214] P. Zang, J.A. Ryan and L.T. Bryndzia, Environ. Sci. Technol., 31 (1997) 2673.
[215] A.P. Davis, in"Encyclopedia of Surface and Colloid Science" (A. Hubbard, ed.), p. 440-449, Marcel Dekker, New York, 2002.
[216] H.B. Bradl, in"Encyclopedia of Surface and Colloid Science" (A. Hubbard, ed.), p. 373-384, Marcel Dekker, New York, 2002.
[217] H.B. Bradl, J. Colloid Interface Sci., 277 (2004)1.
[218] B. Gworek, Plants and Soil, 143 (1992) 71.
[219] C.L. Rugh, S.P. Bizily and R. B. Meagher, in "Phytoremediation of Toxic metals – Using Plants to clean up the Environment" (J. Raskin and B.D. Ensley, eds.), p. 151-169, John Wiley & Sons, New York, 2000.
[220] M.P.de Souza, E.A.H. Pilon-Smith and N. Terry, in "Phytoremediation of Toxic metals – Using Plants to clean up the Environment" (J. Raskin and B.D. Ensley, eds.), p. 171-190, John Wiley & Sons, New York, 2000.
[221] J.B. Robinson and O.H. Tuovinen, Microbiol. Rev.e 48(198) 95.
[222] O.A. Summers, Annu. Rev. Microbiol.e 40 (1986) 607.
[223] T. Tezuka and K. Tonomura, J. Biochem., 80 (1976) 79.
[224] T.P. Begley, A.E. Walts and C.T. Walsh, Biochemistry, 25 (1986) 7186.
[225] A.C.P. Heaton, C.L. Rugh, N.-J. Wang and R.B. Meagher, J. Soil Contam., 7 (1998) 497.
[226] M.J. Blaylock and B.R. James, Plant Soil, 158 (1994) 1.
[227] J.E. Leggett and E. Epstein, Plant Physiol., 31 (1956) 222.
[228] M.P. Arvy, J. Exper. Bot., 44 (1993) 1083.
[229] G.L. Dilworth and R.S. Bandurski, Biochem. J., 163 (1977) 521.
[230] T. Adler, Sci. News, 150 (1996) 42.
[231] S. Dushenkov and Y. Kapulnik, in "Phytoremediation of Toxic metals – Using Plants to clean up the Environment" (J. Raskin and B.D. Ensley, eds.), p. 89-106, John Wiley & Sons, New York, 2000.
[232] M. Falbo and T.E. Weaks, Econ. Bot., 44 (1990) 40.
[233] F.E. Dierberg, T.A. de Busk and N.A. Goulet, Jr. in" Aquatic Plants for Water Treatment and Resource Recovery" (K.R. Reddy and W.H. Smith, eds.), p. 497-504, Magnolia Publishing, Orlando, 1987.
[234] S.C. Mo, D.S. Choi and J.W. Robinson, J. Environ. Sci. Health, A24 (1989) 135.
[235] H.J. Dittmer, Am. J. Bot. 24 (1937) 417.
[236] V. Dushenkov, P.B.A. Nanda Kumar, H. Motto and I. Raskin, Environ. Sci. Technol., 29 (1995) 1239.
[237] D.E. Salt, I.J. Pickering, R.G. Prince, D. Gleba, S. Dushenkov, R.D. Smith and I. Raskin, Environ. Sci. Technol., 31 (1997) 1636.

[238] I. Raskin, R.D. Smith and D.E. Salt, Curr. Opin. Biotechnol., 8 (1997) 221.
[239] C.L. Rugh, H.D. Wilde, N.M. Stack, D.M. Thompson, A.O. Summers and R.B. Meagher, Proc. Nat. Acad. Sci. USA, 93 (1996) 3183.
[240] T. McIntyre, in "Phytoremediation" (D.T. Tsao, ed.), p. 97-123, Springer, Berlin, Heidelberg, 2003.
[241] J. Vangronsveld, J. Colpaert and K. van Tichelen, Environ. Pollut., 94 (1996) 131.
[242] J. Schnoor, L.A. Licht, S.C. McCutcheon, N.L. Wolfe and L.H. Carreira, Environ. Sci. Technol., 29 (1996) 318.
[243] H. Black, Innovations, 103 (1995) 1106.
[244] D. Glass, The 1998 United States Market for Phytoremediation, D. Glass and Associates, Inc., Needham, MA, 1998.
[245] D.D. Genske, Urban Land – Degradation, Investigation and Remediation, Springer, Berlin, Heidelberg, 2003.
[246] R.K. Tucker and J.A. Shaw, in "Phytoremediation" (D.T. Tsao, ed.), p. 33-42, Springer, Berlin, Heidelberg, 2003.
[247] V. Boyd, Environ. Protect., 5 (1996) 38.

Index

A
AAS, 32, 33
Abrasion, 168
Absorption, see AAS, 30, 32-34, 36
Achromobacter sp., 237
Activated carbon, 167, 179, 192, 202
Additives
    for cement, 230, 231, 233, 235
    for gasoline, 114
    for glass formation, 174
    for feed, 110
    for flocculation, 197
    for fuel, 135
    for soil stabilisation, 201, 202, 203
    for stabilisation, 191
    for suspensions, 173
Admicelle, 186
Adsorption
    mechanisms, 48
    models, 59
    of cadmium, 99, 100, 101
    of chromium, 105, 106
    of cobalt, 143
    of copper, 108, 109
    of lead, 112, 113
    of manganese, 116, 117
    of mercury, 119
    of molybdenum, 124
    of nickel, 127
    of zinc, 140, 141
Aerobic microorganisms, 237
Aerosols, 15,
Agrostis capillaries, 229
Alkaline binders, 202
Allelopathy, 241
Alluviation, 6
Alyssum lesbiacum, 244
Alyssum murale, 244
Alyxia rubricaulis, 242
Alyssum sp., 128
Alyssum tenium, 244

Anaemia, 86, 89, 143, 144
Anaerobic microorganisms, 238
Analytical procedures
    for detection of heavy metals, 28
Anglesite, 114
Anhydrite, 4
Annelida, 103
Anthonomus grandis, 96
Antiknock agent, 116
Aoellanthus biformifolius, 242
Apatites, 222, 223, 224
Aporrectadea longa, 103
Aquifer, 12, 13, 18
Aquifuge, 13
Aquitard, 13
Arabidopsis thaliana, 248
Argentite, 146
Arsenic
    chemical and physical character of, 93
    ecotoxicological effects of, 96
    sources and applications of, 94
Arsenopalladinite, 133
Arsenopyrite, 3
Artificial ground freezing, 170
Asphyxiation, 129
Asthma, 89, 139
Astralagus racemosus, 242
Ataxia, 88, 124
Atmosphere, 1, 15, 16, 23
Atmospheric deposition, 16
Atomic absorption spectrometry, 32, 33
Atomic emission spectroscopy, 36
Attractive potential, 196
Attrition, 167
Avena sativa, 229
Azotobacter, 127

B
Bald eagle, 124
Bangladesh, 86, 98
Basalt, 7, 8, 132
Bauxite, 12
Beer's law, 34
Bentonite, 170
Beryl, 3
Bioaccumulation, 85, 87-90, 91, 92
Bioavailability, 87-90
Biobarriers, 188, 190
Biofilms, 267
Biological methods
    for remediation of heavy metals, 165

Biomagnification, 122
Biopolymer trenching, 188
Biosolids, 21
Biosorption, 186
Blindness, 123
Blue-green algae, 126, 144
Boll weevil, 95
Bored-pile walls, 172
Braggite, 132
Brass, 110
Brassicaceae, 128
Braunite, 117
Bronchitis, 97, 108
Bronze, 110, 148
Bushveld, 128, 131

C

Cacodylic acid, 96, 97
Cadmium
    chemical and physical character of, 98
    ecotoxicological effects of, 103
    sources and applications of, 101
Caisson drilling, 189
Calcite, 4, 5, 60, 61
Carbon monoxide dehydrogenase, 128
Casing, 189
Cassiterite, 148
Catalytic converter, 128, 132-135, 137-139
Cathode lamp, 33, 35
Cement, 217
Cepaea nemoralis, 103
Cerebral palsy, 123
Cerussite, 114
Chalcopyrite, 13, 108
Charge injection devices, 38
Charge-coupled devices, 38
Chemical remediation techniques, 191
China, 85, 114
Chlorinated hydrocarbons, 15
Chlorite, 11
Chlorosis, 103, 128, 142, 147
Chromium
    chemical and physical character of, 104
    ecotoxicological effects of, 106
    sources and applications of, 107
Chromite, 3, 106, 131
Cinnabar, 3, 57, 121
Classification, 167
Clay fraction, 8
Clay minerals, 8-11
Coal ash, 95, 131

Coal combustion, 23
Coarse fraction, 166, 170
Cobalt, 143
Cobaltite, 143
Coccidiosis, 93
Codling moths, 95
Coking, 102, 122
Colloidal systems, 194
Colloids, 101, 119
Concrete, 218
Conglomerate, 5
Conjunctivitis, 97
Contact dermatitis, 89, 108, 128
Cooperite, 132
Copper
    chemical and physical character of, 108
    ecotoxicological effects of, 111
    sources and applications of, 110
Coronadite, 113
Critical flocculation concentration, 198
Critical micelle concentration, 186
Crooksite, 146
Cydia pomonella, 95
Czerny-Turner configuration, 38

D

Darcy`s law, 13
Deep soil mixing, 233
Dental amalgams, 121
Dental fillings, 121
Depigmentation, 101
Depressants, 167
Dermatitis, 89, 97, 108, 128, 143
Destruction
    of agglomerates, 168
Detection limits
    for AAS, 35, 36
    for EDXRF, 45
    for ICP-MS, 40
    for microprobe analysis, 46
    for PIXE, 46
    for XRF, 41
Dibutyl Sn, 148
Dicoma niccolifera, 117
Differentiation, 2, 3
Diffusive double layer, 59-73, 195
Dimethyl Hg, 120
Discus rotundatus, 103
Disintegration
    of particles, 168
Disodium methanearsonate, 96

DLVO theory, 195
Dolostone, 5
DTPA, 216
Dunite, 131
Dynamic Tests, 212

E
EDTA, 216, 242, 243
Eel, 138
Effusive rocks, 2-4
Eichhornia crassipes, 248
Electrode reactions, 178
Electrokinesis, 165
Electrokinetic techniques, 176
Electrolyte content, 170
Encapsulation, 170
Electromigration, 177
Electron microprobe, 32, 46
Electroosmosis, 177
Electrophoresis, 177
Electroremediation, 179
Electrothermal atomisation, 35
Eluviation, 6
Emission, 15, 16, 17, 23, 24
Encapsulation, 170
England, 102
Erythrite, 143
ETAAS, 35
EXAFS, 58, 56-61, 81, 82
Excitation, 36, 37, 42, 45
Ex situ treatment
 of contaminated soil, 230

F
Fe oxyhydroxides, 11
Ferrimolybdenite, 125
Ferrooxidase, 111
Filter velocity, 13
Filtration, 167
Fine particle fraction, 166
Flame photometry, 32, 33
Flocculation chemicals, 197
Flocculation, 167, 194
Flow-around tests, 215
Fluorimetry, 32, 33
Fluoropyromorphite, 225
Fly ash, 219
Forsterite, 3
Froodite, 132
Frozen walls, 173
Funnel and gate systems, 182

G
Galena, 3, 114
Ganges
 basin, 97
 River, 97
Gangrene, 98
Gastroenteritis, 97
Glycosuria, 108
Goethite, 4,
 sorption onto, 60-62
 sorption of Hg onto, 53, 57, 59
 sorption of Pb onto, 53
 surface, 58
Gold, 47, 133
Granodiorite, 131
Graphite furnace, 35
Gravel filters, 188
Gravel, 4, 5
Groundwater, 12-15
Gyprock, 5
Gypsum, 4, 5

H
Halite, 4, 5
Haumaniastrum roberti, 242
Hausmannite, 117
Hemimicelle, 186
Hemimorphite, 142
Hepatic degeneration, 128
Hepatopathy, 98
High pressure jets, 167
High pressure liberation unit, 167
Hordeum vulgare, 229
Hornsilver, 144
Hornworm, 95
Hutchinsonite, 146
Hydraulic fracturing, 188, 190
Hydraulic gradient, 13
Hydraulic remediation, 180
Hydrocotyle umbellata, 248
Hydrocyclone, 167
Hydrogenase, 128
Hydrophilic colloids, 194
Hydrophobic colloids, 194
Hydrophilic particles, 169
Hydrophobic particles, 169
Hydroxyapatite, 184, 223, 224, 226, 230
Hyperaccumulators, 242
Hyperkeratosis, 98
Hypocuprosis, 126
Hypogonadism, 143

I
ICP-AES, 39
ICP-MS, 39
ICP-OES, 32
Illite, 10, 11, 13
Ilsemanite, 125
Immission, 17
Inner-sphere adsorption, 49-52, 58
In situ treatment
   of contaminated soil, 230
Interaction energy, 195
Intrusive rocks, 2
Ion exchange, 193
Ion exchangers, 193
Iridium, 128
isomorphic substitution, 1, 10, 11
Isopoda, 103
Itai-itai, 102

J
Japan, 95, 102, 123
Jet Grouting, 172
Jinzu River, 102
Jordisite, 125
Joule heating, 175

K
Kaolin, 10
Kaolinite, 10, 11
Keratosis, 97, 126
Klebsiella, 127
Kupferschiefer, 131
Kyushu Island, 123

L
Laurite, 132
Leaching procedures, 211
Lead
   chemical and physical character of, 111
   ecotoxicological effects of, 115
   sources and applications of, 114
Lemna minor, 248
Leptinotarsa decemlineata, 95
Leptospermum scoparium, 107
Liberation, 166, 168, 169
Limestone, 5, 12
Liquid to Solid ratio, 211
Livingstonite, 121
Lorandite, 146
Lumbricus terrestris, 103

M
Macroencapsulation, 200
Magma, 1-4
Magmatic rocks, 1, 120
Magnetite, 2, 120
Manduca quinquemaculata, 95
Manganese
   chemical and physical character of, 115
   ecotoxicological effects of, 118
   oxyhydroxides, 120
   sources and applications of, 117
Manganese Nodule, 131
Manganite, 117
Mass spectra, 40
Mass spectrometer, 39-40, 45
Matrix, 6
Melanosis, 101
Membrane filter processes, 198
Mental retardation, 98
mer genes, 247
Mercury
   chemical and physical character of, 119
   ecotoxicological effects of, 122
   sources and applications of, 121
Metacinnabar, 121
Metallogenum sp., 12
Metamorphic rocks, 1, 5, 102
Metamorphosis, 5
   contact, 5
   kinetic, 5
Methylation
   of Hg, 120
Methyl Hg, 123
Methyl coenzyme M reductase, 128
Mexico
   Gulf of, 24
Micro X-ray absorption analysis, 32, 45, 46
Microanalytical methods, 32, 45
Microencapsulation, 200
Micronutrients, 111
Microwave heating, 175, 176
Minamata Bay, 123
Minamata disease, 123
Mineral processing
Mineral surfaces, 11
Mining, 22
Mixed-layer clay, 11
Mollusca, 103
Molybdate, 124, 125
Molybdenite, 4, 125

Molybdenosis, 126
Molybdenum
    chemical and physical character of, 124
    ecotoxicological effects of, 126
    sources and applications of, 125
Monobutyl Sn, 148
Monomethyl Hg, 120
Monosodium methanearsonate, 96
Montmorillonite, 10, 11

N
Nanofiltration, 194
Necrosis, 103, 128
Neoplasms, 98
Neuropathy, 88, 89, 97
Neutralization
Nickel
    chemical and physical character of, 126
    ecotoxicological effects of, 128
    sources and applications of, 127
Nicotiana tabacum, 248
Niggliite, 132
Nitrate reductase, 126
Nitrogenase, 126
Norilsk, 128, 131
Nutrients, 239

O
Olivine, 2, 131
Oniscus asellus, 103
Organotin compounds, 148
Osmium, 128
Osprey, 123
Osteomalachia, 103
Osteoporosis, 103
Outer-sphere adsorption, 49-51, 53, 58, 70

P
PAH, 15
Palladium, 128
Parent rock, 6, 17, 18
Paris Green, 93, 95, 97
Parkinson's disease, 88, 119
Permeability coefficient, 13
Permeability, 25
Pedogenesis, 6
Pegmatite, 114
Pentlandite, 127
Peridotite, 3, 128, 131
Permeable reactive barrier systems, 179-186

Permeable walls, 180
Pesticides, 19
Phase transfer, 192
Phaseolus vulgaris, 229
Phosphate treatment
    of contaminated soil, 230
Phosphatic fertilizers, 18
Phosphorite, 4
Phyllanthus serpentinus, 242
Physical remediation techniques, 165
Phytoextraction, 241
Phytofiltration, 248
Phytoremediation
    advantages of, 249
    limitations of, 241
    mechanisms of, 241
Phytostabilization, 247
Phytovolatilization, 247
Pigments, 102, 103, 108, 115, 126
PIXE, 32, 46
Plasma heating, 175
Plasma, 175
Platinum group elements, 128
    bioaccumulation of, 137
    chemical and physical character of, 129
    emission by car catalytic converters, 134
    in environmental matrices, 15
    sources and applications of, 131
    transformation of, 137
Pneumonia, 109
Pneumonitis, 129
Polychromator, 35, 38, 39
Polycyclic aromatic hydrocarbons, 15
Polymer microencapsulation, 221
Portland cement, 217
Potarite, 133
Powellite, 126
Pozzolans, 219
Precipitating agents, 192
Precipitation, 184, 192
Proton induced X-ray fluorescence, 32, 46
Pseudomonas sp., 191, 237
Psoriasis, 94
Pulmonary eosinophilia, 129
Pulmonary oedema, 129
Pulp, 122
Pump-and-treat technology, 180
Pyrite, 3, 4, 12, 18, 22
Pyrolusite, 63, 118
Pyromorphite, 222, 223

Pyrrhotite, 128

Q
Quartz, 2, 5, 6, 12, 61

R
Raphanus sativus, 229
Reactor technologies, 183
Repulsive potential, 196
Reverse osmosis, 191
Rhizobium, 127, 145
Rhizodegradation, 241
Rhizosphere, 266, 267
Rhodium, 129, 141
Rhodochrosite, 118
Rhodonite, 118
Rhodospirulum, 127
Rock
    magmatic, 1
    metamorphic, 1, 5, 6
    sedimentary, 1, 4, 5, 18
Rock cycle, 1
Rock salt, 4, 5
Root nodules, 127
Rotation drums
Rowland circle, 38, 39
Ruthenium, 129

S
Sand, 4, 5, 82, 101, 117, 132
Sandstone, 5
Saturated zone, 13
Schulze-Hardy rule, 198
Secondary ion mass spectrometry, 32
Sedimentites, 4, 5
    biogenic, 4
    chemical, 4
    klastic, 4
Seed dressing, 122
Sequential extraction, 31, 32, 215
Serpentine, 132
Sewage Effluents, 19
Sewage sludge, 19, 96, 102, 103, 108, 111, 115, 119, 122, 126, 128, 138, 143
Shale, 5
Shear stress, 167
Sheet Pile Walls, 172
Sieves, 167
Sieving, 167
Silene vulgaris, 244
Siltstone, 5

Silver, 145
SIMS, 32
Single Extraction Tests, 212
Slimicides, 122
Slurry Walls, 170
Smaltite, 144
Smectite, 10
Smithsonite, 143
Sodium molybdate, 126
Soil, 1, 6, 7-10
    flow sheet, 166
    process steps, 166
Soil air, 6
Soil amendments, 246
Soil constituent, 7-12
Soil electric conductivity, 175
Soil formation, 6
Soil fractions, 166, 167
Soil horizons, 7
Soil matrix, 6
Soil microorganisms, 267
Soil organic material, 7
Soil washing, 165
Soil water, 6
Solidification, 200
Sols, 194
Sorghum bicolor L. Moench, 229
Sorghum vulgare L. Moench, 229
Sorghum vulgare var. *sudanense*, 229
Sorption processes, 47, 48
Sources of heavy metals
    natural, 1, 15-17
    anthropogenic, 1, 17-25
Spectra, 33, 34, 41, 58, 83
Spectrophotometry, 32, 33
Sperrylite, 132, 133
Sphalerite, 3, 4, 74, 143
Sphere
    Area, 167
    Volume, 167
Stabilization/Solidification, 200
    Evaluation of S/S processes, 209
    Field applications of S/S processes, 230
    Technology description of S/S processes, 216
Stenotaphrum secundatum, 229
Sudbury, 129, 131
Sulfite oxidase, 127
Surface sorption complexes, 47
Surface Waters, 12
Sutera fodina, 107

Syphilis, 94

T
Tellurides, 146
Tetrachloroethylene, 14
Tetraethyl lead, 116, 120
Tetramethyl lead, 120
Thallium, 147, 148
Thermoplastic encapsulation, 221
Thin Walls, 171
Thiobacillus ferrooxidans, 12
Thiol groups, 123
Thlaspi caerulescens, 242
Thlaspi calaminare, 242
Thlaspi sp., 129
Tin, 89, 148
Toluene, 14
Transmission, 16
Tributyl Sn, 149
Trichloroethylene, 14
Turmaline, 2

U
Unsaturated zone, 13
Uranitite, 3, 4
Uranium, 185
Uranyl ion, 185
Urban Environments, 23
Urease, 129
Uric acid, 127

V
van der Waals force, 195
vesicular-arbuscular mycorrhizal fungi, 238
Vitrification, 174
Volatilization, 16

W
Washcoat, 133, 138
Wash water
   treatment of
Waste deposition, 25
Waste incineration, 17
Weathering, 1, 2, 6
   chemical, 4
   physical, 4
   products, 8
Wulfenite, 126
Wurzite, 143

X
XANES, 57
Xanthine oxidase, 127
XAS, 47, 56, 57
Xenobiotics, 14
X-ray absorption fine structure, 49, 57, 58
X-ray absorption near edge structure, 49, 58
X-ray absorption spectroscopy, 47, 56
XSW, 49
Xylene, 14

Z
Zea Mays, 229
Zeolites, 186
Zinc
   chemical and physical character of, 140
   ecotoxicological effects of, 143
   sources and applications of, 142
Zircon, 2, 3